Excel® for Chemists

Excel® for Chemists
A Comprehensive Guide

E. Joseph Billo

WILEY-VCH

New York • Chichester • Weinheim • Brisbane • Singapore • Toronto

E. Joseph Billo
Department Of Chemistry
Boston College, MA 02167-3860

This text is printed on acid-free paper.

Trademarks: Macintosh is a registered trademark of Apple
Computer, Inc. Windows is a registered trademark of
Microsoft Corporation.

Library of Congress Cataloging-in-Publication Data is available.
ISBN 0-471-18896-4

Printed in the United States of America
10 9 8 7 6 5 4 3

SUMMARY OF CONTENTS

CONTENTS

PREFACE

Most chemists deal with numbers on a daily basis. They record, calculate, summarize, graph, and report numerical data. Much of this work is done with the aid of a spreadsheet program on a personal computer. Many chemists use spreadsheet programs to *record* data in tabular form, but few have learned to take advantage of the tremendous *scientific calculating power* that is contained within the current versions of these programs. The aim of this book is to show you, a professional chemist, how to use the premier spreadsheet program, Microsoft Excel, to handle chemical calculations, from the relatively simple to the highly complex.

For example, you may need to

• calculate the percentages of carbon, hydrogen, nitrogen, oxygen, and other elements in a newly synthesized compound in order to compare the results of an elemental analysis with the theoretical values

• test various rate laws for a chemical reaction to see which equation best fits the observed data

• create a chart of the concentration of the acid-base forms of a new radiopharmaceutical as a function of pH, to illustrate the species distribution near pH 7

• resolve a UV spectrum into its individual Gaussian components in order to obtain the absorbance contribution of a shoulder peak

• apply linear regression to tensile strength data of polymer samples, to determine the effect of composition and molding conditions

• calculate a binding constant for a host-guest complex from the shift of NMR line position with changes in concentration of the guest molecule

• perform non-linear least-squares curve fitting to obtain the pK_a values of a polyprotic acid from a titration curve

Microsoft Excel can perform all these calculations, and more. You may have access to commercial software programs designed for some of these situations, but often you'll find that these programs don't handle the data you want to treat, or the model you want to fit, in exactly the right way. My purpose in writing this book is to demonstrate that it's relatively easy to "program" Excel to perform the calculations or other data manipulation needed for your specific application. Furthermore, if you use a range of commercial programs to perform data analysis, you'll have to learn (and remember) the commands and idiosyncrasies of each program.

This book is divided into four parts. Part I covers the basics of spreadsheet operations — entering data, cutting and pasting, formatting, creating charts, and so on. Part II shows how to use Excel's wide range of worksheet functions to perform sophisticated chemical calculations, how to create macros to automate spreadsheet tasks or to carry out repetitive calculations, and how to customize menus or toolbars to suit your own particular needs. Part III covers mathematical techniques that are particularly useful in a spreadsheet environment — matrix mathematics, numerical differentiation and integration, basic statistics, graphical and numerical methods of analysis — and shows how you can apply them easily using Excel. Part IV applies the techniques introduced in Parts I, II and III to a wide range of chemical problems.

The intent of this book is not simply to provide a series of templates that can be applied to particular situations (although there are lots of useful spreadsheet templates, macros and other tools on the disk that accompanies this book), but to show how you can create your own spreadsheets or macros to solve completely different chemical problems.

ACKNOWLEDGMENTS

Lev Zompa, University of Massachusetts-Boston, for spectrophotometric data used in Chapter 19.

Ross Kelly, Boston College, and Steve Bell, ICI Australia, for NMR data used in Chapter 20.

Allan D. Waren, Cleveland State University, for discussion about the Solver algorithms, and Edwin Straver, Frontline Systems Inc., for information about the inner workings of the Solver.

Dick Stein, University of Massachusetts-Amherst, and Stan Israel, University of Massachusetts-Lowell, for guidance on polymer databases.

Kavitha Srinivas, Boston College, for guidance about statistics.

Kenneth Kustin, Brandeis University, and Richard Haack, G. D. Searle Inc., Skokie IL, for reading the manuscript and offering helpful comments.

Barbara Goldman, executive editor, Camille Pecoul Carter, managing editor, Brenda Griffing, copy editor and Perry King, associate editor, electronic services, for their assistance and guidance during the publishing process.

My wife, Joanne, for encouragement and patience during the two years it took to write this book.

E. Joseph Billo
Chestnut Hill, Massachusetts

BEFORE YOU BEGIN

MACINTOSH AND WINDOWS VERSIONS OF EXCEL

This book is intended both for users of Excel for the Macintosh and for users of Excel for Windows. There are very few differences between the Mac and PC versions of Excel; although I'm a Macintosh user, I've tried to provide even-handed treatment to users of either type of computer. As you read through this book you'll see illustrations taken from Excel for the Macintosh and from Excel for Windows.

The small differences that *do* exist between Mac and Windows versions of Excel are mostly in the keystrokes that are used to perform some Excel operations. I've "piggybacked" these different instructions within a particular section. For example, in the sections on array formulas, you'll read "to enter an array formula, press COMMAND+ENTER (Macintosh) or CONTROL+SHIFT+ ENTER (Windows)". These keystroke differences are also listed in a table in Appendix G.

In the rare cases of instructions that are markedly different depending on whether you are using Excel for Windows or Excel for the Macintosh, I've placed those instructions in separate sections.

VERSIONS 4.0, 5.0 AND EXCEL FOR WINDOWS 95

This book covers versions 4.0, 5.0 and the latest version of Excel, Excel for Windows 95 (Excel 7.0). I know that there are lots of PC users who are still getting along fine with Excel 5.0. Similarly there are lots of Mac users who are still using version 4.0 instead of version 5.0. The majority of worksheet functions, menu commands, toolbuttons and dialog boxes are identical or near-identical in all three versions. Unless it's specifically stated otherwise, you can follow the instructions no matter which version you're using.

Some important changes were made when Excel 5.0 replaced Excel 4.0, but the changes made in going from Excel 5.0 to Excel 7.0 were smaller (and not all those changes were improvements, in my opinion). In the chapters that follow, I'll mention procedures to follow if you're using Excel 4.0 or Excel 5.0; when I refer to Excel 5.0, I mean either Excel 5.0 or 7.0. I'll mention Excel 7.0 specifically only when the instructions apply uniquely to Excel 7.0. Again, where the procedures for Excel 4.0 and the newer versions are different, I've usually put the procedures in separate sections.

(As this book was being completed, the next version of Excel, Excel 97, is about to appear. I've provided a preview of some of the new features in Excel 97 in Appendix I.)

TYPOGRAPHIC CONVENTIONS

As you read through this book, you'll see several different fonts and capitalization styles within the text. Here are the conventions that I've used.

- Names of keyboard keys are in ALL CAPS: TAB, SHIFT, CONTROL, OPTION, SHIFT, COMMAND, RETURN. (In Windows, the key is CTRL, but CONTROL is used for both Windows and Macintosh.)

- Menu headings and menu commands are in boldface type: **File**, **Format**, **Delete**....

- Dialog box titles and options are in Title Case: "The Rename Sheet dialog box...", "... press Cancel".

- Occasionally, menu commands and dialog box options are combined for clarity and conciseness: "... use **Paste Special** (Values)...".

- Cell references are in Geneva font: "In cell A9 ...".

- Worksheet functions and macro functions are in Geneva: SUM, ACTIVATE.

- General (i.e., placeholder) arguments in functions or in text are in Geneva italic; required arguments are in bold italic: LINEST(***known_y's, known_x's***, *const, stats*).

- Specific arguments in functions or in text are in Geneva, not italic: ACTIVATE(SourceSheet), "... to copy the SourceSheet, you must ....".

- Visual Basic statements are in Geneva; VBA reserved words are bold: **For** Counter = Start **To** End **Step** Increment.

SPECIAL FEATURES IN THIS BOOK

This book has a number of features that you should find useful and helpful. There are over 50 **Excel Tips** to simplify and improve the way you use Excel. For example:

*Excel Tip.   To **Fill Down** a value or formula to the same row as an adjacent column of values, select the source cell and double-click on the Fill Handle.*

Throughout the book you'll see **"How-To" Boxes** that outline, in a clear and systematic manner, how to accomplish certain complex tasks. For example:

To Link Worksheet Text to a Chart

1. The worksheet must be open.

2. The chart must be activated. If the chart is an embedded chart, you must activate it by double-clicking on it.

3. With the chart as the active document, select the chart text item (Chart Title, X-axis Title, Y-axis Title or Unattached Text) by selecting it by using the **Chart** menu or by double-clicking on the chart text item if it already exists in the chart.

4. In the formula bar, type
   =worksheet_name!absolute_cell_reference,

   e.g., ='Voltammetric curve'!$A$7. If the worksheet name contains spaces, it must be enclosed in single quotes.

5. Alternatively, you can have Excel supply the worksheet name and cell reference. In the formula bar, type "=". Activate the worksheet by clicking on it or by selecting it from the **Window** menu. The external reference to the worksheet will appear in the formula bar. Select the desired cell in the worksheet. The absolute reference to the cell will appear in the formula bar.

THE DISKETTE

The diskette that accompanies this book is in PC format; the document names are ones that are compatible with Excel for Windows. Most Macintosh computers can open and read Excel files in PC format, using the utility program PC Exchange. If you have trouble, please contact John Wiley's tech support system at (212) 850-6194.

The diskette contains most of the worksheets that are discussed in the book. They are in Excel 4.0 format, so that they can be opened using either Excel 4.0, Excel 5.0 or Excel 7.0. A complete list of all files on the diskette, with short descriptions, is in Appendix H.

INSTALLATION INSTRUCTIONS

The files are included on the diskette in chapter directories within the main software directory CHEM-XL. You may install the files to your computer by dragging the CHEM-XL folder to your hard drive. If you are running on a Windows system on a PC, you can also load the files by using the installation program provided on the diskette. To use the program, type a:install in the **Run** option of the **File** menu or double-click on the INSTALL.EXE file in File Manager or other file management program. Follow the instructions on the installation screens to load the files to your selected target.

PART I

THE BASICS

1

WORKING WITH EXCEL

This chapter covers the basics of working with Excel: navigating around the worksheet, entering values and formulas, and formatting and editing a worksheet. If you are an experienced Excel user, you can probably skip this chapter; however, there may be a few useful tips in this chapter for even experienced users.

THE EXCEL DOCUMENT WINDOW

An Excel workbook or worksheet is a *document* that appears in its own *document window*. (Documents of other types will be discussed later.) Although you can have several documents open at the same time, and can see them all

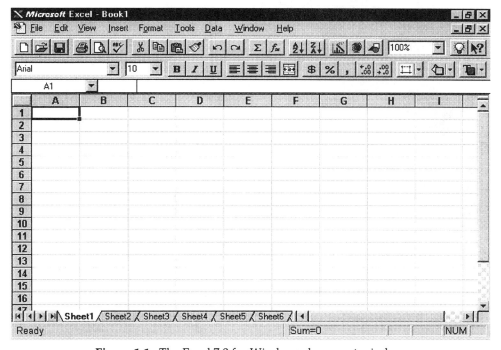

Figure 1-1. The Excel 7.0 for Windows document window.

displayed on the screen simultaneously, only one document can be the *active document*. The Excel 5.0[*] workbook usually contains 16 worksheets; only one worksheet in the workbook can be the active document.

An Excel worksheet consists of 256 columns (labeled A, B, C, ..., IV) and 16,384 rows (1, 2, 3, ...). The rows and columns define *cells* (A1, H27, etc.) which constitute the worksheet. Information can be entered into a cell from the keyboard after the cell has been selected with the mouse pointer. Data can also be entered into a cell, or many cells, by calculation. The Excel 7.0 for Windows document window is shown in Figure 1-1. Your screen may not show the same number of rows or columns, depending on your monitor.

Reading from the top down you'll see the *application title bar*, the *menu bar* (with **File**, **Edit**, **View,** etc. menus), the *toolbars* (with Bold, Italic and Alignment tools, for example), the *formula bar*, the rows and columns of cells, the *sheet tabs* (your screen may not show the same number of sheet tabs, depending on your monitor), the *horizontal scroll bar* and, at the bottom, the *status bar*. The formula bar contains the *cell reference area* (displaying the cell reference of the currently selected cell) and the editing area. As you enter values at the keyboard, they appear in the editing area of the formula bar. When you begin to type an entry, the Enter and Cancel buttons appear.

The Excel 5.0 for the Macintosh document window, shown in Figure 1-2, is almost identical (although the menu bar and toolbar are somewhat different).

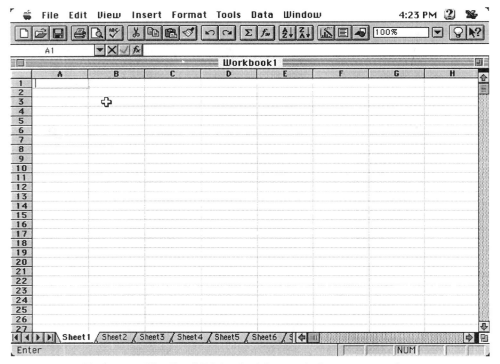

Figure 1-2. The Excel 5.0 for the Macintosh document window.

* References to Excel 5.0 apply to both Excel 5.0 and Excel 7.0 unless stated otherwise.

With Excel 5.0, Microsoft introduced the TipWizard, the lightbulb icon near the right end of the Standard toolbar. When the lightbulb icon lights up, Excel has a tip that suggests a different or better way to accomplish the task you are doing. Click on the TipWizard icon to see the list of tips. Like Macintosh System 7's Balloon Help, the TipWizard soon becomes superfluous. In Chapter 9 you'll learn how to get rid of the TipWizard.

Excel 5.0 also introduced ToolTips; if you place the tip of the mouse pointer on one of the toolbuttons, a yellow ToolTips box appears, describing the button's function. You can deactivate ToolTips by choosing **Toolbars...** from the **View** menu and de-selecting the View ToolTips check box.

The Excel 4.0 menu bar and toolbars (not shown) differ somewhat from their Excel 5.0 counterparts. The Excel 4.0 for the Macintosh menu bar contains **File, Edit, Formula, Format, Data, Options, Macro** and **Window** menus; the **File, Edit, Format** and **Window** menus are similar in versions 4.0 and 5.0, but most commands that are in the Excel 4.0 **Formula**, **Options** and **Macro** menus are located in the Excel 5.0 **Tools** menu.

MOVING OR RE-SIZING DOCUMENTS (EXCEL FOR WINDOWS)

To change the size of a workbook or worksheet, click and drag any of its borders or corners; the mouse pointer changes shape when you click on a border or corner. You can adjust the document to any size you desire. If you click on the Minimize button (the open square containing an "underline" symbol in the upper right corner of the document) the document will be minimized so that only the title bar is visible. To restore it to its full size, click the Maximize button, the open square in the upper right corner of the title bar .

To change the position of a document within the Excel window, click on the title bar and drag the document. It can even extend off-screen.

MOVING OR RE-SIZING DOCUMENTS (EXCEL FOR THE MACINTOSH)

To change the size of a workbook or worksheet, click and drag in the Size box, the open square in the lower right corner of the document. You can adjust the document to any size you desire. To restore it to its full size, click the Maximize button, the open square in the upper right corner of the title bar, or anywhere in the title bar.

To change the position of a document within the Excel window, click on the title bar and drag the document.

NAVIGATING AROUND THE WORKSHEET

You can move around a worksheet either by means of the mouse or by using keystrokes.

Use the arrows in the *vertical* and *horizontal scroll bars* (the gray bars on the right edge and at the bottom of the window) to scroll through the worksheet. A single click of the mouse on an arrow moves the worksheet one row or column. The position of the *scroll box* (the white square in the gray bar) indicates the position of the window relative to the worksheet. You can also scroll through the

Table 1-1. Keys for Cursor Movement*

Arrow keys	Moves left, right, up, down one cell
RETURN	Moves down one cell
TAB	Moves right one cell
HOME	Moves to the beginning of a row
END	Moves to the end of a row
PAGE UP	Moves to the top of the window
PAGE DOWN	Moves to the bottom of the window
CONTROL+→	Moves to the first occupied cell to the right in the same row that is either preceded or followed by a blank cell
CONTROL+←	Moves to the first occupied cell to the left in the same row that is either preceded or followed by a blank cell
CONTROL+↑	Moves to the first occupied cell above in the same column that is either preceded or followed by a blank cell
CONTROL+↓	Moves to the first occupied cell below in the same column that is either preceded or followed by a blank cell

*On the Macintosh, use either CONTROL+(key) or COMMAND+(key).
The COMMAND key is the ⌘ key.

worksheet by clicking on an arrow and holding down the mouse button, by dragging the scroll box with the mouse, or by clicking in the gray space on either side of the scroll box.

Table 1-1 lists keystroke commands for cursor movement.

SELECTING A RANGE OF CELLS ON THE WORKSHEET

You can select a range of cells on the worksheet in several ways:

- Click on the cell in one corner of the range, hold down the mouse button and drag to the cell in the opposite corner of the range. The range of cells will be highlighted. The size of the selection, e.g., $10R \times 3C$, is displayed in the Reference box of the formula bar.

- Select the cell in one corner of the range, move to the cell in the other corner of the range, hold down the SHIFT key and select the cell in the opposite corner of the range. The range of cells will be highlighted.

- Select a complete row or column of cells by clicking on the row or column heading. The row or column will be highlighted.

SELECTING NON-ADJACENT RANGES

To select non-adjacent ranges, select the first range, then hold down the CONTROL key (Windows) or the COMMAND key (Macintosh) while selecting

	A	B	C	D
3				
4	t	[A]	[B]	[C]
5	0	1.0000	0.0000	0.0000
6	5	0.7788	0.2101	0.0111
7	10	0.6065	0.3537	0.0398
8	15	0.4724	0.4474	0.0802
9	20	0.3679	0.5041	0.1281
10	25	0.2865	0.5334	0.1801
11	30	0.2231	0.5428	0.2341
12	35	0.1738	0.5380	0.2882
13	40	0.1353	0.5233	0.3413
14	45	0.1054	0.5020	0.3927
15	50	0.0821	0.4763	0.4416
16				

Figure 1-3. Selecting non-adjacent ranges.

the second range. Both cell ranges will be highlighted (Figure 1-3).

To extend the range of a cell selection you just made, hold down the SHIFT key, select the last cell in the selection and drag to include the additional cells.

SELECTING A BLOCK OF CELLS

A *block* of cells is a range of cells containing values and bounded by empty cells. To select cells within a block, select a cell at a boundary of the block (at the top, bottom or side of the block). Move the mouse pointer over the edge of the selected cell until the pointer changes to the arrow pointer (Figure 1-4A). Hold

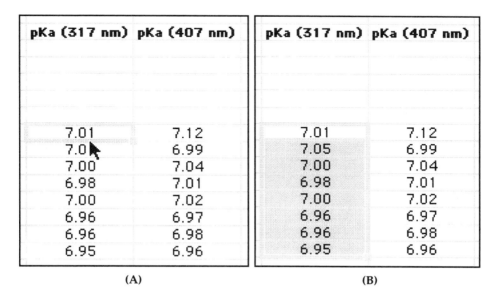

pKa (317 nm)	pKa (407 nm)		pKa (317 nm)	pKa (407 nm)
7.01	7.12		7.01	7.12
7.0	6.99		7.05	6.99
7.00	7.04		7.00	7.04
6.98	7.01		6.98	7.01
7.00	7.02		7.00	7.02
6.96	6.97		6.96	6.97
6.96	6.98		6.96	6.98
6.95	6.96		6.95	6.96

(A)   (B)

Figure 1-4 . Using the Drag-and-Drop pointer: (A) selecting a cell edge and (B) selecting a block by clicking.

down the SHIFT key and double-click on the bottom edge of the selected cell to select all cells in the column from the top to the bottom of the block, as shown in Figure 1-4B. You can select cells from top to bottom, from bottom to top, from left to right or from right to left within a block. You can also select multiple columns or rows in the same way.

ENTERING DATA IN A WORKSHEET

To enter a value in a worksheet cell, select the cell with the mouse pointer, which appears as a large open cross when it passes over cells. Clicking on the desired cell highlights it, indicating that this is the *active cell*, the cell where you can now enter a value. As you type in a value, the characters will appear in the formula bar (Excel 4.0) or in both the formula bar and the active cell (Excel 5.0). You can complete the entry in several ways.

- Press the Enter button ☑ in the formula bar. The cell remains selected.

- Press the RETURN key or the ENTER key. This moves the selection to the cell below (although you can change the default option so that the selection is not moved).

- Use the mouse pointer to select a different cell. This is not a good habit to get into, though, because it has a completely different and unwanted effect if you are entering a formula rather than a value.

To cancel the entry and revert to the original contents of the cell, press the Cancel button ☒ or the ESC key.

Excel Tip. *To enter the same value in a range of cells, select the range of cells, type the value, then press COMMAND+RETURN, COMMAND+ENTER, CONTROL+RETURN or CONTROL+ENTER (Macintosh) or CONTROL+ ENTER (Windows).*

ENTERING NUMBERS

Excel has a remarkable ability to recognize the format of the value that you have entered: as a number, a percent, a debit value, as currency, in scientific notation, as a date or time, or even as a fraction. The number will be displayed in the cell in the proper format, but the number equivalent of the value will appear in the formula bar. Figure 1-5 illustrates the number formats recognized by Excel.

If you enter a fraction less than 1, such as 1/3, it will be interpreted as a date ("3-Jan"). To prevent Excel from converting the fraction to a date, add a zero and a space (0 1/3). The zero indicates that the entry is a number, and the value will appear in the formula bar as 0.333333333333333.

HOW EXCEL STORES AND DISPLAYS NUMBERS

Excel can accept numbers in the range from $\pm1E-307$ to $\pm9.99999999999999E+307$.

	A	B	C	D
1	Type	As Entered	As Displayed in Cell	As Displayed in Formula Bar
2	percent	15%	15%	0.15
3	scientific	2e-3	2.00E-03	0.002
4	currency	$50	$50	50
5	currency	$20000	$20,000	20000
6	fraction	2 5/8	2 5/8	2.625
7	debit	(5000)	-5000	-5000
8	date	7/4	4-Jul	7/4/1997*
9	date	8-3-38	8/3/38	8/3/1938
10	time	4:30	4:30	4:30:00 AM
11	time	16:00	16:00	4:00:00 PM
12	time	4 p	4:00 PM	4:00:00 PM
13		* enters current year		

Figure 1-5. Number formats recognized by Excel.

Excel stores numbers with 15-significant-figure accuracy. These are displayed in the formula bar and used in all calculations, no matter what number formatting has been applied. Thus the fraction 1/3 appears in the formula bar as 0.333333333333333, and π as 3.14159265358979. Fifteen-significant-figure accuracy is roughly equivalent to double precision in other computer systems.

Excel switches between floating point and scientific for best display of values. The formula bar can display numbers up to 21 characters, including the decimal point. Thus 1E-19 entered on the keyboard will appear as 0.0000000000000000001 (21 characters) in the formula bar, while 1E-20 will appear as 1E-20. Similarly, 1E20 appears as 100000000000000000000, while 1E21 appears as 1E21. Since a total of 21 characters can be displayed, the number of significant figures determines the magnitude of a number less than 1 that can be displayed in non-E format in the formula bar. Thus 1.2345E-15 appears as 0.0000000000000012345, while 1.23456E-15 is displayed as 1.23456E-15.

ENTERING TEXT

If you enter text characters (any character other than numbers, the decimal point, or the characters +, -, *, /, ^, $, %) in a cell, Excel will recognize the entry as text. For example, the entry Chestnut Hill MA 02167-3860 is a text entry. A cell can hold up to 255 characters of text. You can distinguish text entries from number entries in the following way: in a cell that has not been alignment-formatted (e.g., left, centered, right, etc.), text entries are left-aligned, numbers right-aligned. Of course, if you format the alignment of a cell to be right-aligned, its value will be right-aligned whether the value is a number or text.

Sometimes it is necessary to enter a number as a text value. To do this, begin the entry with a single quote.

ENTERING FORMULAS

Instead of entering a number in a cell, you can enter an equation (called a *formula* in Microsoft Excel) that will calculate and display a result. Usually formulas refer to the contents of other cells by using *cell references*, e.g., a reference to a cell, such as A2, or a reference to a range of cells, such as B5:B12. The value displayed in a cell containing a formula will be automatically updated if values elsewhere in the worksheet are changed. Formulas can contain values, arithmetic operators and other operators, cell references, the wide range of Excel's worksheet functions, and parentheses.

The rules for writing formulas (the *syntax*) are as follows:

- A formula must begin with the equal sign (=).
- The *arithmetic operators* are addition (+), subtraction (-), multiplication (*), division (/) and exponentiation (^). Other types of operator are described in Chapter 3.
- Parentheses are used in the usual algebraic fashion to prevent errors caused by the *hierarchy of arithmetic operations* (multiplication or division is performed before addition or subtraction, for example).

Excel Tip. Formulas that return the wrong result because of errors in the hierarchy of calculation are common. When in doubt, use parentheses.

Some examples of simple formulas:

=A1+273.15	Adds 273.15 to the value in cell A1
=A2^2+13*A2-5	Evaluates the function $x^2 + 13x - 5$ where the value of x is stored in cell A2.
=SUM(B3:B47)	Sums the values contained in cells B3 through B47
=(-C3+SQRT(C3^2-4*C2*C4))/(2*C2)	Finds one of the roots of the quadratic equation whose coefficients a, b and c are stored in cells C2, C3 and C4 respectively.

Excel formulas are discussed in much greater detail in Chapter 3.

ADDING A TEXT BOX

You can add visible comments or other information to a worksheet by typing them into one or more worksheet cells. Another way to add comments, in a much more flexible form, is by using a *text box*.

To create a text box, press the Text Box toolbutton [▤] . The mouse pointer will change to a crosshair. Position the crosshair pointer where you want to place the text box, and click and drag to outline it (the text box can be moved and

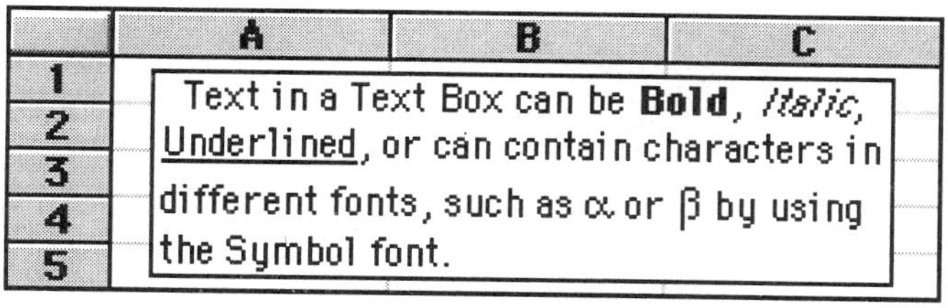

Figure 1-6. A text box.

sized later). An empty text box will be displayed with a blinking text cursor. Type the desired text within the box.

Text-box input has many features of a simple word processor: you can **Cut**, **Copy** or **Paste** text, make individual portions of text Bold, Italic or Underlined, use different font styles, etc., as shown in Figure 1-6. The text within the box can be formatted with the alignment toolbuttons or with the Alignment command.

To move a text box, click the mouse pointer anywhere within the text box and drag it to its new position. To re-size a text box, select it (black handles will appear), then place the mouse pointer over one of the black handles and click and drag to move the border of the box. If you hold down the CONTROL key while dragging, the text box will align with the cell gridlines.

ENTERING NOTES

You can attach comments to a cell, for documentation purposes, in the form of a *text note*. Text notes are not displayed on the worksheet; a small square in

the upper left corner of the cell   ▪   indicates that the cell contains a note.

To add a text note to a cell, choose **Note...** from the **Formula** menu (Excel 4.0) or from the **Insert** menu (Excel 5.0). Enter the text of the note in the Cell Note dialog box (Figure 1-7), then press the Add or OK buttons. A list of other notes in the sheet is also displayed. After the text has been entered, you can enter the same note in another cell by selecting the cell or changing the reference in the Cell box and pressing Add. To delete a note, choose the note from the Notes in Sheet list, or select the cell containing the note, or type the cell reference in the Cell box, then press the Delete button.

To view a text note, double-click on the cell (Excel 4.0), or select the cell, then choose **Options** from the **Tools** menu and check the InfoWindow box (Excel 5.0). The note is displayed in the Cell Note dialog box (Excel 4.0) or the InfoWindow (Excel 5.0).

Excel 7.0 introduced a new way to view cell notes: when the mouse pointer is moved over a cell that displays a cell note indicator, the cell note text appears in a small box similar to a ToolTip.

Note indicators are not printed when you print a worksheet. You can turn screen display of Note indicators on/off by choosing **Workspace...** from the **Options** menu (Excel 4.0) or **Options** from the **Tools** menu (Excel 5.0) and checking/unchecking the Note Indicator box.

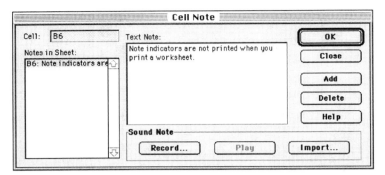

Figure 1-7. The Cell Note dialog box.

EDITING CELL ENTRIES

You can edit cell entries in one of two ways — either in the formula bar or by using the Edit Directly in Cell feature (Excel 4.0 does not have this feature). When you select a cell that contains an entry, the contents of the cell appear in the formula bar. As soon as you begin to enter a new value the old value disappears. To make minor editing changes in the old entry, place the mouse pointer in the text at the point where you want to edit the entry. The mouse pointer becomes the vertical insertion-point cursor. You can now edit the text in the formula bar using the **Copy, Cut, Paste** or **Delete** commands or keys. Complete the entry using the Enter button in the formula bar.

To use Excel 5.0's Edit Directly in Cell feature, double-click on the cell. The text can now be edited in the cell in the same way as in the formula bar.

Excel 5.0 permits you to format individual characters in a cell using Bold, Italic, Underlined, etc., or with different fonts.

EXCEL'S MENUS: AN OVERVIEW

In Excel 5.0 for Windows or Macintosh, and Excel 7.0 for Windows, the worksheet menu bar has the following pull-down menus: **File, Edit, View, Insert, Format, Tools, Data, Window** and **Help.** In Excel 4.0 for the Macintosh the menus in the menu bar are **File, Edit, Formula, Format, Data, Options, Macro** and **Window.** The **File, Edit, Format** and **Window** menus are discussed in this chapter. Commands in other menus will be discussed in later chapters.

The way in which a command appears in a menu provides information about its form or availability:

- A menu command with an ellipsis (...), such as **Save As...**, indicates that the command opens a dialog box to obtain user input.

- Many Excel 5.0 menus contain submenus, indicated by the ▶ symbol at the right edge of the menu.

- Some menu commands are dimmed, i.e., appear as gray characters, when the menu command is unavailable. Others appear on the menu only when they are available.

- Some menu commands change the text of their command, depending on circumstances. For example, in Excel 4.0, if you use **Select Print Area** to select a worksheet area to be printed, the command changes to **Remove Print Area** so that you can Undo the area selection later.

- Some menu commands are preceded by a check mark if the choice has been selected. Select the command again to remove the selection.

SHORTCUT MENUS

Excel also provides "context-sensitive" *shortcut menus*. If you press the right mouse button (Windows) or press COMMAND+OPTION (Macintosh) while you select a worksheet element with the mouse pointer, a menu is displayed containing commands that apply to the selection. For example, if you select a row or column while holding down the right mouse button, a shortcut menu containing editing and formatting commands appears.

MENU COMMANDS OR TOOLBUTTONS?

Most menu commands can be carried out by using toolbuttons. Toolbuttons are more convenient; they often combine a whole series of actions — menu selection plus dialog box options — into a single click of the mouse button.

Some buttons mentioned in this chapter don't appear on either the Excel 4.0 or Excel 5.0 Standard toolbar. To make them available for use, you can have more than one toolbar displayed at one time, or you can customize a toolbar (see Chapter 9).

OPENING, CLOSING AND SAVING DOCUMENTS

Most menu commands for managing documents are in the **File** menu. For the most part, the menu is similar to the **File** menu in other Windows or Macintosh applications, with **New**, **Open...**, **Close**, **Save**, **Save As...**, **Page Setup...**, **Print Preview...**, **Print...** and **Exit** (Windows) or **Quit** (Macintosh) commands. **Save Workspace...** (Excel 5.0) and **Save Workbook** (Excel 4.0) are commands specific to Excel.

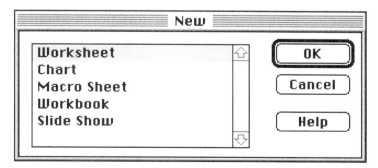

Figure 1-8. The Excel 4.0 for the Macintosh New dialog box.

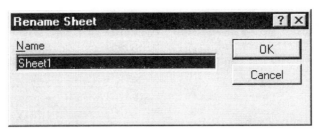

Figure 1-9. The Rename Sheet dialog box.

OPENING OR CREATING WORKSHEETS OR WORKBOOKS

Use the **Open...** command to locate and open an existing document; use **New** to create a new document. In Excel 4.0, the **New...** command opens a dialog box listing five document types (Figure 1-8): worksheet, chart, macro sheet, workbook or slide show (slide shows are not discussed in this book).

In Excel 5.0, **New** does not open a dialog box but instead immediately creates a workbook containing 16 blank worksheets. Additional worksheets, charts, macro sheets or modules can be added to the workbook by means of the **Insert** menu.

In Excel 7.0, **New...** displays a dialog box in which you have a choice of opening either a new worksheet or any of the built-in or user-created template sheets.

To open an existing workbook or worksheet from the desktop, simply double-click on it. This will open the document (and will start Excel as well if it wasn't already running). If you start Excel first, it will open a new blank worksheet (Excel 4.0) or a new workbook (Excel 5.0).

EXCEL 5.0 WORKBOOKS

The default Excel 5.0 workbook contains 16 worksheets, but you can add or remove sheets.

You select a worksheet in an Excel 5.0 workbook by clicking on its sheet tab. Scroll through the whole range of sheet tabs by using the arrow buttons to the left of the sheet tabs: .

When you create a new workbook, the sheet tabs have the names Sheet1, Sheet2, etc. To rename a sheet, double-click on the sheet tab. The Rename Sheet dialog box (Figure 1-9) will appear and you can enter a more descriptive name, as, for example, in Figure 1-10.

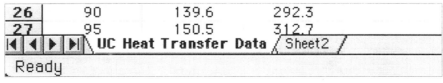

Figure 1-10. Descriptive sheet names are helpful.

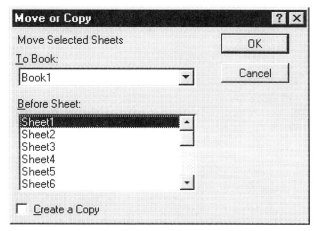

Figure 1-11. The Excel 5.0 Move or Copy Sheet dialog box.

USING MOVE, COPY OR DELETE SHEET (EXCEL 5.0)

Three Excel 5.0 commands permit you to add or remove sheets from a workbook. The **Delete Sheet** command in the **Edit** menu permanently removes the active sheet from the workbook. To add sheets to a workbook, use the **Worksheet** command in the **Insert** menu. You can also insert chart or macro sheets. To move or copy sheets within a workbook, or from one workbook to another, use the **Move or Copy Sheet...** command (Figure 1-11) in the **Edit** menu.

CREATING AND USING WORKBOOKS (EXCEL 4.0)

As you'll learn in Chapter 10, an Excel document can be linked to other Excel documents. A document that gets data from another document is called a *dependent* document, while a document that contributes values to other documents is called a *supporting* document. (A document that does not rely on or contribute to any other document is called an *independent* document.) If you open a dependent document that refers to a document that is not open, you will get a "Update references to unopened documents?" message or a "This document establishes links. Update links?" message. To avoid getting these messages, always open supporting documents first. Excel 4.0 provides the option of arranging your worksheets in a *workbook*. If you bind related documents in an Excel 4.0 workbook, opening the workbook opens all documents and avoids "Update?" messages.

USING CLOSE OR EXIT/QUIT

You can **Close** a document either with the **Close** command from the **File** menu or, more conveniently, by using the Close button on the document title bar. You will be asked if you want to save changes.

If you hold down the SHIFT key while you pull down the **File** menu, the **Close** command becomes **Close All**. That way you can close all open Excel documents at once.

When you use the **Exit** command (Windows) or the **Quit** command (Macintosh), you close all open documents (you will be asked if you want to save changes) and then exit from Excel.

USING SAVE OR SAVE AS...

When you **Save** a newly-created worksheet, workbook, chart etc., the Save dialog box will prompt you to assign a name to the document. In Excel 5.0 for Windows, document names can have up to 8 characters; no spaces are allowed. Excel for Windows automatically appends a three-letter filename extension, e.g., .XLS, to identify the file format type. In Excel for the Macintosh, names can have up to 31 characters, upper- or lowercase, and spaces are allowed. Any keyboard character except the colon (:) can be used in a filename.

Excel 7 allows long filenames, essentially the same as Excel for the Macintosh.

You can use **Save As...** to create a backup copy of a worksheet by giving the copy a different name.

THE TYPES OF EXCEL DOCUMENT

Excel uses a variety of file formats for its documents. Table 1-2 lists the filename extensions used by Excel for Windows. Excel for the Macintosh assigns a specific name or appends a descriptor to some, but not all, of the file types.

USING SAVE WORKSPACE... (EXCEL 5.0)

You can use **Save Workspace** if you have to interrupt your work, when you have several workbooks open at once. The **Save Workspace** command saves the current configuration of the workspace. Excel for the Macintosh proposes the filename RESUME. When you open the RESUME file, the workbooks will be restored to their former size and position on screen.

USING DELETE... (EXCEL 4.0)

You can **Delete** documents from the desktop, hard drive or floppy without exiting Excel. But beware — the document doesn't go into the Trash, where you

Table 1-2. Excel for Windows Filename Extensions and Macintosh Equivalents

Windows filename extension		Macintosh equivalent
.XLA	Add-in macro	Appends "Add-In" to filename
.XLB	Toolbar configuration	Filename is "Excel Toolbars (n)"*
.XLC	Excel 4.0 chart	No identifier
.XLM	Excel 4.0 macro	No identifier
.XLS	Excel 4.0 worksheet	No identifier
.XLS	Excel 5.0 workbook	No identifier
.XLT	Template	No identifier
.XLW	Workspace configuration	Proposes "RESUME" as filename

* n = 4 or 5; indicates Excel version.

can recover it, nor can you reverse your action with the Undo button. Before using this command, be sure you really want to lose the document forever.

Excel 5.0 does not have the **Delete...** command.

PRINTING DOCUMENTS

Menu commands for printing documents are located in the **File** menu.

USING PAGE SETUP...

Use **Page Setup** to choose Portrait or Landscape orientation (Figure 1-12), adjust margins (Figure 1-13A), add or delete page numbers, or change or delete header or footer (Figure 1-13B). In Excel 5.0, the default header and footer, which will be printed on every page unless you choose otherwise, are "Sheetname" and "Page #", respectively. To enter a different header or footer, choose the Header/Footer tab in the Page Setup dialog box; it includes list boxes with a wide range of built-in formats for header and footer. You can also create your own custom headers or footers by pressing the Custom Header... or Custom Footer... button. The Header or Footer dialog box (they are identical) enables you to enter filename, sheetname, page number, date or time information. For example, you can date-stamp a worksheet by pressing the Date button; Excel inserts the custom header/footer text & [Date] in the selected position of the header or footer.

In Excel 4.0 you enter header or footer text by using the Header or Footer dialog boxes (Figure 1-14), which are virtually identical to the Excel 5.0 Custom Header/Footer dialog box. The codes for filename, page number, date, etc. are slightly different from those used in Excel 5.0 (Excel inserts Page &P when you press the Page button, for example) but are fairly self-explanatory. The default header is &F, which means that the filename will be printed at the top of the page. The default footer is Page &P.

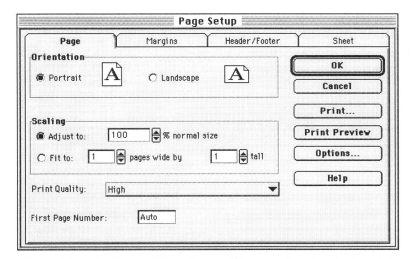

Figure 1-12. The Excel 5.0 Page Setup dialog box showing the Page tab.

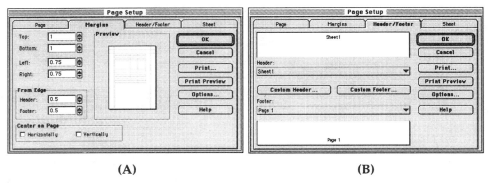

(A)   **(B)**

Figure 1-13. The Page Setup dialog box, showing (A) the Margins tab and (B) the Header/Footer tab.

To squeeze the maximum amount of worksheet information on a single page, you may want to decrease the margin widths. The default margin values are 0.75 inch left and right and 1 inch top and bottom. If you set the margins to zero, the header and footer information will still be printed, usually right on top of data in your worksheet, so delete the header and/or footer information by choosing "(none)" from the list box (Excel 5.0) or highlighting it and deleting the text (Excel 4.0).

You can elect to Print Row And Column Headings and/or Cell Gridlines by choosing the Sheet tab. If you de-select Cell Gridlines, they will not be printed but they will still be displayed on the screen. If you turn off screen display of gridlines by choosing the **Display** command from the **Options** menu (described later), the Cell Gridlines check box will also be cleared in the Page Setup dialog box, and gridlines will not be printed.

You may need to use Print Black and White if your worksheet uses color. Colors may be printed as various patterns by your printer; to remove the patterns and produce text in cells in black and white, check the Print Black and White Cells box.

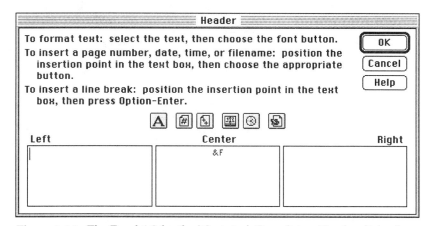

Figure 1-14. The Excel 4.0 for the Macintosh Page Setup Header dialog box.

USING **PRINT PREVIEW**

In addition to showing what your finished worksheet will look like when printed, **Print Preview** is useful in other ways. If you preview your worksheet and then return to the document window, page breaks will be displayed on the worksheet, as dashed lines. They will assist you in adjusting column widths, for example, before printing.

USING **PRINT...**

If you choose the **Print** command and simply press the OK button, Excel will print the rectangular array of sheets that includes all filled cells. Use **Print Preview** before printing; the total number of pages to be printed will be displayed in the status bar. This will tell you whether you can print the whole worksheet, or whether you need to specify a range of pages to be printed.

If you choose **Print Preview** or **Page Setup**, Excel displays the automatic page breaks as dashed lines in the worksheet. You can also insert a forced page break if you want to print a portion of a worksheet page. To insert a horizontal page break, select an entire row as if you were going to insert a row. Then choose **Insert Page Break** from the **Options** menu (Excel 4.0) or **Page Break** from the **Insert** menu (Excel 5.0). The page break will be inserted immediately above the row selected. Excel displays forced page breaks as dashed lines that are heavier than the dashed lines used to indicate automatic page breaks. A forced vertical page break is inserted in a similar fashion; the page break is inserted immediately to the left of the selected column. If you want to insert both a vertical and a horizontal page break, select a single cell within the worksheet; the page breaks will be immediately above and to the left of the cell.

PRINTING A SELECTED RANGE OF CELLS IN A WORKSHEET

In Excel 4.0, to print a selected range of cells within a worksheet, first select (highlight) the range to be printed, then choose **Select Print Area** from the **Options** menu. The range to be printed will be indicated by Page Break lines. To cancel the area selection, select the whole document by clicking the mouse pointer on the row/column heading box in the upper left corner (where the row and column headings intersect); then choose **Remove Print Area** from the **Options** menu.

In Excel 5.0, to select a range of cells to be printed, first choose **Page Setup** from the **File** menu, and choose the Sheet tab. Click in the Print Area text box to select it. Now select the range of cells that you want to print (you can move the dialog box out of the way if necessary), and press the OK button. To cancel an area selection, delete the text within the Print Area text box.

The process for selecting or removing a Print Area was more convenient to use in Excel 4.0 than in Excel 5.0. (**Select Print Area** returned as a standard menu option in Excel 7.0.) In subsequent chapters you'll learn how to install the Set Print Area toolbutton or to create a new menu command in Excel 5.0 to **Select Print Area** or **Remove Print Area.**

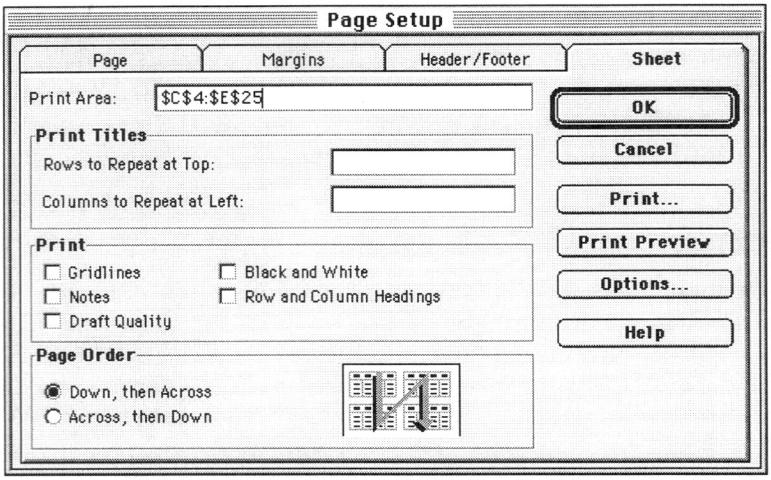

Figure 1-15. Designating an area to be printed in Excel 5.0 for the Macintosh.

If the Print Area you selected requires more than one page, you can choose **Page Setup** and change the value in the Reduce/Enlarge box to less than 100%. Sheets printed with values less than about 60% are difficult to read, though. To obtain the appropriate reduction value automatically, after you've selected the area to be printed, choose the Page tab and press the "Fit to 1 pages wide by 1 tall" button.

PRINTING ROW OR COLUMN HEADINGS
FOR A MULTI-PAGE WORKSHEET

If you are printing a multi-page worksheet, you can duplicate row or column headings automatically on each printed page. In Excel 5.0, choose **Page Setup** from the **File** menu, and choose the Sheet tab, as in Figure 1-15. Select the Rows to Repeat at Top or the Columns to Repeat at Left text box by clicking the cursor in it. Now select the range of cells that you want to print on every page as a title (you can move the dialog box out of the way if necessary). Then click the OK button. The headings will appear at the top or left of each printed page.

In Excel 4.0, choose **Set Print Titles...** from the **Options** menu. When the Set Print Titles dialog box appears (Figure 1-16), select the area containing the row or column headings to be printed.

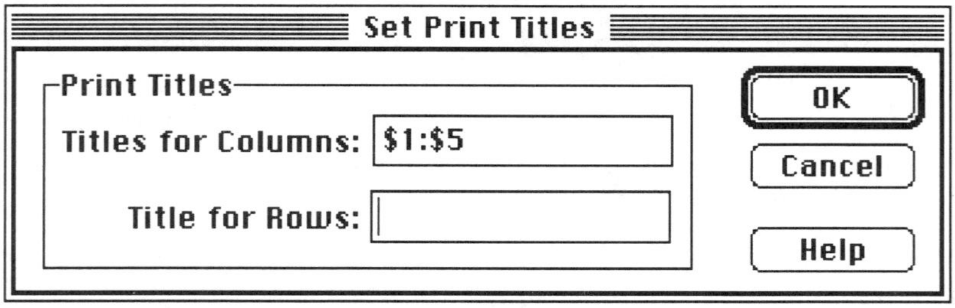

Figure 1-16. The Excel 4.0 for the Macintosh Set Print Titles dialog box.

EDITING A WORKSHEET

INSERTING OR DELETING ROWS OR COLUMNS

To insert a row of blank cells, click on a row heading, which selects an entire row. Then choose **Insert** from the **Edit** menu (Excel 4.0), or choose **Rows** from the **Insert** menu (Excel 5.0). A new column will be inserted above the row you selected. (Insertion occurs above or to the left of the selected row or column, respectively.) Multiple rows or columns can be inserted in a similar fashion, by selecting as many rows or columns as you want to insert.

To insert an extra cell or cells within a row or column, select the cell range above or to the left of which you want to insert cells, then choose **Insert....** (Note that **Insert** has become **Insert...** because the Insert dialog box will be displayed.) Excel usually makes a pretty good guess as to whether the cells should be shifted to the right or down in order to make the proper insertion, but always check to make sure. Then click OK in the dialog box (Figure 1-17).

When inserting partial rows or columns, take care that other parts of the worksheet do not become misaligned.

Complete rows or columns are deleted in a similar manner. Click on the row or column header to select the rows or columns to be deleted, then choose **Delete** from the **Edit** menu. To delete cells *within* a row or column, select them with the mouse, then choose **Delete....** You will be asked whether to move the cells up or to the left. When deleting partial rows or columns, take care that other parts of the worksheet do not become misaligned.

USING **CUT, COPY** AND **PASTE**

Single cells, ranges of cells, or whole rows or columns can be copied or cut from the worksheet and inserted or pasted into other locations. In general the destination must be the same size as the copied or cut cells. First, select the cell or range of cells that you wish to copy or cut. Then choose **Copy** or **Cut** from the **Edit** menu, or press the Copy 🖺 or Cut ✂ toolbutton. A *marquee* (a dashed line) will appear around the selected cells and a copy of the cells is placed on the Clipboard. Next, select the destination range. You can now transfer the copy to the destination either by choosing **Paste** from the **Edit** menu, by pressing the

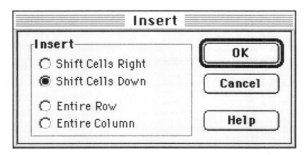

Figure 1-17. The Insert dialog box.

Paste toolbutton 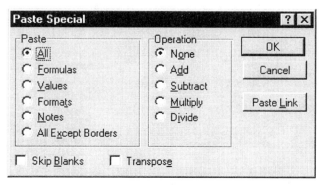 , or by pressing either the PASTE function key or the ENTER key.

> *Excel Tip.   Instead of selecting a destination range that is the same size as the copied or cut range, you can select a single cell, then **Paste**. The selected cell will be the upper left corner of the pasted range of cells.*

You can also **Copy** or **Cut** text in the formula bar and **Paste** it in a worksheet cell. Select the text to be copied or cut, then press the Copy or Cut key or choose the appropriate command from the Edit menu. Complete the operation by clicking the Enter button in the formula bar. Then **Paste** in the desired cell.

> *Excel Tip.   Use the ESCAPE key to cancel a **Copy** or **Cut** operation.*

USING **PASTE SPECIAL**...

When you **Copy** a cell and **Paste** it, Excel transfers the cell's contents, format and Note. You can choose to transfer only some of these cell attributes by using **Paste Special**. The Paste Special dialog box (Figure 1-18) permits you to paste only Formulas, Formats or Notes. In addition, you can convert formulas to constants by choosing Values.

> *Excel Tip. You can use the Paste Values or Paste Formats toolbuttons instead of the **Paste Special** menu command. See Chapter 9 for instructions on how to make these buttons available.*

If you press one of the Operation buttons in the Paste Special dialog box the value in the destination cell will be added to, subtracted from, multiplied by or divided by the value in the copied cell.

If the cell in either the source or destination contains a formula, then the formula will be enclosed in parentheses and joined to the contents of the destination cell by the arithmetic operator. Relative references in the source will be changed in the same way as in a normal **Paste** operation. You can also **Copy**

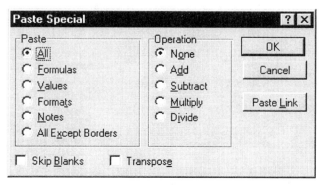

Figure 1-18. The Paste Special dialog box.

	A	B	C	D
1	pH	6.00	6.20	etc.
2	Abs	0.903	0.861	etc.

(A)

	F	G
1	pH	Abs
2	6.00	0.903
3	6.20	0.861
4	etc.	etc.

(B)

Figure 1-19. Rows and columns transposed. (A) Before using **Paste Special** (Transpose) (B) After using **Paste Special** (Transpose).

cells that contain formulas and press both the Values button and one of the Operation buttons to either Add, Subtract, Multiply or Divide.

If you check the Skip Blanks check box, only non-blank cells in the source will be pasted.

USING PASTE SPECIAL TO TRANSPOSE ROWS AND COLUMNS

If data in the source range are arranged in rows, you can convert the data to columnar format, or vice versa, as shown in Figure 1-19.

First **Copy** the cells, then select a cell or range in which you want the transposed values to be placed. Choose **Paste Special** and check the Transpose box, then press OK.

USING CLEAR...

When you choose **Clear...** from the Excel 4.0 **Edit** menu, the Clear dialog box is displayed. If you press the Formats button, for example, you can remove formats from selected cells. Pressing *either* the All button or the Formulas button will delete cell entries (Figure 1-20).

In Excel 5.0, choosing the **Clear** command displays a submenu with four options: All, Formats, Contents, Notes.

Excel Tip. The easiest way to clear a range of cells is to use the Erase tool ✐*. Use the Clear Formats button* $✐ *to remove only formatting from a cell. See Chapter 9 for instructions on how to make these buttons available.*

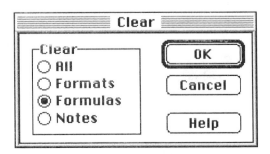

Figure 1-20. The Excel 4.0 for the Macintosh Clear dialog box.

USING **INSERT PASTE**...

If you **Copy** a cell or cells, then click on the **Edit** menu, the **Insert** menu selection becomes **Insert Paste.** You must select a cell range the same size as the copied one, or you'll get a "Copy and paste areas are different shapes" error message. If you forget the shape of the copied area, you can go back and **Copy** it again and make a point to remember it. If you decide that it's easier to insert new rows or columns first, then **Copy** and **Paste**, you'll have to clear the clipboard before **Insert Paste** becomes **Insert.** You can do this either (i) by selecting an empty cell and pressing DELETE, (ii) by choosing **Clear**... from the **Edit** menu, then exiting from the dialog box by pressing the Cancel button or, most conveniently, (iii) by pressing ESC.

TO COPY, CUT OR PASTE USING DRAG-AND-DROP EDITING

You can also **Copy**, **Cut** or **Paste** using Excel's "Drag-and-Drop" method. With this method you **Cut** and **Paste** a selection by using only the mouse pointer.

To Cut and Paste by Using Drag-and-Drop

1. Select the range of cells to be moved.

2. Position the mouse pointer over a border of the selection (top, bottom or side). The mouse pointer will change to an arrow.

3. Drag the selection toward the desired position. The border of the selection will be indicated as you drag it (Figure 1-21).

4. Position the selection as desired and release the mouse button.

To **Copy** the selection instead of cutting, hold down CONTROL (Windows) or OPTION (Macintosh) while dragging. A small plus sign will appear near the arrow pointer.

To **Insert** the selection, hold down the SHIFT key while dragging. The insertion point of the selection will be indicated by a horizontal or vertical bar as you drag (Figure 1-22).

Excel Tip. *To use this method, the Cell Drag and Drop box must be checked in the Workspace Options dialog box. Choose* **Options** *from the* **Tools** *menu and choose the Edit tab (Excel 5.0) or choose* **Workspace** *from the* **Options** *menu (Excel 4.0).*

USING **COLUMN WIDTH**... AND **ROW HEIGHT**...

When Excel creates a blank worksheet, all rows are the same default height, all columns the same width. You can change the width of columns, or the height of rows, to improve the appearance of a worksheet, or to eliminate wasted space so that you can get more information on a single page. You can also hide rows or columns by reducing their height or width to zero. The data they contain will be still there, but hidden.

	A	B	C	D
1	t, sec	A(obsd)		
2	0.2	0.0047		
3	0.6	0.0129		
4	1.0	0.0163		
5	1.4	0.0188		
6	1.8	0.0201		
7				
8				
9				
10				
11				
12				
13				
14				
15				

Figure 1-21. Cutting and Pasting cells using Drag-and-Drop editing.

	A	B	C	D
1	t, sec	A(obsd)		
2	0.2	0.0047		
3	0.6	0.0129		
4	1.0	0.0163		
5	1.4	0.0188		
6	1.8	0.0201		
7				
8				
9				
10				
11				

Figure 1-22. Inserting cells using Drag-and-Drop editing.

To change the width of a cell or column, choose **Column Width...** from the **Format** menu. You can enter the new width of the column; one unit corresponds to the width of one character of the current font.

You can also change the column width by using the mouse pointer. Place the cursor on the separator bar between column headings, on the right of the column whose width you want to change. The cursor changes to a double-headed arrow: ↔. Click the mouse button and drag to the right or left to change column width. The column width is displayed in the reference area of the formula bar as you drag the separator bar.

Excel Tip.   To adjust several columns at a time to the same width, select the columns by clicking and dragging in the headers. Now perform the column width adjustment with the mouse pointer on any of the selected columns. When you release the mouse button, all of the columns will have the adjusted width.

Column widths can also be changed by using the handy Best Fit check box in the Column Width dialog box. You can adjust the column width to fit a single selected cell or to be the best fit to the widest entry in a whole column.

Excel Tip.   You can also get a "best fit" simply by double-clicking on the row or column separator. To adjust several rows or columns at once, select them by clicking and dragging in the headers, then double-click on any row or column separator.

DUPLICATING VALUES OR FORMULAS IN A RANGE OF CELLS

Use Excel 4.0's **Fill Down** or **Fill Right** to duplicate a value or formula in one cell into a range of cells. Select the cell whose value you want to duplicate, and drag down or to the right to select the range of cells where you want the value duplicated. Then choose **Fill Right** or **Fill Down**.

In Excel 5.0, choosing the **Fill** command displays a submenu with the options **Down**, **Right**, **Up**, **Left** and **Across Worksheets**.

If the cell contains a number or a text label, the value will be duplicated in the rest of the cells. If the cell contains a formula, the formula will be copied into the selected cells, except that Microsoft Excel uses *relative referencing* when formulas are copied. For example, if cell A2 contains the formula =A1+1, and **Fill Down** is used to copy the formula into a range of cells below cell A2, the formula copied into cell A3 will be =A2+1, and so on.

Cell references are adjusted when you **Insert** or **Delete** rows or columns, too. If you **Insert** a new column to the left of column A in the preceding example, the formula in cell B2, which used to be cell A2, now reads=B1+1.

To use the **Across Worksheets** option, you must have selected multiple sheets beforehand. Select adjacent or non-adjacent sheets in the same way you select cells. To select an adjacent range of worksheets click on the tab of the first sheet, then hold down the SHIFT key and click on the tab of the last sheet in the range. To select non-adjacent sheets, hold down the CONTROL key (Windows) or the COMMAND key (Macintosh) while selecting. When you choose **Across Worksheets** from the submenu, the Fill Across Worksheets dialog box (Figure 1-23) will appear; you can choose to fill Contents, Formats or both.

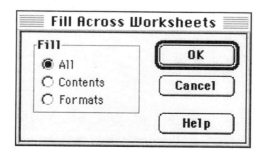

Figure 1-23. The Fill Across Worksheets dialog box.

ABSOLUTE, RELATIVE AND MIXED ADDRESSING

Sometimes relative referencing can cause problems. To keep the address of a cell fixed when you use the **Fill** commands, precede both its letter and number designation by dollar signs, e.g., $B$1. Thus the formula =A1+$B$1 in cell A2, when filled down, yields =A9+$B$1 in cell A10 (Figure 1-24). You will find this *absolute cell addressing* useful if you wish to use numerical constants in formulas.

Occasionally it is useful to use *mixed references*. A *relative reference* in a formula, such as A1, becomes B1, C1, etc., as you **Fill Down** a formula into cells below the original formula. An *absolute reference* such as $A$1 remains $A$1 as you **Fill Down**. A *mixed reference* is a reference such as A$1 or $A1; the row or the column designation, respectively, will remain constant when you **Fill Down** or **Fill Right**.

RELATIVE REFERENCING WHEN USING COPY AND CUT

If you **Copy** and **Paste** a formula, its references will be transferred using relative referencing. Thus, if you **Copy** the formula =A1+1 from cell A2 and **Paste** it in cell A10, the formula in cell A10 will be =A9+1. If you **Copy** the formula from cell A2 and **Paste** it in cell C2, the formula in cell C2 will be =C1+1. (This is probably not the formula you want.)

On the other hand, if you **Cut** the formula in cell A2 and **Paste** it anywhere in the worksheet, it will still be the formula =A1+1.

Thus the difference (with respect to cell references) between **Copy** and **Paste** and **Cut** and **Paste** is that **Cut** adjusts relative references so that they still refer to the original cells, while **Copy** does not adjust relative references, with the result that they refer to different cells.

The best way to copy a formula to a different row and column without altering relative references is to **Copy** it from the formula bar, click the Enter box, then **Paste** in the destination cell.

	A	B
1	1	0.001
2	=A1+$B$1	
3	=A2+$B$1	
4	=A3+$B$1	
5	=A4+$B$1	
6	=A5+$B$1	
7	=A6+$B$1	
8	=A7+$B$1	
9	=A8+$B$1	
10	=A9+$B$1	

	A	B
1	1	0.001
2	1.001	
3	1.002	
4	1.003	
5	1.004	
6	1.005	
7	1.006	
8	1.007	
9	1.008	
10	1.009	

Figure 1-24. Two views of the same worksheet, showing formulas (left) and values (right). The formula in cell **A2** has been Filled Down into **A3:A10**.

USING AUTOFILL TO FILL DOWN OR FILL RIGHT

Excel's AutoFill feature lets you **Fill Down** or **Fill Right** simply by using the mouse pointer. To use AutoFill, select a cell by clicking on it. You will see a small black square on the lower right corner of the selected cell. Position the mouse pointer exactly over the small black square (the *fill handle* or AutoFill handle). The mouse pointer becomes a small black cross. Click and drag in the usual way to select a range of cells. If the cell contains a formula, it will be duplicated to the rest of the range just as if you had used **Fill Down** or **Fill Right**. If the cell contains a number or a text label, the value will be duplicated in the rest of the cells. With AutoFill you can also **Fill Up** and **Fill Left**.

Excel Tip. To **Fill Down** *a value or formula to the same row as an adjacent column of values, select the source cell and double-click on the Fill Handle.*

USING AUTOFILL TO CREATE A SERIES

AutoFill provides an additional feature: you can use it to create a series. There are three ways to create a series. For example, to create the series 1, 2, 3, ... in column A, you can either:

- Enter the value 1 in cell A1, enter the formula =A1+1 in cell A2 and then use **Fill Down** to create the series. You can then use **Copy** and **Paste Special** (Values) to convert the formulas to values.

- Use **Series**... from the **Data** menu (Excel 4.0) or from the **Fill** submenu of the **Edit** menu (Excel 5.0). With **Series** (Figure 1-25) you enter the start value, the end value and the increment.

- Use AutoFill. This is by far the simplest, most convenient method. If you select a cell containing a number formatted as a date or time, or a text label containing a number, and use AutoFill to fill a range of cells, AutoFill creates a series using the selected cell as the starting value.

  For example, if you select a cell displaying `30-Jan` and use AutoFill to **Fill Right**, the series `30-Jan  31-Jan  1-Feb  2-Feb` will be created. The value of the series being entered in the active cell is displayed in the Reference box of the formula bar as you move the AutoFill handle. If you select two or more cells, AutoFill will create a series based on the cells you select, as shown in the second and third examples of Figure 1-26.

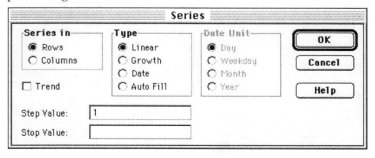

Figure 1-25. The Series dialog box.

	A
1	January

	B
1	1
2	3

	C
1	Compound 1
2	Experimental
3	Calculated
4	

After selecting as shown above, **Fill Down** using AutoFill to get:

	A
1	January
2	February
3	March
4	April
5	May
6	June
7	July
8	August
9	September
10	October
11	November
12	December

	B
1	1
2	3
3	5
4	7
5	9
6	11
7	13
8	15
9	17
10	19
11	21
12	23

	C
1	Compound 1
2	Experimental
3	Calculated
4	
5	Compound 2
6	Experimental
7	Calculated
8	
9	Compound 3
10	Experimental
11	Calculated
12	

Figure 1-26. Some examples of the use of AutoFill.

Excel Tip.   To use AutoFill to fill a cell range by copying the same cell contents to all the cells in the range (in other words, to prevent AutoFill from creating a series), hold down the CONTROL key as you position the black cross pointer over the Fill Handle. A small plus sign will appear to the right of the black cross pointer. Click and drag in the usual way to fill rather than create a series.

FORMATTING WORKSHEETS

You can use commands from the **Format** menu to control the way values are displayed and to control other features of the worksheet.

USING NUMBER... TO FORMAT NUMERICAL DATA

Choose **Cells...** from the **Format** menu and choose the Number tab (Excel 5.0) or choose **Number...** from the **Format** menu (Excel 4.0) to display numerical data in virtually any format. Excel has a wide range of built-in number formats. For example, selecting a cell and then choosing 0.00 from the format list box will display the value in the cell to two decimal places. You can also display values

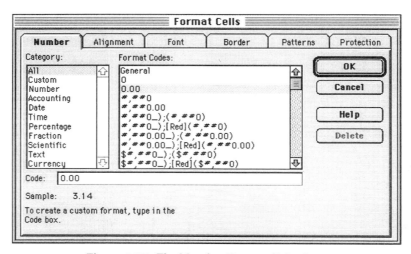

Figure 1-27. The Number Format dialog box.

as percent or in exponential notation, as indicated in Figure 1-27. For the meaning of the built-in number format code symbols, see Table 1-3 or "Number Format Codes" in Excel's On-Line Help.

CUSTOM NUMBER FORMATS

You can create your own custom number formats. If you want to display numbers to four decimal places, type 0.0000 in the input box at the bottom. The new format will be stored in the list of formats so that you can apply it to other cells.

Table 1-3 lists the formatting symbols you can use to create custom formats.

To format a column of social security numbers, enter the format 000-00-0000, which will display the number as e.g., 012-54-5842. To format a column of telephone numbers, use the custom format (###) ###-####. This will format a cell entry such as 6175523619 in the format (617) 552-3619. The format #.??????? was used to format the table of atomic weight values shown in Figure 1-28, so that they are aligned on the decimal. (Note that, since the format contains seven ? symbols and the atomic weight of Na has only six digits to the right of the decimal point, there is an additional space to the right of the number.)

You can create some fairly sophisticated number formats. For example, the format $#.0,, (dollar sign, number sign, decimal, zero, comma, comma) formats financial entries rounded to millions, with one decimal; the value 21180000 is displayed as $21.2.

Excel Tip. *You can use number formatting to add units to a number value. For example, the format #" g" appends the grams unit g to a number value; the value 50 is displayed as 50 g, as shown in Figure 1-29.*

Table 1-3. Number Formatting Symbols

#	Placeholder for digit.
0	Placeholder for digit. Displays an extra zero if the number has fewer digits than the number of zeros specified in the format.
?	Placeholder for digit. Same as #, except that numbers are aligned on the decimal point. Also used when formatting a number as a fraction.
,	Thousands separator (if used with #, 0 or ?). Used alone, it rounds and truncates to the thousands place (millions place if two commas are used, etc.)
%	Converts to percent .
E	Converts to scientific format. Use E- to include sign with negative exponents only, E+ to include sign with both positive and negative exponents.
/	Converts to a fraction. Usually used in the form # ??/?? or ##/##. The number of ? or # symbols determines the accuracy of the display.
"*text*"	Text characters can be included in a format by enclosing them in double quotes. Some special formatting characters, including $ - ( ) : comma and decimal point, do not need to be enclosed in quotes.
@	Text placeholder. If the cell contains a text entry, the text is displayed in the format where the @ symbol appears.
[RED]	Displays the characters in the cell in red. You can also use [BLUE], [GREEN], [YELLOW], etc.

As you saw earlier, Excel recognizes the format of data typed into cells. If you enter a value such as 3/23/93 and then choose **Number...** from the **Format** menu, you'll see that Excel has recognized the value as a date in the m/d/yy format. Now choose the format d-mmm-yy from the menu and observe the sample display at the bottom of the dialog box. The data is displayed as 23-Mar-93. Create your own custom date formats by using the year, month and day formats listed in Table 1-4. Day or month formats can have one-, two-, three- or four-letter formats; year formats can have either two- or four-letter formats. For example, the number format dddd, mmmm d, yyyy applied to a date entered as 8/3/38 will display Wednesday, August 3, 1938.

	A	B
1	H	1.00797
2	O	15.9994
3	Na	22.989768
4	S	32.066

Figure 1-28. Values aligned on the decimal point by using the ? formatting symbol.

Figure 1-29. Units added to a value by means of number formatting.

VARIABLE NUMBER FORMATS

Different number formats can be applied to positive, negative, zero and text values entered into a cell. A complete format consists of four sections, separated by semicolons, for positive, negative, zero and text values, respectively. If only one number format is specified, it applies to all values. If two number formats are specified, then the first one applies to positive numbers and zero, the second to negative numbers. The text format must be the fourth section in the number format. For example, the format $#,###;[Red]$#,### formats positive amounts in black, Excel's default color, and negative amounts in red.

CONDITIONAL NUMBER FORMATS

Conditional number formats can be created by using the syntax [*condition, value*] *format statement*. *Condition* is one of the symbols <, >, =, >=, <=, <>; *value* may be any number. Format statement may be any built-in or custom format. For example, the number format [>1] "Number too large" accepts any input less than 1 but otherwise issues an error message.

Several conditions may be combined using semicolons. The number format

[>999]#.##,,%;###" ppm"

displays the analytical results 110 and 21560 ppm as 110 ppm and 2.16%, respectively. The number format

[>999.9]#," L";[>0.5]#" mL";0.##" mL"

displays the volumes 25678, 12 and 0.37 as 26 L, 12 mL and 0.37 mL, respectively.

Table 1-4. Date Formatting Symbols

d	Displays the day as a number without leading zeros (1-31)
m	Displays the month as a number without leading zeros (1-12)
dd	Displays the day as a number with leading zeros (01-31)
mm	Displays the month as a number with leading zeros (01-12)
ddd	Displays the day as an abbreviation (Sun-Sat)
mmm	Displays the month as an abbreviation (Jan-Dec)
dddd	Displays the day as a full name (Sunday-Saturday)
mmmm	Displays the month as a full name (January-December)
yy	Displays the year as a two-digit number, e.g., 97
yyyy	Displays the year as a four-digit number, e.g., 1997

USING THE NUMBER FORMATTING TOOLBUTTONS

You can also format number values in cells by using the number formatting toolbuttons shown below.

+.0 / .00	Displays an additional decimal place.
.00 / +.0	Increases the number of decimal places.
%	Formats the number in percent style, with no decimal places.
$	Formats the number in currency style, with 2 decimal places.
,	Formats a number with commas and 2 decimal places.

Excel Tip.   There isn't a toolbutton to format number values in exponential format.   Apply exponential format conveniently by using the shortcut key sequence CONTROL+SHIFT+(^).

FORMATTING NUMBERS USING "PRECISION AS DISPLAYED"

To permanently change *all values* stored on a worksheet to their displayed values, use the Precision as Displayed option.   Once this command has been invoked, you can't restore the original values.

To apply Precision as Displayed, choose **Options** from the **Tools** menu and choose the Calculation tab (Excel 5.0), or choose **Calculation...** from the **Options** menu (Excel 4.0).   Check the Precision as Displayed box, (shown in Figure 15-10) then press OK.   Because this is an irreversible change, Excel asks you to confirm the change,

To change only a *selected range of values* to "precision as displayed", use the FIXED worksheet function (see "Text Functions" in Chapter 3).

USING **ALIGNMENT...**

The **Alignment** command provides a number of formatting options for the alignment of values in cells,.   Choose **Cells...** from the **Format** menu and choose the Alignment tab (Excel 5.0) or choose **Alignment...** from the **Format** menu (Excel 4.0).   To use the Center Across Selection option, select the columns across which you want the text to be centered, as indicated in Figure 1-30.   (The text must be in the leftmost column of the selected cells.)   There are option buttons for both horizontal and vertical alignment (Figure 1-31).   The Vertical orientation options are useful if you want to add a text label to a narrow data column.

	A	B	C	D	E
1	Centered across selected columns				
2					

Figure 1-30. Using Center Across Selected Columns.

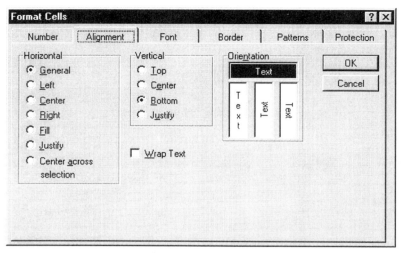

Figure 1-31 The Alignment dialog box.

Alternatively, for the most common horizontal alignment options you can use the alignment toolbuttons [≡][≡][≡][+a+] to format cells left, centered, right or centered across selected cells.

You can also format text entries in cells so that the text wraps and is displayed in more than one line (Figure 1-32). In Excel 5.0, text can be both aligned vertically and wrapped; in Excel 4.0, Wrap Text can be applied only to horizontal text entries.

When you select a text box the Alignment dialog box has a slightly different form. The Automatic Size check box, when selected, fits the borders to the text.

USING **FONT**...

The Font tab in the Format Cells dialog box (Excel 5.0) allows you to format individual cells in any of the installed fonts (Figure 1-33). In addition, you can format individual characters in various font styles or sizes, or as strikethrough, superscript or subscript characters. This is a big improvement over Excel 4.0, where the **Font** command in the **Format** menu allows only the complete contents of a cell to be formatted. By using Font you can include Greek letters, for example, in text labels.

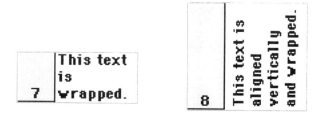

Figure 1-32. Examples of using Wrap Text

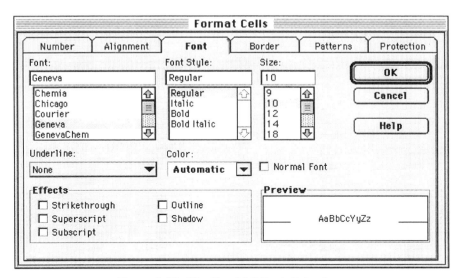

Figure 1-33. The Excel 5.0 Font dialog box.

THE ALTERNATE CHARACTER SET

There is another way to enter some useful characters. From the keyboard, you can enter symbols that are in the so-called *alternate character set*. In Excel for the Macintosh these characters are obtained by pressing OPTION+(key) or OPTION+SHIFT+(key). The characters produced may be different for each different font. The ones shown in Figure 1-34 are obtained with the Geneva font, the default font for Excel for the Macintosh.

In Excel for Windows, the characters are obtained by holding down the ALT key and typing the 4-digit ASCII code for the character, *using the numeric keypad*. The range of useful characters obtainable in this way is rather limited. Table 1-5 gives some characters for Macintosh; for characters obtained using Arial, one standard font for Excel 5.0 for Windows, you would use the ASCII codes shown in parentheses in Table 1-5.

If you use a different font, you'll have to experiment to see what alternate characters are produced.

ENTERING SUBSCRIPTS AND SUPERSCRIPTS IN EXCEL 5.0

Excel 5.0 introduced the capability of entering subscripts and superscripts in text labels, by means of choosing the Font tab in the Format Cells dialog box and checking the Subscript or Superscript check box (see Figure 1-33). However, to subscript a number in a chemical formula requires the following sequence of actions after the text with numbers and letters has been typed in the standard

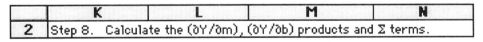

Figure 1-34. Special characters in Excel for the Macintosh by using the OPTION key.

Table 1-5. Some Useful Alternate Characters*

ß	OPTION+s	(0223)		∂	OPTION+d	
Δ	OPTION+j			μ	OPTION+m	(0181)
π	OPTION+p			Π	OPTION+P	
Σ	OPTION+w			Ω	OPTION+z	
÷	OPTION+/	(0247)		±	OPTION+(plus)	(0177)
≈	OPTION+x			≠	OPTION+=	
«	OPTION+\	(0171)		»	OPTION+SHIFT+\	(0187)
≤	OPTION+(,)			≥	OPTION+(.)	
√	OPTION+v			∫	OPTION+b	
∞	OPTION+5			Å	OPTION+A	(0197)
°	OPTION+SHIFT+8	(0176)				

*For Macintosh. Four-digit ASCII codes in parentheses are for Windows.

font: (i) double-click on the cell to actuate the Edit Directly in Cell mode, (ii) select the character(s) to be subscripted, (iii) choose **Cells...** from the **Format** menu, (iv) choose the Subscript option from the Format Cells dialog box, (v) click the OK button. Formatting the numbers in, for example, the text C12H22O11 requires that operations (ii) through (v) be performed three times. Chapter 7 illustrates a macro that does this formatting automatically. The macro can be assigned to a custom button on Excel's standard toolbar; after typing a chemical formula, clicking the button automatically formats the text as a chemical formula. Chapter 9 provides instructions on how to assign a macro to a toolbutton. A more advanced version of this macro is provided on the diskette that accompanies this book. The advanced version formats text in a cell, a range of cells, text in a text box or text in a chart label; it is for Excel 5.0 (Windows or Macintosh).

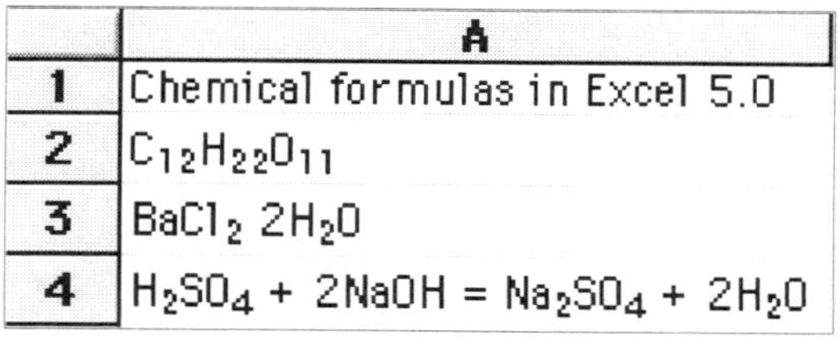

	A
1	Chemical formulas in Excel 5.0
2	$C_{12}H_{22}O_{11}$
3	$BaCl_2\ 2H_2O$
4	$H_2SO_4 + 2NaOH = Na_2SO_4 + 2H_2O$

Figure 1-35. Subscripts in Excel 5.0.

	A
1	Chemical formulas in Excel 4.0
2	$C_{12}H_{22}O_{11}$
3	$[Co(NH_3)_5Br]^{2+}$
4	$Ni(ClO_4)_2 \cdot 6H_2O$

Figure 1-36. Sub- and superscripts in Excel 4.0 by using a custom font.

ENTERING SUBSCRIPTS AND SUPERSCRIPTS IN EXCEL 4.0 FOR MACINTOSH

Excel 4.0 makes no provision for subscripts or superscripts — a real drawback for chemists. Unless you're satisfied with typing e.g., H2SO4 in your worksheets and charts, you've probably wondered how to produce text with subscripted numbers in your worksheets.

One way to produce subscripts or superscripts is to employ a custom font designed for the purpose. You can create a custom font using a software program such as Fontographer (Macromedia, Inc., 600 E. Townsend St., San Franciso CA 94103). I created a font based on the Geneva font, the default font for Excel for the Macintosh. It provides sub- and superscript digits and plus and minus signs. Subscript numbers can be entered by pressing OPTION+(number key), superscript numbers by pressing OPTION+SHIFT+(number key). The centered dot character is produced by pressing OPTION+SHIFT+(period). (Note that if you adopt thisapproach you will no longer be able to enter the characters ∞, • or ± from the alternate character set.)

Install the custom font in the way that is appropriate for your system. Then, to produce chemical formulas or other text with subscript or superscript numbers, simply choose **Font** from the **Format** menu before typing, and choose the custom font. Figure 1-36 shows some examples of text labels produced by using such a font.

USING BORDER... AND PATTERNS...

The Border tab in the Format Cells dialog box allows you to place a border around one or more sides of a selected cell or range. This is useful if you want to emphasize comments, instructions or values. The Patterns tab is used to change the background color or pattern of cells.

The Border option is often used to underline headings, or in a sheet in which the gridlines have been removed, to create a custom form. Figure 1-37 illustrates a portion of a template sheet produced in this way. To remove gridlines, choose **Options** from the **Tools** menu and choose the View tab (Excel 5.0), or choose **Display...** from the **Options** menu (Excel 4.0), then uncheck the Gridlines checkbox.

The built-in template sheets provided with Excel 7 ("Spreadsheet Solutions") are good examples of the use of Border and Patterns to create custom forms.

BOSTON COLLEGE				DATE		6/5/1996	
OFFICE OF THE UNIVERSITY REGISTRAR							
CHESTNUT HILL, MASSACHUSETTS 02167				ACADEMIC YEAR		1996	
				SEMESTER		96S	
COURSE			CH 222 01 INTRO/INORGANIC CHEM				
INSTRUCTOR			BILLO, E JOSEPH				
MEETING PLACE							

STUDENT'S NAME	ID NUMBER	SCHOOL	CLASS	MAJOR	CREDIT	GRADE

Figure 1-37. Using Border to create a custom report form.

ADDING "RICH" TEXT TO SPREADSHEETS IN EXCEL 4.0

Excel 5.0 provides the capability of formatting individual words or characters in cells, via the Edit Directly in Cells feature. You can produce the same effect in Excel 4.0 by creating a text box and positioning it to coincide exactly with one or more cells. Here's how to achieve the results shown in Figure 1-38.

Adding "Rich" Text to Spreadsheets in Excel 4.0

1. Create a text box by clicking on the Text Box toolbutton and moving the crosshair cursor to place the text box anywhere on the worksheet that is convenient.

2. Type your text in the text box. You can use Bold, Italic, Underline and different fonts and font sizes within the text box.

3. Click on the text box to select it. Size it to fit and align with worksheet cells by holding down CONTROL while you move the handles.

4. Click on the text box to select it. Choose **Patterns...** from the **Format** menu. In the Patterns dialog box, choose Style = Solid line, Color = Gray, Weight = lightest, to duplicate the appearance of the gridlines.

Data on this worksheet from E. J. Billo *et al.*, *Inorg. Chim. Acta* **1993**, *210*, 71.

Figure 1-38. "Rich text" in Excel 4.0 by using a text box.

PROTECTING DATA IN WORKSHEETS

Sometimes you'll want to protect data in a worksheet, either from changes by other users, or changes entered accidentally by yourself.

USING CELL PROTECTION

There are two ways you can protect the contents of an Excel document. You can lock cells within a document so that they cannot be changed, or you can hide cells so that it is impossible to view the cell contents. Use the Protection tab of the **Cells...** command in the **Format** menu (Excel 5.0) or the **Cell Protection...** command in the **Format** menu (Excel 4.0) to set and complete this protection status.

The process for doing this is somewhat complicated. First you select cells to be locked and/or hidden, and set their status using the **Protection** command. Then you put the status into effect by choosing the **Protect Sheet...** command from the **Tools** menu (Excel 5.0) or the **Protect Document** command from the **Options** menu (Excel 4.0).

Before you begin, it's important to know that when a new worksheet is opened, the status of *all* cells in the document is "Locked". Thus if you select a range of cells to be locked, and set the status using **Protect Sheet...**, you will find that all cells in the document are locked.

To lock only a limited range of cells in a document (as you will most often want to do), you must first set the status of all the cells in the document to "Unlocked" and then select the range of cells that you want to be locked.

To Lock a Range of Cells in a New Document

1.   Select all cells in the document by clicking on the row/column heading box in the upper left corner of the worksheet.

2.   Choose **Protection** from the **Format** menu, uncheck the Locked option in the Cell Protection dialog box, and press the OK button.

3.   Select the range of cells that you want to protect. Choose **Protection** again and check the Locked option.

4.   Choose **Protect Sheet...** from the **Tools** menu (Excel 5.0) or **Protect Document** from the **Options** menu (Excel 4.0). The Protect Sheet dialog box will appear. You can enter a password if you wish. If you merely want to prevent yourself from making accidental changes, no password is necessary. If you want to protect the document from changes by others, you need a password; make sure that you will be able to retrieve it when you need it.

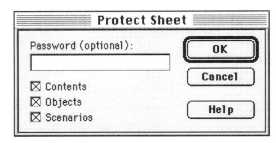

Figure 1-39. The Excel 5.0 Protect Sheet with Password dialog box.

CONTROLLING THE WAY DOCUMENTS ARE DISPLAYED

Use the **Window** menu to switch between one Excel document and another. All open documents are listed at the bottom of the **Window** menu; the active document is indicated with a check mark.

Use **Hide** to hide a worksheet or macro sheet. Most commonly you'll use **Hide/Unhide** with macro sheets. A macro sheet is still "active" even when it's hidden.

VIEWING SEVERAL WORKSHEETS AT THE SAME TIME

Although only one worksheet can be the active document, Excel provides a number of ways to examine data in several different worksheets, or different areas of the same worksheet, at the same time.

USING ARRANGE...

If you have more than one document open, you can view them all simultaneously in a number of ways. One way is to re-size and move the documents so that the desired part of each can be seen at the same time. Another way is to use **Arrange...**. When you choose **Arrange...** from the **Window** menu, Excel displays the Arrange Windows dialog box (Figure 1-40).

If you have two documents open, you can arrange them horizontally (one above the other) or vertically (side by side). (The active document will be on top or on the left, respectively.) If you have created a separate chart document from

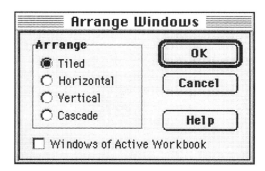

Figure 1-40. The Arrange Windows dialog box.

data in a worksheet, **Arrange**... provides a convenient way to work with a worksheet and observe changes in the associated chart. With **Arrange...**, chart documents are reduced in size so that the whole chart appears in the window; worksheet documents are not reduced in size. Figure 1-41 illustrates a worksheet/chart combination displayed using the **Arrange** (Vertical) option.

With three open documents, the **Tiled** option arranges the documents with the active sheet occupying the left half of the screen; the other two sheets each occupy one-quarter of the screen, one above the other. With four documents **Tiled**, each occupies one-quarter of the screen. Click on any document to make it the active sheet. Double-click anywhere on the solid border between the windows to undo the arrangement.

DIFFERENT VIEWS OF THE SAME WORKSHEET

As your worksheets get larger and more complicated, it becomes impossible to view all of a single worksheet at one time, or even all cells in one row or column at one time. Excel provides several convenient ways to display separate portions of a single worksheet on the screen at the same time, so that you can view one part while entering or changing data in another part.

USING **NEW WINDOW**

When you choose **New Window**, a second window of the active document is opened. You can then re-size and move the windows so that the desired parts of the worksheet can be seen at the same time. This is useful if you want to **Cut** or **Copy** several cell ranges and then **Paste** them into another area of a worksheet, but the two areas of the worksheet are far apart.

Excel Tip.   You can set different display options for the two windows. Display values in one window and formulas in another to see the effect of changes.

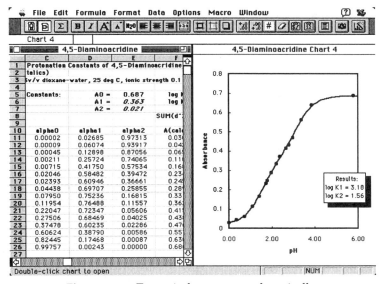

Figure 1-41. Two windows arranged vertically .

USING SPLIT SCREENS

To use the **Split** feature to split a document window horizontally into two windows, select an entire row as if you were going to insert a row. Then choose **Split** from the **Window** menu. This creates a split in the window, above the selected row, with each part of the window displaying the active document. Each part of the document now has its own scroll bar, and you can scroll one part of the document while the other part remains fixed. A vertical split is accomplished in the same way.

You can also split the document window by placing the mouse pointer on the black rectangle at the left end of the horizontal Scroll Bar or at the top of the vertical Scroll Bar, then click and drag the rectangle.

The document window can be split both horizontally and vertically, by selecting a single worksheet cell, then choosing **Split**, as illustrated in Figure 1-42.

To remove splits, choose **Remove Split** from the **Window** menu, or slide the split box back to its original position.

Excel Tip. To remove a split from a window, it's not necessary to slide the split box back to its original position at the top or left hand side of the scroll bar. Just place the pointer on the split box and double-click.

USING FREEZE PANES

Freeze Panes can be used to create a similar split document window, but the upper or left part of the window is fixed and cannot be scrolled. Split panes are useful to display fixed row or column headings (or both) while scrolling through the rest of the worksheet.

To use the **Freeze Panes** feature to split a document window horizontally into two windows, select an entire row as if you were going to insert a row. Then choose **Freeze Panes** from the **Window** menu.

	A	E	F	G	H	I	J	K
1	**Grade Sheet**							
2		**Hour Exams**						
3	**Name**	**#1**	**#2**	**#3**	**Oral report**	**Paper**	**Final Exam**	**Total**
4								
7	FERREIRO, Kathy	24	16	32	20	45	52	63.0
8	GANGE, Eric	28	13	43	20	40	51	65.0
9	GREALEY, John	22	14	40	17.5	40	56	63.2
10	HAPPERSBACH, Bill	28	12	30	17.5	45	59	63.8
11	HOGAN, Derek	37	17	37	20	50	60	73.7
12	LAROZI, Patrick	27	12	38	20	45	58	66.7

Figure 1-42. A document with a two-way split.

SAVING DIFFERENT VIEWS OF A WORKSHEET

Use the **View Manager...** command from the **View** menu to save different display or print settings of a worksheet. You give each different view a name, e.g., global, summary. The settings and their associated names are stored in the worksheet. You can then choose any of the views by name, in order to display or print the information.

The View Manager is an add-in, a separate software package. To save memory, it may not automatically be opened whenever you start Excel. You can open an add-in by choosing **Add-Ins...** from the **Tools** menu (Excel 5.0) or from the **Options** menu (Excel 4.0), then checking the appropriate box for the add-ins you desire. You may have to run the Microsoft Excel Setup program to install the add-in on your hard disk.

USEFUL REFERENCES

Microsoft Excel 5 for the Macintosh Step by Step, Microsoft Press, Redmond, WA, 1994.

John Walkenbach, *Excel for Windows 95 Bible*, IDG Books Worldwide, San Mateo, CA, 1995.

2

CREATING CHARTS:
AN INTRODUCTION

Nothing can be as helpful as displaying data in graphical form. With Excel you can quickly and easily create a chart, simply by selecting the data to be plotted and choosing the way you want the data to be displayed; Excel does the rest. In this chapter you'll learn the basics of creating Excel charts.

ONLY ONE CHART TYPE IS USEFUL FOR CHEMISTS

Excel 5.0 provides a gallery of 15 chart types — bar charts, column charts, line charts or pie charts, among others. Since Excel originated as a financial tool, most of the chart types are those that are useful for displaying financial and related information — a bar chart to show sales figures for each business quarter, a line chart to show stock values each day over a one-month period, etc. Only one kind of chart, the X-Y or scatter plot, is of general usefulness for displaying scientific data. It is the only one in which numeric values are used along both axes. All other charts plot the numeric y values vs. equal increments on the X-axis and use the x values only for labels (called *categories* by Excel). The line chart, which appears to be a form of X-Y chart, is actually only a bar chart of y values with the y values connected by straight lines.

CREATING A CHART

There are two ways to create a chart: either as a separate document (Excel 4.0) or separate chart sheet (Excel 5.0), or as a chart *embedded* in a worksheet, so that you can see both the data and the chart at the same time. This feature is useful if you want to see how a curve changes as you change its parameters. As you change the values in worksheet cells, the chart will update automatically. Also, it is convenient to be able to save a spreadsheet and one or more charts in a single document.

CREATING AN EMBEDDED CHART USING THE CHARTWIZARD

You can use the ChartWizard tool ▨ to create an embedded chart. Later, if you wish, the embedded chart can be made a separate chart document. To use the ChartWizard, first select the data series to be plotted. The data can be in rows or columns. If the two data series are not adjacent, hold down the

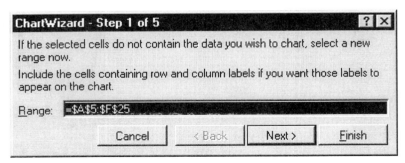

Figure 2-1. The first ChartWizard dialog box.

CONTROL key (Windows) or COMMAND key (Macintosh) while selecting the separated rows or columns of data. Then click the ChartWizard tool. The mouse pointer becomes a crosshair; in Excel 5.0 it also displays the characteristic miniature chart bars seen on the ChartWizard button. Click and drag the crosshair pointer to outline the area on the spreadsheet where you want the chart to be located. When you release the mouse button, the first of a series of five dialog boxes will appear.

> *Excel Tip. If you hold down the SHIFT key while outlining the chart area with the mouse pointer, the embedded chart will be a square.*

The first ChartWizard dialog box (Figure 2-1) lets you check that the correct data has been selected. The selected range appears in the dialog box and can also be seen on the sheet. You can change the data range by typing in the Range text box or by selecting with the mouse, or you can click the Cancel button or proceed to the next step with the Next button. You can also go directly to the finished chart by clicking the Finish button; this capability will be useful only if you want to create a chart in the default format (usually a column chart, but you can change the default format).

The second and third dialog boxes allow you to select the desired chart format. The second Excel 5.0 dialog box (Figure 2-2) corresponds to the **Gallery** menu in the **Chart** menu bar (Excel 4.0). Click on the desired chart type with the mouse button or press the indicated key for the selection. (You'll be creating an X-Y chart, so press S for Scatter plot.) The second and subsequent dialog boxes also provide the Cancel and Back buttons, in case you want to change what you've already selected.

The third dialog box (Figure 2-3) provides a range of possibilities for each chart type. For the X-Y chart type, there are six chart formats: format #1 uses data markers with no connecting line, format #2 uses both data markers and a connecting line, format #3 provides gridlines, format #4 is a semi-log plot, format #5 is a log-log plot and format #6 interpolates between data points to produce a smooth curve. Format #2 is a common selection, but a chart created in one chart type or format can easily be converted into another. Select the chart format by clicking on it with the mouse or by pressing the appropriate number key.

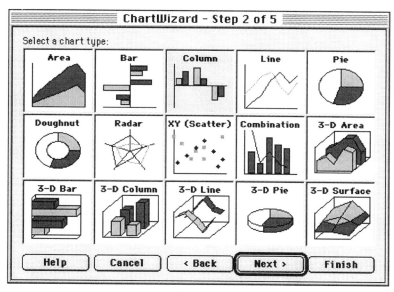

Figure 2-2. The Gallery of chart types displayed by Excel 5.0's ChartWizard dialog box.

The fourth dialog box (Figure 2-4) shows you what the completed chart will look like. At this point you can change the way Excel assigned the chart data, etc. Clicking the Next button takes you to the fifth and final dialog box (Figure 2-5), where a title, chart axis labels, or a legend can be added. (All of these, and many other changes, can be added or removed after the chart has been completed.)

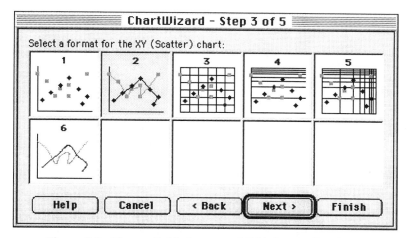

Figure 2-3. The available formats of X-Y chart in Excel 5.0.

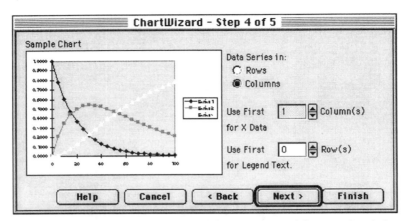

Figure 2-4. The fourth dialog box provides a preview of the completed chart.

When you click on the final OK button, the completed chart appears on the worksheet in the area that you selected. To move the chart, place the mouse pointer anywhere inside the chart and click and drag the chart to its new location. To change the size of the chart, first select it by clicking on it. When selected, an embedded chart has black "handles" at the sides and corners. Use these to change the size of the chart. Click on one of the handles, then drag to move that edge or corner inward or outward.

CREATING A CHART AS A SEPARATE CHART SHEET (EXCEL 5.0)

To create a chart as a separate sheet in a workbook, first select the chart data. Then choose **Chart** from the **Insert** menu and choose **As New Sheet** from the submenu. The ChartWizard dialog box will be displayed and you then proceed as above.

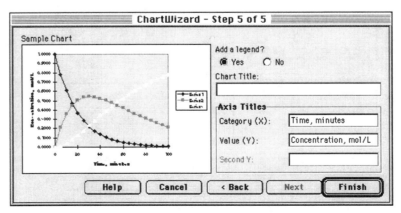

Figure 2-5. The final ChartWizard dialog box.

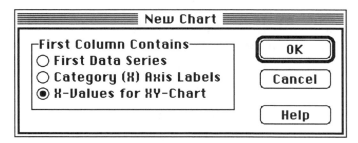

Figure 2-7. The Excel 4.0 for the Macintosh New Chart dialog box.

COPYING AN EMBEDDED CHART TO A SEPARATE CHART SHEET (EXCEL 5.0)

Insert a separate chart sheet in the workbook by choosing **Chart** from the **Insert** menu and then choosing **As New Sheet** from the submenu. You'll see a "nonsense" chart, depending on what cells had been selected on the worksheet. Select the chart by clicking near the border (black "handles" will appear) and delete it, leaving a completely blank chart sheet. Now select the worksheet containing the embedded chart, double-click on the embedded chart to activate it, and **Copy**. Select the chart sheet and **Paste**.

CREATING A CHART AS A SEPARATE DOCUMENT (EXCEL 4.0)

To create a chart as a separate Excel 4.0 document, first select the chart data. Then choose **New** from the **File** menu, choose Chart in the dialog box and click the OK button, as in Figure 2-6. Then, in the New Chart dialog box, choose "X-Values for XY-Chart" (Figure 2-7). When you click the OK button, your X-Y chart will be displayed as a separate document, which can be named and saved in the usual manner.

COPYING A CHART TO A WORKSHEET (EXCEL 4.0)

To create an embedded chart from a separate chart document, select the chart area by clicking on it, or choose **Select Chart** from the **Chart** menu. (To select the chart with the mouse pointer, click in the area between the chart border and the edge of the document window.) The selected area will be

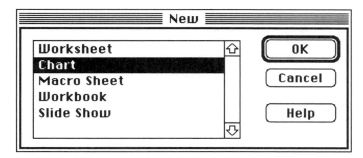

Figure 2-6. The Excel 4.0 for the Macintosh **New** dialog box.

indicated by the appearance of "handles". Choose **Copy** from the **Edit** menu. A marquee will appear around the chart. Now activate the worksheet window. Select a worksheet cell in the area where you want the chart to be located, and **Paste**; the selected cell determines the position of the upper-left corner of the embedded chart. You can now move or re-size the chart.

FORMATTING CHARTS: AN INTRODUCTION

Excel scales and formats your chart automatically. It does a good job, but usually there is plenty of room for improvement. Excel provides a wide range of tools for modifying a chart. A few of these are discussed in this chapter; for more details, see Chapter 11 ("Advanced Charting Techniques").

If the chart is an embedded chart, you must "activate" it before it can be modified. To activate an embedded chart, double-click on it. An embedded chart in Excel 4.0 becomes a separate chart document, i.e., it is displayed in a separate window; in Excel 5.0, the chart remains displayed on the worksheet, but with a wide patterned border.

To add text to the chart, choose **Titles...** from the **Insert** menu (Excel 5.0) or **Attach Text** from the **Chart** menu (Excel 4.0). The resulting dialog box (Figure 2-8) lets you select the chart elements for which to provide titles: the chart title or a text label for the X- or Y-axis. When you choose the Chart Title option, for example, and then click the OK button, a text box with black "handles" will appear on the chart. You can then type the desired title; when you press Return or click the Enter button, the title text will appear on the chart. Excel wraps the text if it is too long to fit on one line.

Occasionally you will need to change the axis scales. Excel always creates axis scales that include zero, and this may not always be suitable, as in Figure 2-9, where the wavelength data ranged from 350 nm to 820 nm.

Figure 2-8. The Titles dialog box.

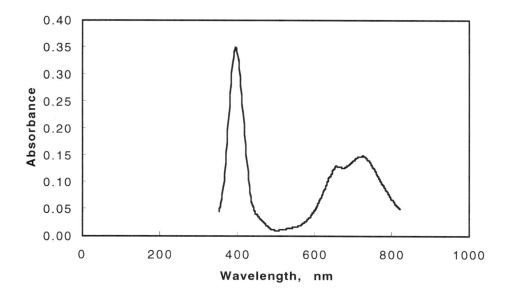

Figure 2-9. An X-Y chart as originally created by Excel.

To modify the X- or Y-axis scale, first select the axis by clicking on it. White "handles" will appear at the ends of the selected axis. Then choose **Selected Axis...** from the **Format** menu and choose the Scale tab (Excel 5.0), or choose **Scale** from the **Format** menu (Excel 4.0). Enter new values for Maximum and Minimum, as in Figure 2-10.

Format Axis				
Patterns	**Scale**	Font	Number	Alignment

Value (X) Axis Scale
Auto
☒ Minimum: `0`   (**OK**)
☒ Maximum: `1000`   (Cancel)
☒ Major Unit: `500`
☒ Minor Unit: `100`   (Help)
☒ Value (Y) Axis
   Crosses At: `0`

☐ Logarithmic Scale
☐ Values in Reverse Order
☐ Value (Y) Axis Crosses at Maximum Value

Figure 2-10. The Format Axis dialog box.

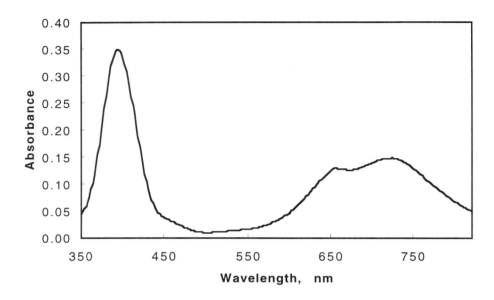

Figure 2-11. The chart after adjustment of the scale of the X-axis.

Figure 2-11 shows the same chart after the horizontal axis scale has been adjusted.

PART II

ADVANCED SPREADSHEET TOPICS

3

CREATING
ADVANCED WORKSHEET FORMULAS

This chapter shows you how to use Excel's wide range of worksheet functions to construct sophisticated worksheet formulas. In addition, you'll learn how to use *named references* in formulas, which makes constructing complicated formulas an easier and more error-free task, and how to use arrays in worksheet formulas. As well, you'll learn techniques of formula editing and worksheet troubleshooting.

THE ELEMENTS OF A WORKSHEET FORMULA

A worksheet formula consists of *operators* and *operands*. Worksheet operators are either arithmetic, text, comparison or reference operators. Operands may be values, references, names or functions.

OPERATORS

The *arithmetic operators* for addition, subtraction, multiplication, division and exponentiation are familiar ones and have already been mentioned in Chapter 1. In addition to these operators there are text, reference and comparison operators.

There is only one *text operator*: the & (ampersand) symbol. It is used to concatenate text, or text and variables. For example, if cell G256 contains the value 1995 and cell A1 contains the formula ="Chemical Inventory for "&G256, then cell A1 displays Chemical Inventory for 1995.

The *comparison operators* compare two values and produce a logical result, either TRUE or FALSE. The logical operators are the following: = (equal to), > (greater than), < less than), >= (greater than or equal to), <= (less than or equal to) and <> (not equal to). Note that >=, <= and <> must be typed as shown here, or Excel will not recognize them as operators and will produce an "Error in formula" message. In worksheet formulas, FALSE is equivalent to zero and TRUE to any non-zero value.

There are three *reference operators*: the range operator (colon), the union operator (comma) and the intersection operator (space). The *range operator* produces a reference that includes all the cells between and including the two references, e.g., G3:L3. The *union operator* produces a reference that includes the two references, e.g., G3,L3. More than one union operator can be used in a single reference. The *intersection operator* produces a reference to the cells common to

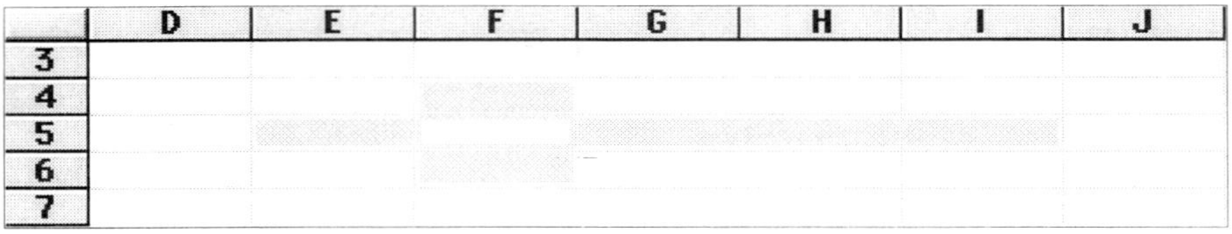

Figure 3-1. A single cell reference ($F$5) produced by the intersection operator.

two references. For example, the reference $F$4:$F$6 $E$5:$I$5 refers to cell $F$5, as illustrated in Figure 3-1. The intersection operator is particularly useful when used with *named references* (see "Using Create Names" later in this chapter).

ABSOLUTE, RELATIVE AND MIXED REFERENCES

Cell references can be absolute, relative or mixed. A *relative reference* in a formula, such as A1, becomes B1, C1, and so on, as you **Fill Down** a formula into cells below the original formula. An *absolute reference* such as $A$1 remains $A$1 as you **Fill Down**. Use *mixed reference* style if you want to control how the reference is changed as a formula is duplicated using **Fill Down** or **Fill Right**. A mixed reference is a reference such as A$1 or $A1; in the first case the row designation will remain constant when you **Fill Down** or **Fill Right**, in the second case the column designation will remain constant. For example, the reference $D2 becomes $D3 when duplicated to the cell in the next row down, but remains $D2 when duplicated to the cell in the next column to the right.

You can type the reference in the form you want, or you can use the **Reference** command in the **Formula** menu (Excel 4.0 only). To change the reference style, first select the reference, then choose **Reference**. The reference format is changed in the sequence $A$1, A$1, $A1, A1. The **Reference** command appears in the **Formula** menu only after you begin typing or editing in the formula bar (only when the Cancel and Enter boxes appear), and it is dimmed unless you type an equal sign.

> *Excel Tip. You can also use COMMAND+T (Macintosh) or F4 (Windows) to toggle between cell reference types. Select the cell reference by double-clicking on it in the formula bar (or just put the insertion point cursor anywhere in the reference), then press F4 to cycle through the formats. If you are typing a formula, you can use COMMAND+T after typing the cell reference; Excel converts the reference to the immediate left of the insertion point.*

ENTERING WORKSHEET FORMULAS

There are a number of useful techniques that you can use to enter worksheet formulas or to edit formulas that you have entered in worksheet cells.

Type formulas in lowercase to detect typographical errors. When you enter a formula, Excel converts functions and cell references to uppercase. If you type the formula =offset(d1,5,1), Excel will convert it to =OFFSET($D$1,5,1) when you enter the formula, but if you type "ofsett" instead of "offset" Excel won't recognize it and will display the error message #NAME?. When you examine the formula, you'll easily see that the incorrect function name remained in lowercase letters.

If complicated formulas contain terms identical to those used in other cells, you can **Copy** that part of the formula and **Paste** it into the new formula. Here's one method: before beginning to type the new formula, select the cell containing the formula you want to copy. In the formula bar, highlight the part of the formula you want to copy, **Copy** it, then click the Check box in the formula bar. Now select the cell where you want to type the new formula and **Paste** (or begin to type the new formula until you reach the part that you've copied, then **Paste** in the formula fragment).

> *Excel Tip.   To select (highlight) a word or reference for editing in the formula bar, double-click on it.*

USING NAMES INSTEAD OF REFERENCES

A name can be applied to a cell, a range of cells, a value or a formula. Most often you'll use names for cell references. Using names makes it easier to create and to decipher complex formulas. For example, the formula

=pKa+LOG(base/acid)

is easier to understand than the formula

=$B$1+LOG($B$2/$B$3)

In Excel 5.0, the **Name** submenu of the **Insert** menu contains several commands for working with names: **Define...**, **Paste...**, **Create...** and **Apply....** You will probably find **Define...** and **Create...** most useful. Use **Define...** to assign a single name to a cell or range; use **Create...** to create names for several cells or ranges, based on row and/or column titles. In Excel 4.0, the analogous commands (**Define Name...**, **Create Names...**, **Apply Names...** and **Paste Name...**) are located in the **Formula** menu.

USING DEFINE NAME

To assign a name to a cell reference, first select the cell or range. Then choose **Define...** from the **Name** submenu (Excel 5.0) or **Define Name...** from the **Formula** menu (Excel 4.0) to display the Define Name dialog box (Figure 3-2). The absolute reference of the selected cell will appear in the Refers To box. Excel will usually propose a name in the Name box, using text from the selected cell or from the cell immediately above or to the left of the selected cell. Edit the name if necessary, then press OK.

The first character of a name must be a letter. Subsequent characters can be letters or numbers or the period or underline character. Spaces are not allowed (the space character is the intersection operator). Use a period or underline character instead, or dispense with spaces altogether. Excel will substitute an underline character for a space in any name that it proposes based on text in worksheet cells. Names that look like references (e.g., A1), will not be accepted by **Define Name.**

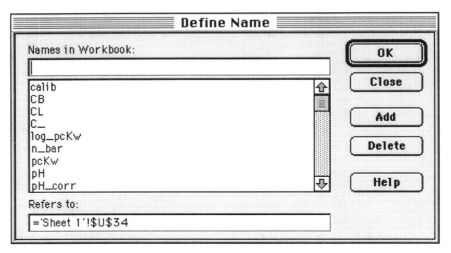

Figure 3-2. Excel 5.0 Define Name dialog box.

Instead of a reference, you can type a numeric value or a formula in the Refers To box.

DELETING NAMES

The Names in Workbook box lists all names that have been assigned in the workbook, even if they are no longer used or valid. If you have removed an unwanted name by deleting the cell, row or column in which it was located, the reference to that name in the Refers To box will be #REF!. Use the Delete button to delete unwanted or invalid names from the list.

USING CREATE NAMES

You'll find **Create Names** very useful if you have worksheets with constants or other values arranged in a table format, as in Figure 3-3.

To assign the names V_0, C_acid and C_base to the cells B1, B2 and B3, select A1:B3 (the cells to be named and the cells containing the names). Choose **Name** from the **Insert** menu and choose **Create...** from the submenu (Excel 5.0) or choose **Create Names...** from the **Formula** menu (Excel 4.0). Excel displays the Create Names box; in Figure 3-4 the Left Column check box is checked, indicating that Excel proposes to use text in the cells in the left column for names. Press the OK button. The names will be assigned to the appropriate cells.

	A	B
1	V_0	50.00
2	C_acid	0.002056
3	C_base	0.1381

Figure 3-3. A table of constants and labels.

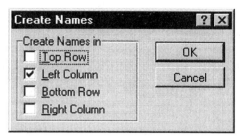

Figure 3-4. The Create Names dialog box.

	B	C	D	E	F	G
4	Initial volume, mL				50.00	(V_0)
5	Initial Concentration of aedach.3HBr, M				0.002056	(CL)
6	pH correction factor C				-0.11	(C_)
7	Concentration of titrant NaOH, M				0.1381	(CB)
8	Buret calibration factor				0.990	(calib)
9	pcKw				13.78	(pcKw)

Figure 3-5. An example of selecting cells for Create Names.

Names can be enclosed in parentheses for clarity in the worksheet, as in Figure 3-5. The parentheses are ignored in creating the names. Here the cells F4:G9 are selected and Excel proposes to use names in the Right Column. Equal signs and colons are also ignored, as in Figure 3-6.

If the data table is a two-dimensional one, as in Figure 3-7, cells are referenced both by row and by column. Excel proposes the row and column titles as variable names, as shown in the Create Names dialog box of Figure 3-8.

Excel will apply the name Emax to the range F4:I4, the name band1 to the range F4:F6, etc. The intersection operator can then be used to identify the named variables. For example, band3 A_0 refers to cell H5.

	F	G	H	I
4		Protonation constants:		
5		logK1H =	10.42	
6		logK2H =	9.74	
7		logK3H =	8.21	
8		logK4H =	5.44	
9				

Figure 3-6. Another example of selecting cells for Create Names.

	E	F	G	H	I	J
3		band1	band2	band3	band4	
4	max	29.25	22.65	18.56	10.11	
5	A_0	1.12	0.15	0.87	1.57	
6	s	1.60	1.58	1.36	1.99	
7						

Figure 3-7. A two-dimensional data table.

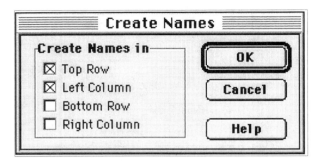

Figure 3-8. The Create Names dialog box.

USING APPLY NAMES

Use **Apply Names** (Figure 3-9) if you created a spreadsheet with formulas using cell references and want to replace the cell references with names. First, use **Define Name** or **Create Names** to assign names to the references. Then choose **Apply Names**. The names that you have assigned will be shown in the list box. The Ignore Relative/Absolute box should usually be checked. Select a name from the list and press OK. All cells containing the reference will be replaced by the name.

> *Excel Tip.* *You can select more than one name at a time from the list box. In Excel for Windows, press the CONTROL key and then click on the names you want to select. In Excel for the Macintosh, press the COMMAND key and click on the names you want to select.*

USING PASTE NAME

The **Paste Name** command allows you to select a variable name from the list of all names in the worksheet and **Paste** it into a cell or formula. There doesn't seem to be any advantage to this command over simply typing in the name.

CHANGING A NAME

You can easily change all occurrences of a name in a spreadsheet. For example, to change the name assigned to the cell containing the concentration of an acid from CL to C_acid, choose **Replace...** from the **Edit** menu (Excel 5.0) or

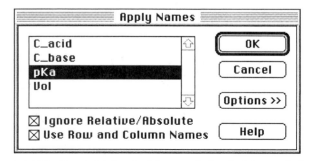

Figure 3-9. The Excel 4.0 for the Macintosh Apply Names dialog box.

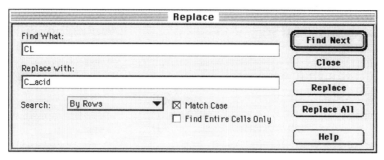

Figure 3-10. The Replace dialog box.

from the **Formula** menu (Excel 4.0). The Replace dialog box (Figure 3-10) asks you to specify the search text and the replacement text. You should select Look at Part, or the only cells that will be modified will be those that contain the search text and nothing else.

The Match Case box is not checked when the dialog box is initially displayed, but because this default case can often cause problems, it's usually a good idea to check it. In the example of Figure 3-10, where CL is changed to C_acid, unless the Match Case box is checked, all occurrences of the letters "cl" will be replaced, including in text entries such as $NaClO_4$.

Although all occurrences of the name are changed in the worksheet, the name definition in the Define Name box is not changed, so you'll have to change it there also. Until you do so, all formulas containing the changed name will display the #NAME? error value. Note that when you change CL to C_acid in the Define Name dialog box, the old name CL will remain in the list until you delete it.

Excel 4.0 provides the Name Changer add-in macro, which changes the name both in the sheet and in the Define Name dialog box.

WORKSHEET FUNCTIONS: AN OVERVIEW

Even though Excel is primarily a business tool, it provides a wide range of functions that are useful for scientific calculations. There are over 300 worksheet functions, organized in eleven categories: Engineering, Financial, Date & Time, Information, Math & Trig, Statistical, Lookup & Reference, Database, Text, Logical and Direct Data Exchange/External functions. This chapter provides examples using selected worksheet functions from the Math & Trig, Statistical, Logical, Date & Time, Text and Lookup & Reference function categories. Excel's Engineering functions (none of which are discussed in this book) are available in Excel 4.0 only through the Add-In Macros feature; they include functions to perform conversions from one number system to another (e.g., decimal to hexadecimal) and functions to operate on complex numbers.

Appendix A lists selected worksheet functions in the Database, Date & Time, Information, Logical, Lookup & Reference, Math & Trig, Statistical, Lookup & Reference, and Text categories. (Engineering, Financial and Direct Data Exchange functions are not discussed.) Appendix B provides an alphabetical list of these worksheet functions along with the required syntax, some comments on the required and optional arguments, one or more examples, and a list of related functions.

FUNCTION ARGUMENTS

Most worksheet functions require one or more *arguments*: the values that the function uses to calculate a return value or to perform an action. The arguments are enclosed in parentheses following the function name, e.g., SQRT(125) or SUM($F$3:$F$28) or SUBSTITUTE(PartNumber, "-1995", "-1996"). A few functions, such as PI() or NOW(), do not require arguments, but the opening and closing parentheses must still be provided.

Function arguments are either required or optional; required arguments must be included, but optional ones can be omitted. In the following sections of this chapter and in the Appendices, function arguments are shown in italic. When a function is shown with its arguments to illustrate the syntax of the function, the required arguments are shown in bold. For example, the worksheet function LOG(***number***, *base*) returns the logarithm of a number to a particular base. If the optional argument *base* is omitted, the function returns the base-10 logarithm.

Most arguments must be of a particular data type (number, text, reference, array, logical or error). Most argument names indicate the data type that is required, by using the words *number*, *text*, *reference*, etc., or by appending *_num* or *_number*, etc. to the argument name. For example, the syntax of the SUBSTITUTE function is SUBSTITUTE(***text***, ***old_text***, ***new_text***, *instance_num*) The first three arguments must be text, the fourth must be a number. Some functions can operate on arguments of any data type, indicated by the use of *value* as an argument name.

In every case, a function can use a cell reference as an argument, but the cell must contain a value of the correct data type.

MATH AND TRIG FUNCTIONS

Excel's mathematical and trigonometric functions (59 of them in Excel 5.0) include functions that correspond to the following "calculator" functions: $\sqrt{x}$, log x, ln x, Σ, e^x, π, sin x, cos x, tan x, as well as many others. See Appendix A for a complete listing.

FUNCTIONS FOR WORKING WITH MATRICES

Excel provides functions for the manipulation of arrays or matrices: TRANSPOSE(***array***) returns the transpose of an array, MDETERM(***array***) returns the matrix determinant of an array, MINVERSE(***array***) returns the matrix inverse of an array, MMULT(***array1***, ***array2***) returns the matrix product of two arrays and SUMPRODUCT(***array1***, ***array2***, ...) returns the sum of the products of corresponding array elements. These functions are discussed more fully in Chapter 14.

STATISTICAL FUNCTIONS

Excel 5.0 provides 71 statistical functions, including functions that return the mean, median, maximum and minimum values, average deviation, standard deviation, variance, n^{th} quartile, rank and many others. Many of these functions are described in Appendices A and B.

A statistical function of considerable use for chemists is LINEST (*linear estimation*). It returns the least-squares regression parameters of the linear function that best describes a data set. LINEST is discussed in detail in Chapter 16.

LOGICAL FUNCTIONS

Logical functions allow you to use different formulas in a cell, depending on the values in other cells. The logical functions provided with Excel are IF, AND, OR and NOT. The latter three are almost always used in combination with the IF function, as are the comparison operators.

THE IF FUNCTION

The syntax of the IF function is IF(*logical_test*, *value_if_true*, *value_if_false*). *Logical_test* is an expression that evaluates to either TRUE or FALSE, e. g., B3<C3, SUM(F3:F28)<>0, etc. Since FALSE = 0 and TRUE = any non-zero value, the formula =IF(G27, *value_if_true*, *value_if_false*) tests whether G27 is non-zero or zero.

If *value_if_true* or *value_if_false* are omitted, the IF function returns TRUE or FALSE in place of the missing expression. To avoid this, use a null string ("") instead of omitting the expression.

A common use of an IF function is to prevent the display of error values when values are missing or inappropriate. In the table shown in Figure 3-11, the percentage change in freshman chemistry enrollment from one year to the next is calculated using the following formula in cell C4:

=100*(B4-B3)/B3

but if the formula is filled down to cells where both of the operands are missing, as in Figure 3-11, the formula returns the #DIV/0! error message.

If the formula is replaced by

=IF(B4<>0,100*(B4-B3)/B3,"")

the calculation is not performed for cells in which the operand in column B is missing, as shown in Figure 3-12. Note that the formula =IF(B4,100*(B4-B3)/B3,"") is equivalent.

	A	B	C
1		Enrollment	
2	Year	Chem I	% change
3	1984	363	
4	1985	328	-10
5	1986	358	9
6	1987	285	-20
7	1988	257	-10
8	1989	255	-1
9	1990	255	0
10	1991	329	29
11	1992	414	26
12	1993	481	16
13	1994	495	3
14	1995	536	8
15	1996	522	-3
16	1997		-100
17	1998		#DIV/0!
18	1999		#DIV/0!

Figure 3-11. A worksheet displaying error values.

	A	B	C
1		Enrollment	
2	Year	Chem I	% change
3	1984	363	
4	1985	328	-10
5	1986	358	9
6	1987	285	-20
7	1988	257	-10
8	1989	255	-1
9	1990	255	0
10	1991	329	29
11	1992	414	26
12	1993	481	16
13	1994	495	3
14	1995	536	8
15	1996	522	-3
16	1997		
17	1998		
18	1999		

Figure 3-12. Display of error values suppressed by using an IF function.

NESTED IF FUNCTIONS

IF functions can be nested, for example by using a second IF function for *value_if_false*. The following formula performs different operations depending on whether C3-B3 is positive, zero or negative.

=IF(C3-B3>0, C3-B3, IF(C3-B3=0, C3, "unable to calculate"))

Up to seven IF functions can be nested.

Nested IF functions allow you to calculate, in one row or column, values that require different formulas depending on the value in one or more different cells. Otherwise you'd have to enter a different formula for each separate case and manually select the cells in which it was to be entered. For example, in the titration of a weak acid with a strong base, four different equations for the pH of the solution are required: at $V = 0$, $pH = \sqrt{C_{HA}K_a}$; before the equivalence point, $pH = pK_a + \log(C_A/C_{HA})$; at the equivalence point, $pH = 14 - \sqrt{C_{HA}K_b}$ and after the equivalence point, $pH = 14 - pOH$. You can calculate the pH of the solution by entering different formulas into cells in a column labeled "pH", as follows: for the initial point of the titration, =-LOG(SQRT(C_acid*10^-pKa); for the remaining points before the equivalence point, =pKa+LOG(A/HA); at the equivalence point, =14-(-LOG(SQRT(A*10^-(14-pKa)))); beyond the equivalence point, =14-(-LOG((C_base*V_base-V_0*C_acid)/V_tot)).

The following formula, using nested IF statements, calculates the pH using a single equation (the formula is entered all in one cell, of course).

=IF(V_base=0,-LOG(SQRT(C_acid*10^-pKa)), IF(HA>0,pKa+LOG(A/HA),

IF(C_base*V_base>V_0*C_acid,14-(-LOG((C_base*V_base-

V_0*C_acid)/V_tot)), 14-(-LOG(SQRT(A*10^-(14-pKa))))))))

Here, combining the calculations into a single expression makes for a more compact spreadsheet, and doesn't require you to decide which cells require which formula. The downside is that it's a pretty complicated formula. If you are relatively new to Excel, you'll probably find it easier to break such calculations up into parts, each in a different row or column of your worksheet. See "Creating 'Megaformulas'" later in this chapter.

AND, OR AND NOT

The AND and OR functions are similar to the comparison operators — they produce a logical result, either TRUE or FALSE, and are almost always used in conjunction with IF. AND and OR can take up to 30 arguments. AND(*logical1*, *logical2*,...) returns TRUE if all of its logical arguments are TRUE; OR(*logical1*, *logical2*,...) returns TRUE if at least one of its logical arguments is TRUE.

EXAMPLE. The following formula (all in one cell, of course) calculates the pK_a values of a diprotic weak acid from the pH and the parameter n-bar (symbolized by $\bar{n}$ in printed equations; see, e.g., Chapter 20), using one of two different formulas, one if n-bar is between 1.2 and 1.8, the other if n-bar is between 0.2 and

0.8; otherwise the formula returns "".

=IF(AND(n_bar>1.2,n_bar<1.8),pH+LOG((n_bar-1)/(2-n_bar)),
IF(AND(n_bar>0.2,n_bar<0.8),pH+LOG((n_bar)/(1-n_bar)),""))

The NOT function reverses the logical value of its argument. For example, in the macro statement

=IF(NOT(ISERROR(GET.NAME("'GLOBAL.XLM'!Chem.Formula.Convert"))),GO
TO(exit))

the function GET.NAME returns the error #N/A if the argument does not appear in the list of names in the Define Name dialog box. Thus in the above example NOT(ISERROR(logical_expression) returns TRUE if the argument is present in the list.

You can use OR to test whether a value is equal to one of the values in an array. For example, the formula

=OR(month={"Jan","Feb","Mar","Apr","May","Jun","Jul","Aug","Sep","
Oct","Nov","Dec"})

returns TRUE if month contains the value Oct.

It's better to restrict use of this approach to arrays entered within formulas. If you define an array of months of the year elsewhere and use a formula such as =OR(month=array), you must remember to enter the formula as an array formula, that is, by pressing COMMAND+ENTER (Macintosh) or CONTROL+SHIFT+ENTER (Windows). Otherwise the formula returns FALSE unless month = Jan.

DATE AND TIME FUNCTIONS

Excel records dates and times by means of a serial value. There are two different serial value systems — the 1900 Date System, used by Excel for Windows, and the 1904 Date System, used by Excel for the Macintosh. In Excel for Windows, a date is calculated as the number of days elapsed since the base date, January 1, 1900; in Excel for the Macintosh, from the base date January 1, 1904. Thus in Excel for the Macintosh, July 1, 1996 is represented as the serial value 33785, the number of days elapsed since the base date. Dates can extend to the year 2078 (serial number of 65536 or 2^{16}). Times are represented by the decimal part of the serial number. You can use either *number formatting* or *worksheet functions* to convert these arcane serial values into comprehensible dates.

As you've already seen, dates and times can be entered into worksheet cells using any one of several convenient formats: July 1 can be entered as 7-1, 7/1, July 1, Jul 1 or 1 July, among others. All these date entries produce the date 7/1/xxxx in the formula bar and the displayed date 1-Jul. Excel enters the current year unless a different year is specified. Thus in 1997 "xxxx" would be displayed as 1997.

Times are also recognized by Excel. If you enter 10:00 in a cell, it will be recognized as a time, and 10:00:00 AM will appear in the formula bar. Excel assumes a 24-hour clock (military time) unless you indicate differently. You can use AM/PM or am/pm designations with times. Even "2 p" can be used to enter 2:00 PM in a cell.

> *Excel Tip. Enter the current date in a worksheet by using CONTROL+ (semicolon) (Windows) or COMMAND+(hyphen) (Macintosh); to enter the time use CONTROL+(colon) (Windows) or COMMAND+(semicolon) (Macintosh). The date appears in the format mm/dd/yy but can be formatted otherwise.*

Dates can also be entered using the worksheet functions DATE(*yy,mm,dd*), TODAY() or NOW(). DATE is used to enter the serial value of any date. The function TIME(*hh, mm, ss*) performs the same function for a particular time.

TODAY()and NOW()return the serial value of today's date. TODAY() returns an integer number (date only) while NOW() produces a number with decimal (date and time). Both functions return the serial value at the time the function was entered; it is not updated continuously, but it *is* updated each time the worksheet is recalculated.

There are two additional functions for entering dates and times. DATEVALUE(*text*) and TIMEVALUE(*text*) convert text arguments into date serial values. For example, TIMEVALUE("8:30 PM") returns the value 0.85416667.

The following worksheet functions operate on the date serial value to return a date or time: YEAR(*value*), MONTH(*value*) and DAY(*value*). *Value* can be a serial value, a cell reference or a date as text. MONTH returns a number between 1 and 12, DAY a number between 1 and 31. The text function TEXT can also be used to format a date serial value; any custom date format can be applied (see "The TEXT, FIXED, REPT and VALUE Functions" later in this chapter).

The HOUR(*value*), MINUTE(*value*) and SECOND(*value*) functions are similar to the DAY, MONTH and YEAR functions.

DATE AND TIME ARITHMETIC

If you keep in mind that Excel stores dates and times as serial values, performing date or time arithmetic is simple. For example, in a kinetics experiment you may have a table of times at which data points were recorded at irregular intervals (Figure 3-13). To analyze the data you need the elapsed time from $t_{initial}$. Subtracting the time values from the initial value (using the formula =(A10-$A$9)*1440 in cell B10) yields numbers that are decimal fractions of a day and are converted into minutes by multiplying by 24×60. The expression =MINUTE(A10-$A$9) produces the same result.

If you enter 10 AM in cell A1 of a worksheet, and =A1+3 in cell B1, you may at first be confused when the cell displays simply 10:00 AM. But remember that you've added three days, not three hours; if you apply the date format m/d/yy to the cell, you'll see that you've calculated a date three days from the current

	A	B
1	**Time**	**Minutes**
2	10:01 AM	0
3	10:15 AM	14
4	10:33 AM	32
5	11:00 AM	59
6	12:01 PM	120
7	1:07 PM	186
8	2:15 PM	254
9	3:30 PM	329
10	4:59 PM	418

Figure 3-13. Calculating elapsed times

date. If you change the formula to =A1+3:00, you get an "Error in formula" message. That's because 3:00 is not a numerical value. To do it correctly, use the formula =A1+"03:00". This produces the desired result, 1:00 PM. Excel recognizes that 03:00 is text and evaluates it just as it would if you'd typed it into a cell.

TEXT FUNCTIONS

Excel provides a wide range of worksheet functions that operate on text. You are already familiar with the & operator, to concatenate text or text and values. Most of Excel's text functions select or modify one or more characters within a text string.

THE LEFT, RIGHT, MID AND LEN FUNCTIONS

The LEFT(*text*, *num_characters*) function returns the leftmost character or characters in a text string. For example, LEFT("02167-3860",5) returns 02167. If *num_characters* is omitted, the value 1 is assumed. The RIGHT(*text*, *num_characters*) function is similar. If cell B7 contains a nine-digit number, then RIGHT(B7,4) returns the last four digits of the number.

The syntax of the MID function is MID(*text*, *start_num*, *num_characters*); it returns a specific number of characters from a specified position in a text string. For example, if cell A1 contains H2SO4, then MID(A1,3,1) returns S.

LEN(*text*) returns the number of characters in a text string.

THE UPPER, LOWER AND PROPER FUNCTIONS

Three functions change the case of a text string: the UPPER(*text*) and LOWER(*text*) functions do what their names suggest; the PROPER(*text*) function capitalizes the first letter in each word of a text string, as illustrated in column B of Figure 3-14.

THE FIND, SEARCH, REPLACE, SUBSTITUTE AND EXACT FUNCTIONS

FIND(**find_text, within_text,** *start_at_num*) and SEARCH(**find_text, within_text,** *start_at_num*) are similar. Each returns the position number of *find_text* within the text string *within_text*. FIND is case-sensitive, SEARCH is not. For example, if cell A4 contains toluene, 2-chloro-, the expression FIND(",",A4,1) returns the value 8. Unless the optional starting position is specified, the functions begin at position 1.

The following two functions are complementary: REPLACE(**text, start_num, num_characters**), **new_text**) and SUBSTITUTE(**text, old_text, new_text**, *instance_num*).

REPLACE replaces unspecified characters at a specified position within a text string. Note that its syntax is similar to that of the MID function, except that a fourth argument, *new_text*, is appended. Example: if cell A1 contains the text 1995, REPLACE(A1, 3, 2, "96") returns 1996.

SUBSTITUTE replaces specific characters within a string. For example, if cell A1 contains Et and cell B1 contains (C2H5) then SUBSTITUTE("Et3N", A1, B1) returns the text (C2H5)3N. If the optional argument *instance_num* is specified, only that instance of *old_text* will be replaced. If *instance_num* is omitted, all instances of *old_text* will be replaced.

EXACT(*text1, text2*) returns TRUE if the two strings are identical, FALSE otherwise. EXACT is case-sensitive. Simple comparison of strings is not case-sensitive. For example, the formula =("Name"="NAME") returns TRUE. Use EXACT if you want to make a case-sensitive comparison of two strings.

EXAMPLE. The following formula re-formats a list of names in which the original names, in column A of Figure 3-14, are in the form LAST_NAME,FIRST_NAME; the re-formatted names are in column B.

=PROPER(RIGHT(A1,LEN(A1)-FIND(",",A1))&" "& LEFT(A1,FIND(",",A1)-1))

	A	B
47	ANTONIUS,STEPHEN J	Stephen J Antonius
48	BRUNEL,STEVEN D	Steven D Brunel
49	CARRESTO,KATHY E	Kathy E Carresto
50	LAKLIS,CLAIR L	Clair L Laklis
51	PEDROZO,BENITO A	Benito A Pedrozo
52	SOUSSANE,WALID	Walid Soussane
53	WOODSTOCK,PAUL	Paul Woodstock
54	WYNDLAKE,KEVIN D	Kevin D Wyndlake
55	ZILARIO,J PATRICK	J Patrick Zilario

Figure 3-14. A list of names re-formatted using text functions.

The function FIND(",",A1) returns the position of the comma in the string. The first names and/or initials are obtained using the RIGHT function; the number of characters to be returned is equal to the length of the string minus the position number of the comma. A space is concatenated, then the last name is obtained using the LEFT function. The PROPER function is used to change the case of the string.

THE TEXT, FIXED, REPT AND VALUE FUNCTIONS

TEXT(*value, format_text*) converts a number to text and formats it. It performs the same function, and uses the same symbols, as number formatting using the **Format** menu. For example, if cell A8 contains a nine-digit number, =TEXT(A8,"000-00-0000") formats it as a Social Security number.

FIXED(*number, decimals, no_comma_logical*) formats a number as text with a specified number of decimal places, with or without commas.

REPT(*text, times_number*) repeats text a specified number of times. For example, =REPT("+",10) displays   ++++++++++ in a worksheet cell.

VALUE(*text*) converts a text argument to a number, in the rare instance where this is necessary.

> *Excel Tip.* *When text and values are concatenated using the & operator, the number can no longer be formatted using menu commands or tool buttons. Use the TEXT or FIXED functions instead. For example, if cell B10 contains 41 and cell B11 contains 495, then the formula =B10/B11 returns "0.08282828" and this can be formatted to display "8%". But if the cell contains ="Percent yield = "&B10/B11, it will display Percent yield = 0.08282828282828 and the number can't be formatted. Use the formula ="Percent yield = "&TEXT(B10/B11,"0%") to format the number part of the text.*

THE CODE AND CHAR FUNCTIONS

CODE(*text*) and CHAR(*number*) perform opposite functions. CODE returns the numeric code (either Macintosh character codes or ANSI character codes for Windows) for a single character or the first character in a text string. CHAR returns the character corresponding to the character code. For example, CODE("a") returns 97, CHAR(36) returns $.

CHAR(13) is the carriage return (the ¶ symbol you see when you choose **Show ¶** from the **View** menu in MS Word). Use CHAR(13) to insert a line break into text within a formula. For example, the formula (all in one cell, of course)

="Missing Reports as of "&TEXT(NOW(),"h AM/PM mmm dd, yyyy")

&CHAR(13)&"(X = a report that has not been received"&CHAR(13)&" or was returned for recalculation)"

produces the text displayed in Figure 3-15.

If you use Wrap Text in **Alignment** from the **Format** menu, Excel will decide where to break the text. By using CHAR(13), you make the decision.

> **Missing Reports as of 9 AM Dec. 15, 1994.**
> **(X = a report that has not been received**
> **or was returned for recalculation)**

Figure 3-15. Text with line breaks inserted.

LOOKUP AND REFERENCE FUNCTIONS

There are several functions for obtaining values from a table, based on position or value.

THE OFFSET FUNCTION

OFFSET(*reference*, *rows*, *columns*, *height*, *width*) returns a value offset from a given reference in a one- or two-dimensional range of cells.

For example, Figure 3-16 shows part of a data table of the elements, arranged in order of atomic number. The cells in column A were named *Element* using **Define Name**.

The formula =OFFSET(Element,z,) returns the name of the element whose atomic number is z. It is not necessary to specify the value of *columns*, since the array *Element* is one-dimensional, but the comma must be included to satisfy the syntax of OFFSET.

If the cell range is two-dimensional and both row offset and column offset are specified, then the *height* and *width* of the reference to be selected must be specified.   To select a single cell within the cell range, the formula =OFFSET(reference, rows, columns, 1, 1) must be used.

THE INDEX AND MATCH FUNCTIONS

The function INDEX(*array*, *row_num*, *column_num*, *area_num*) returns a single value from within a one- or two-dimensional range of cells. If the atomic weights of the elements in the preceding example using the OFFSET function are located in the range \$C\$2:\$C\$103, the formula =INDEX(\$C\$2:\$C\$103,z,) returns the atomic weight of the element whose atomic number is z. Non-adjacent selections are permitted; they are handled by *area_num*. See Appendix B for details.

	A	B	C	D
1	**Element**	**Symbol**	**At.Wt.**	**Elec. Config.**
2	Hydrogen	H	1.00797	1s1
3	Helium	He	4.0026	1s2
4	Lithium	Li	6.939	[He] 2s1
5	Beryllium	Be	9.0122	[He] 2s2
6	Boron	B	10.811	[He] 2s2 2p1
7	Carbon	C	12.01115	[He] 2s2 2p2

Figure 3-16. Portion of a data table of the elements.

The function MATCH(*lookup_value*, *array*, *match_type_num*) returns the relative position of a value in a one-dimensional array. If *match_type_num* = 1, MATCH returns the position of the largest array value that is less than or equal to *lookup_value*. The array must be in ascending order. If *match_type_num* = -1, MATCH returns the position of the smallest value that is greater than or equal to *lookup_value*. The array must be in descending order. If *match_type_num* = 0, MATCH returns the position of the first value that is equal to *lookup_value*. The array can be in any order. If no match is found, #N/A! is returned.

For example, in the formula

=MATCH(char,{188,193,170,163,162,176,164,166,165,187},0)

the array is the keycode values of the subscript numbers in the Macintosh custom font GenevaChem that are produced by pressing COMMAND+(number key). The subscript numbers are in the order 0-9. MATCH finds the position of the value in the array that is equal to char, the CODE value of a subscript character in a text string.

THE **VLOOKUP** AND **HLOOKUP** FUNCTIONS

The function VLOOKUP(*lookup_value*, *array*, *index_num*, *match_type_num*) looks for a match in the first column of a two-dimensional array and returns a value offset by *index_num* across the row. *Lookup_value* is the value to be found in the first column of array. An *index_num* of 2 returns a value from column 2 of array. If *match_type_num* is TRUE or omitted, VLOOKUP returns the largest array value that is less than or equal to *lookup_value*. The array must be in ascending order. If *match_type_num* is FALSE, VLOOKUP returns an exact match or, if one is not found, the #N/A! error value. The array can be in any order.

	A	B
137	Ac	[227]
138	Ag	107.870
139	Al	26.9815
140	Am	[243]
141	Ar	39.948
142	As	74.9216
143	At	[210]
144	Au	196.967
145	B	10.811
146	Ba	137.34
147	Be	9.0122
148	Bi	208.980
149	Bk	[247]
150	Br	79.909
151	C	12.01115

Figure 3-17. A portion of an atomic weight table.

This function was improved in Excel 5.0. The *match_type_num* argument is not available in Excel 4.0 (see the following section).

For example, if a spreadsheet contains a list of element symbols and atomic weights in cells A137:B245, a portion of which is shown in Figure 3-17, then the formula =VLOOKUP(symbol,$A$137:$B$245,2,0) returns the atomic weight corresponding to symbol.

The function HLOOKUP(*lookup_value, array, index_num, match_type_num*) is similar to   VLOOKUP.

USING **EXACT** WITH LOOKUP FUNCTIONS IN EXCEL 4.0

In Excel 4.0, the *match_type_num* argument is not available, and the LOOKUP functions will return a value even if an exact match isn't found. Thus it's often necessary to check to make sure that an exact match has been found.

For example, in a sheet that contains student names and exam grades, you can retrieve the exam grade for a particular student by using VLOOKUP to find the student name, then returning the corresponding quiz average. But if VLOOKUP can't find an exact match to name, it uses the largest value less than or equal to name. If a student's name is misspelled or missing from the list, VLOOKUP may return an incorrect value. In the following formula

=IF(EXACT(name,VLOOKUP(name,'94F Exam Grades'!$A$3:$U$151,1)),
VLOOKUP(name,'94F Exam Grades'!$A$3:$U$151,21),"?")

the EXACT function is used to find whether the string name is exactly the same as the best-match value in relative column 1 of the array of names and grades in the reference $A$3:$U$151 in the sheet 94F Exam Grades. If so, the value in relative column 21 is returned; if not, a "?" is returned.

USING PASTE FUNCTION (EXCEL 4.0)

Because Excel provides such a wide range of functions, it is sometime difficult to remember them or to enter their arguments correctly. You can use **Paste Function** (Figure 3-18) from the **Formula** menu to paste a function in a cell, or within a formula that you're typing in the formula bar. When you choose **Paste Function**, you can first select a function category; Excel will display all the functions in that category in the Paste Function box. When you select a function, the function name and syntax appear at the bottom left of the dialog box. The Paste Arguments box should be checked. When you press the OK button, the function will be pasted into the formula in the formula bar, at the insertion point. If you **Paste** a single function into a cell, or begin a formula by pasting a function, Excel automatically inserts the equal sign.

Excel pastes dummy arguments, called *placeholder arguments*, along with the function. The first placeholder argument will be selected (highlighted) so that you can enter a value.

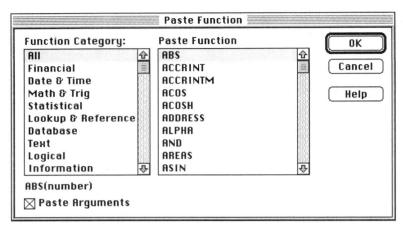

Figure 3-18. The Excel 4.0 Paste Function dialog box.

Excel Tip.   Selecting functions in the Paste Function dialog box: if you type a letter, the first function beginning with that letter is selected. For example, if you type the letter D, the DATE function is selected. You can type several letters in succession to zero in on the function you want. If you type R, the RAND function is selected, but if you type R-O-W (rapidly), you will select the ROW function. If you type a string of letters that doesn't correspond to any function, you'll get a beep.

A SHORTCUT TO A FUNCTION

Most often you'll know what function you want to enter, in which case it's much faster just to type in the function and its arguments. Occasionally you will not be sure of the arguments and their proper order. After typing the equal sign, the function name and the opening parenthesis, press CONTROL+A. This will paste the placeholder arguments and add the closing parenthesis. For example, after entering =LINEST(, press CONTROL+A. The function will be completed and will appear as

```
=LINEST(known_y's,known_x's,const,stats)
```

Known_y's is highlighted, so that you can conveniently enter the desired value. After entering it, double-click on the next argument to select it.

THE FUNCTION WIZARD IN EXCEL 5.0

The Function Wizard, a new feature in Excel 5.0, is actuated by clicking the f_x button or by choosing **Function...** from the **Insert** menu. The Function Wizard does the same thing as **Paste Function...** in Excel 4.0, but provides a little more guidance and information about the arguments required by the function. To enter a function in a worksheet cell or a formula, click the Function Wizard button. The initial dialog box is almost identical to the **Paste Function...** dialog box (Figure 3-19), although there is an additional Function Category, "Most

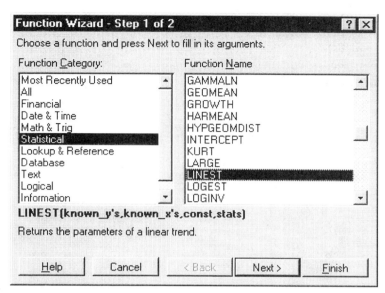

Figure 3-19. The Function Wizard's Step 1 of 2 dialog box.

Recently Used". After selecting the Function Category and then the function to be entered, press the Next button, which activates the Step 2 of 2 dialog box.

The Step 2 of 2 dialog box (Figure 3-20) presents information about each argument as you enter it: a description of the argument, and whether it is required or optional. As you enter each argument, its value is displayed in the text box on the right. Use Tab to move to the next argument. If you need information about a particular argument (the effect of entering either **TRUE** or **FALSE** for a logical argument, for example), press the Help button. When you press Finish, or RETURN or ENTER, the function is entered into the worksheet cell. If you press Finish without entering values for the arguments, Excel will enter the placeholder arguments.

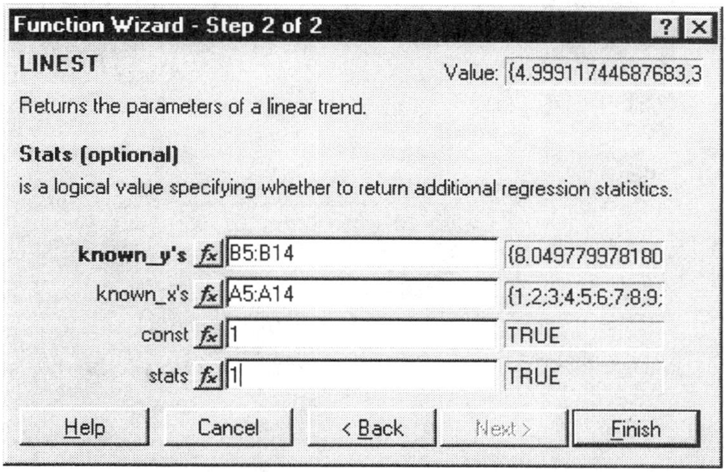

Figure 3-20. The Function Wizard's Step 2 of 2 dialog box.

If you type the function and the opening parenthesis in a worksheet cell, and then press CONTROL+A, you will go directly to the Step 2 of 2 dialog box. You can then enter values for the function's arguments or press the Finish button to paste placeholder arguments in the function.

As you become familiar with the range of functions provided by Excel, you will probably type most of them directly, rather than using the Function Wizard.

CREATING "MEGAFORMULAS"

As you become more experienced in constructing worksheet formulas, you will probably create more and more complicated ones. You can simplify a complicated calculation by performing it in steps, storing the results of intermediate calculations in separate cells of the worksheet. You can even hide the rows or columns containing the intermediate stages of the calculation. But there are advantages to constructing a single "megaformula". You'll use less memory and, more importantly, recalculation of the worksheet will take less time.

You saw some examples of megaformulas earlier in this chapter. A good way to begin to construct a megaformula is to break the calculation into steps and store the results in separate cells. When the formulas are working correctly, you can combine them all in a single cell by copying and pasting. Here's an example: a list of names was imported into Excel from a word processing document. The first few entries are shown in Figure 3-21. You want to create a column in Excel containing the last names, and a second column containing the first names plus initials, if any.

The formulas to accomplish this are the following (the values returned by the formulas are shown in Figure 3-22): in cell B4, use the formula =LEN(A4)-LEN(SUBSTITUTE(A4," ","")) to determine the number of spaces in the text. In cell C4, use =SUBSTITUTE(A4," ","*",B4) to substitute a marker character for the last space in the name. (SUBSTITUTE accepts the optional argument *instance_number* which specifies which instance of *find_text* is to be substituted.) In cell D4, use =FIND("*",C4) to find the location of the marker character, which immediately precedes the last name portion of the string. In cell E4, use =RIGHT(A4,LEN(A4)-D4) to abstract the last name.

	A
4	Jeffrey N. Adams
5	Nicholas Bartlett
6	Cindy A. Bronstein
7	Ming-Hwang Chung

Figure 3-21. Portion of text imported from MS Word.

	A	B	C	D	E
4	Jeffrey N. Adams	2	Jeffrey N.*Adams	11	Adams
5	Nicholas Bartlett	1	Nicholas*Bartlett	9	Bartlett
6	Cindy A. Bronstein	2	Cindy A.*Bronstein	9	Bronstein
7	Ming-Hwang Chung	1	Ming-Hwang*Chung	11	Chung

Figure 3-22. Portion of worksheet to parse test into separate columns .

Finally, combine the formulas: first, **Copy** the formula in cell **B4** from the formula bar (don't include the equal sign) and press the Enter button; then select cell **C4**; in the formula bar, select **B4** in the formula and **Paste** the formula fragment. Repeat the process for cells **D4** and **E4**. The final megaformula is

=RIGHT(A4,LEN(A4)-FIND("*",SUBSTITUTE(A4," ","*",LEN(A4)-
LEN(SUBSTITUTE(A4," ","")))))).

A similar formula is used to abstract the first name plus initial. Finally, **Delete** columns B-D.

A formula can contain up to 1024 characters, so your megaformulas can be quite complicated.

USING ARRAY FORMULAS

In Excel, cell ranges and arrays are essentially identical. They can either be one- or two-dimensional.

A number of Excel's worksheet functions are array functions — they either require arrays (cell ranges) as arguments or return arrays as results. To create a worksheet formula that returns an array result, you must first select a suitable range of cells, with dimensions (R × C) large enough to accommodate the returned array, then type the formula in the formula bar, and finally enter the formula by pressing COMMAND+ENTER (Macintosh) or COMMAND+RETURN (Macintosh) or CONTROL+SHIFT+ENTER (Windows). Excel will indicate that the formula is an array formula by enclosing it in braces and will enter the array formula in all the selected cells.

Even if a worksheet formula returns only a single result to a single worksheet cell, if it contains an array formula you will still need to enter the formula by pressing COMMAND+ENTER (Macintosh) or CONTROL+SHIFT+ENTER (Windows). Worksheet functions such as SUM(*range*) are an exception to this rule.

Arrays can simplify worksheets, as illustrated in the following example, where absorbance values from a first-order rate process are fitted to the equation $A_{calc} = A_0 e^{-kt}$. The worksheet calculates the sum of squares of residuals, which was minimized by changing A_0 and k to obtain the least-squares best fit of the calculated absorbance to the experimental values. The values of A_0 and k obtained in this way were 0.85503 and 0.49537, respectively.

To calculate the sum of squares of the residuals, $\Sigma(A_{obsd}-A_{calc})^2$, you could use the (non-array) approach shown in the worksheet of Figure 3-23.

	A	B	C	D
1				
2	t, min	A(observed)	A(calculated)	(difference)^2
3	0	0.855	0.855	9.00E-10
4	1	0.521	0.521	7.40E-11
5	2	0.317	0.317	2.25E-07
6	3	0.195	0.193	2.40E-06
7	4	0.118	0.118	1.47E-08
8	5	0.070	0.072	3.34E-06
9				
10			Σ(difference)^2 =	5.98E-06
11		equation used in cell D10:		=SUM(D3:D8)

Figure 3-23. Calculating a sum of squares.

The sum-of-squares calculation can also be done by using the array formula shown in the second example (Figure 3-24). Since the SUM function contains the argument (C3:C8-B3:B8) it must be evaluated as an array formula. Remember to use COMMAND+ENTER (Macintosh) or CONTROL+SHIFT+ENTER (Windows); otherwise the #VALUE! error value will be returned.

Using names for the $A_{observed}$ and $A_{calculated}$ ranges simplifies the array formula and makes it much more self-documenting. By using either **Define Name** or **Create Names**, the cell ranges B3:B8 and C3:C8 are defined as A_obsd and A_calc, respectively.

In the two array examples shown in Figures 3-25 and 3-26, the array calculations are carried out just as if separate formulas had been produced using the **Fill Down** or **Fill Right** commands. The values of A_calc and A_obsd that occupy cells in the same row are subtracted.

	A	B	C	D
1				
2	t, min	A(observed)	A(calculated)	
3	0	0.855	0.855	
4	1	0.521	0.521	
5	2	0.317	0.317	
6	3	0.195	0.193	
7	4	0.118	0.118	
8	5	0.070	0.072	
9				
10			Σ(difference)^2 =	5.98E-06
11		equation used in cell D10:	{=SUM((C3:C8-B3:B8)^2)}	

Figure 3-24. Calculating a sum of squares using an array formula.

	A	B	C	D
1				
2	t, min	A(observed)	A(calculated)	
3	0	0.855	0.855	
4	1	0.521	0.521	
5	2	0.317	0.317	
6	3	0.195	0.193	
7	4	0.118	0.118	
8	5	0.070	0.072	
9				
10			$\Sigma(\text{difference})^2 =$	5.98E-06
11	equation used in cell D10:		{=SUM((A_calc-A_obsd)^2)}	

Figure 3-25 Calculating a sum of squares using named arrays.

	A	B	C	D
1				
2	t, min	A(observed)		
3	0	0.855		
4	1	0.521		
5	2	0.317		
6	3	0.195		
7	4	0.118		
8	5	0.070		
9				
10			$\Sigma(\text{residuals})^2 =$	5.98E-06
11	equation used in cell D10:{=SUM((A_0*EXP(-k*t)-A_obsd)^2)}			

Figure 3-26. Calculating a sum of squares without intermediate formulas by using an array formula.

In the final example, shown in Figure 3-26, the column of $A_{calculated}$ values has been eliminated and A_calc is obtained from the formula =A_0*EXP(-k*t).

Array formulas can simplify worksheets and eliminate intermediate formulas.

ARRAY CONSTANTS

In the same way that a worksheet formula can contain a simple constant, e.g., the value 3 in the formula =3*A1+A2, an array formula can contain an *array constant*. An array constant is included in a formula by enclosing the array of values in braces. In this case *you* must type the braces; when entering a completed array formula by pressing COMMAND+ENTER (Macintosh) or CONTROL+SHIFT+ENTER (Windows), Excel automatically provides the braces around the completed formula.

Within the array, values in the same row are separated by commas, and rows of values are separated by semicolons. For example, the array constant $\{2, 3, 4; 3, 2, -1; 4, 3, 7\}$ represents the array $\begin{vmatrix} 2 & 3 & 4 \\ 3 & 2 & -1 \\ 4 & 3 & 7 \end{vmatrix}$.

Array constants can contain numerical or text values, but they cannot contain references. Individual text values in an array constant must be enclosed in quotes.

An array constant is usually incorporated in a worksheet formula, e.g.,
=MATCH(char,{188,193,170,163,162,176,164,166,165,187},0).
Alternatively the array constant can be entered as a named variable in the Define Name dialog box, as shown in Figure 3-27.

EDITING OR DELETING ARRAYS

Since a single formula is entered in all the cells of an array, you can't change or delete part of an array; if you try, you'll get a "Cannot change part of an array" message.

To edit an array formula, simply select any cell in the array. Then edit the formula in the formula bar. When you begin to edit, the braces surrounding the formula will disappear. To re-enter the edited formula, press COMMAND+ENTER (Macintosh) or CONTROL+SHIFT+ENTER (Windows). The formula will be entered into all of the cells originally selected for the array.

Although you can't change part of an array's formulas or values, you can format individual cells. You can also **Copy** values from individual cells of an array and **Paste** them elsewhere.

You can select individual values in an array by using the INDEX function, but this is a "read-only" option. If you need to be able to change individual values in an array, you should enter the values in a range of worksheet cells, rather than as an array.

Excel Tip. *To select an entire array range (in order, for example, to Clear it), select any cell in the array while pressing COMMAND+SHIFT (Macintosh) or CONTROL+"/" (Windows).*

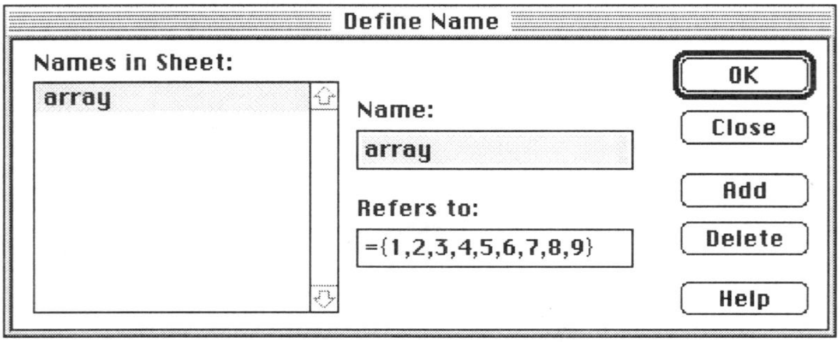

Figure 3-27. An array constant entered as a named variable.

CREATING A THREE-DIMENSIONAL ARRAY

It's possible to create an array with three dimensions by entering an array formula in each cell of a rectangular range of cells. By using an array formula, each cell in the two-dimensional range effectively contains a one-dimensional array. The following example illustrates the use of a three-dimensional array to calculate an "error surface" curve such as the one shown in Figure 11-21. The error-square sum, i.e., the sum of the squares of the residuals, $\Sigma(Y_{obsd} - Y_{calc})^2$ for a one-dimensional array of data points, was calculated for each cell of a two-dimensional array of trial values. The "best" values of the independent variables are those which produce the minimum error-square sum.

The spreadsheet shown in Figure 3-28 contains kinetic data obtained for the reaction sequence $A \rightarrow B \rightarrow C$. The concentration of the intermediate species B was measured at time intervals from 0 seconds to 100 seconds (only a portion of the data table in rows 4 and 5 is shown); the t and $[B]_{obsd}$ values were defined as named ranges t and B_obs. $[B]_{calc}$ values were obtained from the equation

$$[B] = [A]_0 \frac{k_1}{k_2 - k_1} \{e^{-k_1 t} - e^{-k_2 t}\}$$

which can be found in any standard text on chemical kinetics. The error-square sum was calculated for a range of trial values of k_1 and k_2 using an array formula.

	A	B	C	D	E	F	G	H	I	J	K	L
1		Consecutive First Order Reactions (A->B->C)										
2		$[B]_{calc}$ = (k$_1$/(k$_2$-k$_1$))*(EXP(-k$_1$*t)-EXP(-k$_2$*t))										
3		Experimental values										
4		t, sec	0	2	4	6	8	10	12	14	16	18
5		$[B]_{obsd}$	0.002	0.169	0.292	0.387	0.445	0.476	0.498	0.499	0.493	0.479
6												
7		Error sum = $\Sigma([B]_{calc}-[B]_{obsd})^2$										
8		k$_1$ trial values										
9			0.060	0.070	0.080	0.090	0.100	0.110	0.120	0.130	0.140	
10		0.020	0.833	0.829	0.839	0.860	0.889	0.922	0.958	0.995	1.034	
11		0.025	0.505	0.481	0.475	0.483	0.501	0.525	0.553	0.584	0.617	
12		0.030	0.311	0.273	0.257	0.256	0.266	0.284	0.307	0.333	0.361	
13		0.035	0.201	0.154	0.130	0.122	0.127	0.140	0.159	0.181	0.206	
14		0.040	0.144	0.090	0.060	0.048	0.049	0.058	0.074	0.093	0.116	
15	k₂ trial values	0.045	0.121	0.063	0.029	0.014	0.011	0.017	0.030	0.048	0.069	
16		0.050	0.122	0.060	0.024	0.005	0.000	0.004	0.015	0.031	0.050	
17		0.055	0.138	0.074	0.035	0.015	0.008	0.010	0.020	0.034	0.052	
18		0.060	0.164	0.099	0.058	0.037	0.028	0.029	0.037	0.050	0.067	
19		0.065	0.198	0.131	0.090	0.067	0.057	0.057	0.064	0.076	0.092	
20		0.070	0.236	0.169	0.126	0.102	0.092	0.091	0.097	0.108	0.123	
21		0.075	0.277	0.209	0.166	0.142	0.130	0.129	0.134	0.144	0.159	
22		0.080	0.319	0.252	0.209	0.184	0.172	0.169	0.174	0.184	0.197	
23												
24		At minimum:			k$_1$ =	0.10	k$_2$ =	0.05				

Figure 3-28. Spreadsheet implementation of a three-dimensional array.

The array formula entered in cell C10 is:

   {=SUM((B_obs-(C$9/($B10-C$9)*(EXP(-C$9*t)-EXP(-$B10*t))))^2)}

(note the use of mixed addressing), then filled into the range C10:K22 by using AutoFill.

   A 3-D chart of the values is shown in Figure 11-21.

TROUBLESHOOTING THE WORKSHEET

   Inevitably, your worksheet formulas will at times contain errors. Earlier in this chapter you learned some of the ways to prevent or discover errors while entering formulas. But even if a formula is syntactically correct, it may produce an error when it is evaluated. This section provides some techniques and tips for tracking down errors in your spreadsheets.

ERROR VALUES AND THEIR MEANINGS

   Excel displays an error value in a cell if the formula can't be evaluated. The error values, which are #DIV/0!, #N/A, #NAME?, #NULL!, #NUM!, #REF! and #VALUE!, can give you a good idea of what caused the error. The following list is in the approximate order of the frequency with which error values are encountered.

- #DIV/0! is displayed if a formula uses a cell containing zero as divisor, or if the cell is blank.

- #NAME? is displayed when you use a name that Excel doesn't recognize. Most often this occurs when you misspell a name or function, or when you enter a name without having defined it. It will also happen when you enter a text argument in a formula and forget to enclose the text in quotes.

- #REF! is displayed when a formula refers to a cell that has been deleted, or when a worksheet using a custom function is opened when the macro sheet is not open.

- #VALUE! is displayed when the wrong type of argument is used in a function.

- #N/A is displayed when certain built-in functions (especially HLOOKUP, VLOOKUP or MATCH) contain incorrect arguments.

- #NUM! is displayed when a number supplied to a function is not a valid argument, e.g., SQRT(-1).

- #NULL! is displayed if you specify, using the intersection operator, an intersection of two references that do not intersect.

EXAMINING FORMULAS

   When you see an error value displayed in a cell, you'll need to examine the offending formula. There are several things you can do to track down the error. The error value displayed can give you a good idea where to begin. For example, if the error value is #NAME?, you most likely have misspelled a variable name or function, or entered a variable name that has not yet been defined by using **Define Name**.

To track down the source of an error in a lengthy, complicated formula, examine the value of individual references, names, functions or function arguments.

> ***Excel Tip.***   *To view the current value of a variable or function in a statement in the formula bar, highlight it (e.g., by double-clicking) and press F9 (Windows) or COMMAND+ = (Macintosh). The value of the selected portion of the formula will be displayed. Click on the Cancel box in the formula bar or press Undo to restore the statement; otherwise the selected portion of the formula will be permanently replaced by the numerical value.*

FINDING DEPENDENT AND PRECEDENT CELLS

To audit the logic of a complicated worksheet, you may want to find all cells that contain formulas that refer to a given cell (*dependent* cells) or all cells that are referred to by the formula in a given cell (*precedents*). You can search backward or forward, finding the direct precedents, then the cells that are referred to by those cells, and so on.

In Excel 5.0, choose **Auditing** from the **Tools** menu. The **Auditing** submenu (Figure 3-29) allows you to trace precedents, dependents or errors. Figure 3-30 illustrates a typical display when a cell is selected and **Trace Precedents** is chosen.

Trace Precedents
Trace Dependents
Trace Error
Remove All Arrows
Show Auditing Toolbar

Figure 3-29. Excel 5.0 Auditing submenu.

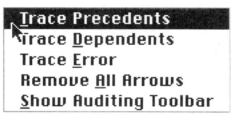

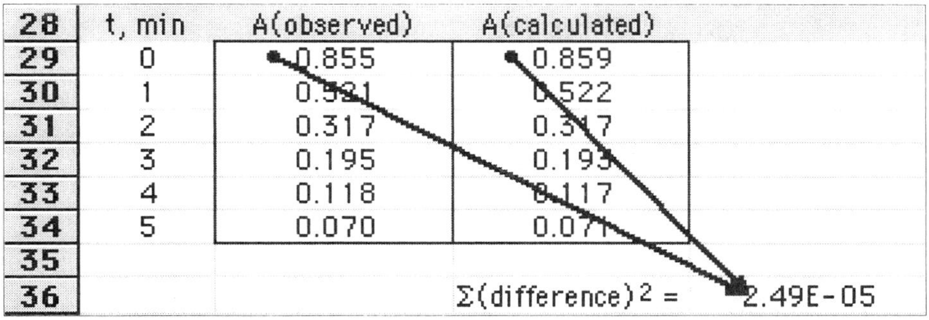

Figure 3-30. Trace Precedents display.

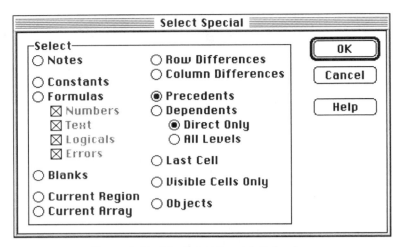

Figure 3-31. The Select Special dialog box.

In Excel 4.0, to find all cells in a worksheet that are direct precedents of a particular cell, choose **Select Special** from the **Formula** menu to display the Select Special dialog box (Figure 3-31). The Direct Only button is pressed; press the Precedents button, then press OK. The cells in the worksheet that are direct precedents will be highlighted. If you pressed the All Levels button, then all cells that are precedents at any level will be highlighted. You can select a range of cells and find the precedents or dependents of all the cells in the range.

Excel examines all cells in the worksheet and highlights all cells in the worksheet that refer to, or are referred to by, a particular cell or cells.

Excel Tip (Excel 4.0 only).   Double-click on a cell to show the direct precedents of that cell.

USING PASTE LIST

The **Paste List** command is useful for worksheet auditing and error tracing. It produces a list of all names used in the worksheet, with their references.

Select the upper-left corner of the range in which you want the list to appear; make sure that nothing will be over-written by the list. Choose **Name...** from the **Insert** menu and choose **Paste...** from the submenu (Excel 5.0) or choose **Paste Names** from the **Formula** menu (Excel 4.0), then press the Paste List button.

Figure 3-32 shows an example of a list of names produced by **Paste List**. Inspection of the list shows that the names log_pcKw and pcKw are duplicates. Of course, it's permissible to assign more than one name to the same reference, but in general it's not good programming practice.

	O	P
12	calib	='Sheet 1'!$G$8
13	CB	='Sheet 1'!$G$7
14	CL	='Sheet 1'!$G$5
15	C_	='Sheet 1'!$G$6
16	log_pcKw	='Sheet 1'!$G$9
17	n_bar	='Sheet 1'!$E$13:$E$43
18	pcKw	='Sheet 1'!$G$9
19	pH	='Sheet 1'!$B$13:$B$43
20	pH_corr	='Sheet 1'!$D$13:$D$43
21	v	='Sheet 1'!$A$13:$A$43
22	V_0	='Sheet 1'!$G$4

Figure 3-32. Using **Paste List** to audit a worksheet.

Sorting the list according to reference makes it easier to find duplications.

USEFUL REFERENCE

Microsoft Excel 5 Worksheet Function Reference, Microsoft Press, Redmond WA, 1994.

4

MACROS: AN INTRODUCTION

In early spreadsheet programs, a macro was simply a string of keystrokes that could be recorded, saved and repeated to automate a simple keyboard operation. In Microsoft Excel, macros are written in a programming language, that provides the capability to perform iterative calculations, or to take different actions based on the results of logic functions.

In Excel 4.0, macros are written in a macro language that uses many of the worksheet functions or menu commands with which you are already familiar. For example, the SUM worksheet function can be used in the same way in a macro as in a worksheet.

With Excel 5.0, Microsoft introduced a new macro language for Excel – Microsoft Visual Basic for Applications. Macros written in the "old" macro language (henceforth called Excel 4.0 Macro Language) can run on either Excel 4.0 or Excel 5.0. Procedures written in Visual Basic can run only on Excel 5.0.

There are advantages to using each system. If you are familiar with Excel's worksheet functions, you have a head start on learning the very similar Excel 4.0 Macro Language. You may already be familiar with the Excel 4.0 Macro Language and may have written macros in it. But as a programming language it's pretty clumsy. Visual Basic is a true programming language, with potential applications across the range of Microsoft products.

EXCEL 4.0 MACROS

An Excel 4.0 macro is a series of statements or commands on an Excel *macro sheet*, which looks much like an Excel worksheet. Macro statements can use any of Excel's worksheet functions, but in addition there are many *macro functions* that can be used only on macro sheets (over 400 macro functions are available in Excel 4.0 Macro Language).

When a macro is "run" or "called", its commands are executed downward in sequential order until a RETURN function is encountered.

A macro may be saved on a separate macro sheet, or on the *Global Macro Sheet* (Excel 4.0) or the *Personal Macro Workbook* (Excel 5.0). Macros saved on a separate macro sheet can be run only when the macro sheet is open. The Global Macro Sheet or Personal Macro Workbook is automatically opened whenever you start up Excel.

Microsoft Excel for Windows gives Excel 4.0 macro sheets the filename extension .XLM.

THERE ARE TWO KINDS OF MACROS IN EXCEL 4 — COMMAND MACROS AND FUNCTION MACROS

In Excel 4.0 you can create two different kinds of macro — function macros and command macros. Although they use many of the same set of macro functions, they are distinctly different.

Function macros augment Excel's library of built-in functions. A function macro is used in a worksheet in the same way as, for example, the SQRT function. It is entered in a single cell of a worksheet, performs a calculation and returns a single result (or an array result) to the cell in which it is located. For example, a custom function macro named ALPHA can be used to calculate α_j, the fraction of an acid-base species in one of its protonated forms H_jX at a particular pH. The function takes three *arguments*, the pH of the solution, the range of pK_a values of the weak acid, and the coefficient j. This function is useful in constructing distribution diagrams, titration curves, etc.

Command macros can automate any of Excel's menu commands. For example, a command macro might be used to open a new worksheet, copy selected ranges of cells from other worksheets and paste them into the new worksheet, format the data in the new worksheet, provide a heading and print the new worksheet. Command macros are not associated with a particular cell of a worksheet; command macros are usually "run" by selecting **Macro...** from the **Tools** menu (Excel 5.0) or **Run...** from the **Macro** menu (Excel 4.0). They can also be run by means of an assigned shortcut key, by being called from another macro, by simply opening or closing a document, or at a predetermined date and/or time.

Both kinds of macro can incorporate decision-making, branching, looping, subroutines and many other aspects of programming languages.

EXCEL 5.0 MACROS

An Excel 5.0 macro is a Visual Basic procedure in a separate *module sheet*. A module sheet does not have cells.

A macro module sheet may be saved as part of the current workbook, which means that it can be run only when the workbook is open, or it may be placed in the *Personal Macro Workbook*, which is automatically opened whenever you start up Excel. This workbook is saved in the Excel Startup Folder in the Preferences folder (Macintosh), or in the XLSTART directory (Windows).

Excel 5.0 can run macros recorded in Visual Basic or in Excel 4.0 Macro Language. Excel 5.0 will automatically record macros in Visual Basic, but you can record new macros in the Excel 4.0 Macro Language if you wish.

THERE ARE TWO KINDS OF MACROS IN EXCEL 5 — SUB PROCEDURES AND FUNCTION PROCEDURES

An Excel 5.0 Sub procedure is equivalent to an Excel 4.0 command macro, and an Excel 5.0 Function procedure is equivalent to an Excel 4.0 function macro.

WHAT'S AHEAD

In the chapters that follow, you'll learn how to create command macros and function macros using the Excel 4.0 Macro Language, and how to create procedures in Excel 5.0 s new Visual Basic for Applications. Appendices C and D provide listings of macro functions for Excel 4.0, which you will find valuable as you create your own macros in Excel 4.0. Appendices E and F provide similar listings of Visual Basic commands.

In Chapters 8 and 9 you'll learn how to create custom menu commands or toolbuttons, and how to assign them to your own macros. In this way you'll be able to create your own custom applications in Excel.

GOOD PROGRAMMING PRACTICES

FIRST, DEFINE THE PROBLEM

As in any type of computer programming, you need to have a clear picture of the task to be solved and the method of solving it. In addition, as in almost every programming venture, the task turns out to be much more complicated than originally envisioned. Especially if the macro is to be used by others, you will need to anticipate and handle errors. Making a macro "bulletproof" often adds more "code" than was in the original macro.

WRITING STRUCTURED MACROS

It's important to get into the habit of writing macros in a structured manner. The best combination of simplicity and structure seems to be a three-column format (Figure 4-1). Column A is reserved for labels, column B for macro statements, and column C for comments. This is identical to the format used in assembly-language programming.

	A	B	C
1	**FixedExpander** by Paul R. Dupuis, ©1993 By the Trustees of Boston		
2	College. All Rights Reserved. Expands a column containing multiple		
3	columns separated by 2 or more spaces and breaks them out into		
4	separate columns.		
5			
6	To use: Select the range containing the text-based table (one column by		
7	any number of rows) and execute the macro 'FixedExpander' by selecting		
8	Run from the Macro menu or pressing CMD-OPT-f on the keyboard.		
9			
10	Names	Formulas	Comments
11	.	=REFTEXT(SELECTION())	Gets reference of range
12		=ECHO(FALSE)	
13		=ROWS(TEXTREF(B11))	

Figure 4-1. Fragment of a structured macro.

Because the Excel worksheet is two-dimensional, there is plenty of latitude for spreading out over the whole worksheet. Each macro is executed downward, but, by employing subroutines, you can place different parts of the same macro side by side. As well, separate macros can exist side by side on the same macro sheet. This feature, combined with Excel's **Edit** capability, makes it all too easy to delete part of an adjacent macro by mistake, when editing. For this reason, it's advisable to use only columns A, B and C for your macros.

DOCUMENTATION IS IMPORTANT

Macros, like any other computer code, need to be maintained (modified/updated/debugged). What is perfectly clear to you now may be incomprehensible in three months. Adequate documentation should be included in every macro.

Documentation can exist at several levels:

- Give the macro a name that indicates its function. (But don't make the name too long to fit in the macro name box.) On the macro sheet, make the name bold for clarity.

- In the first few lines after the title, include a short text description of the purpose and use of the macro. Provide a clear description of the syntax of the macro.

- Within the macro, use variable names rather than cell references to increase clarity.

- If the macro contains subroutines, make their titles bold also.

- In the second line of a subroutine (and subsequent lines if necessary), include a short text description of the purpose and use of the subroutine. Describe any arguments that are passed.

- Use the comments column (column C) for a reminder of the general logic.

- The Cell Note function of Excel can be used to provide more detail, particularly the meaning of complicated expressions.

- For the most complicated, lengthy and important macros, it may be necessary to have separate documentation, especially if the macro is to be used by others.

5

CREATING COMMAND MACROS IN EXCEL 4.0 MACRO LANGUAGE

A command macro can automate any sequence of actions that can be performed by the use of menu commands or keystrokes. Command macros can also carry out much more complicated actions. In this chapter you'll learn how to record a sequence of menu and keyboard commands, how to write macros that incorporate logical branching or iterative looping, how to write interactive macros and how to write macros that run automatically whenever a sheet is opened or closed.

THE STRUCTURE OF A COMMAND MACRO

A command macro consists of a column of Excel statements in a macro sheet, terminated by a RETURN statement (Figure 5-1). Each macro statement is similar to an expression in a worksheet — it begins with an equal sign. Macro statements can use either macro functions, such as COPY, or worksheet functions, such as SUM. When the macro is run, the statements are executed successively until RETURN is encountered. It is permissible to separate blocks of code by blank cells or to intersperse cells containing titles or comments within the macro, for purposes of clarity or documentation. Excel ignores all statements that do not begin with an equal sign. The first cell of the macro should contain the name of the macro.

USING THE RECORDER TO CREATE A COMMAND MACRO

Excel provides the Recorder, a useful tool for creating command macros. When you choose **Record** from the Excel 4.0 **Macro** menu, or **Record New Macro** from the Excel 5.0 **Tools** menu, all subsequent menu and keyboard actions will

	A	B	C
1		MacroName	
2		Macro statements	
3		:	
4		=RETURN()	

Figure 5-1. Structure of a command macro.

be recorded on a macro sheet until you press the Stop Macro button (Excel 5.0) or choose **Stop Recorder** from the **Macro** menu (Excel 4.0). The Recorder is sufficient for creating simple macros, but you can't incorporate logic, branching or looping by using the Recorder. After using the Recorder to create some simple macros, you'll view it as simply a tool to create fragments of macro code for incorporation into more complex macros. Macros that involve only the use of menu or keyboard commands can be created using the Recorder.

> *Excel Tip. Use the Recorder to record a single menu command if the syntax of the corresponding macro function is long or complicated. Then **Copy** it and **Paste** into your macro.*

In Excel 5.0, the Recorder creates Visual Basic commands (it's the default option). You don't have to know anything about Visual Basic in order to record a command macro in Visual Basic. This also provides a good way to gain some initial familiarity with Visual Basic. If you want to record the macro in Excel 4.0 Macro Language, press the Options button in the Record New Macro dialog box, then press the MS Excel 4.0 Macro button.

CREATING A SIMPLE MACRO USING THE RECORDER

As an example of a simple command macro created by using the Recorder, we'll create a macro called *FullPage*. Very often, to squeeze as much information on a single printed worksheet page, it is helpful to eliminate the default margins (Top, Bottom, Left and Right) and the Header and Footer text. This is done using the **Page Setup...** command in the **File** menu.

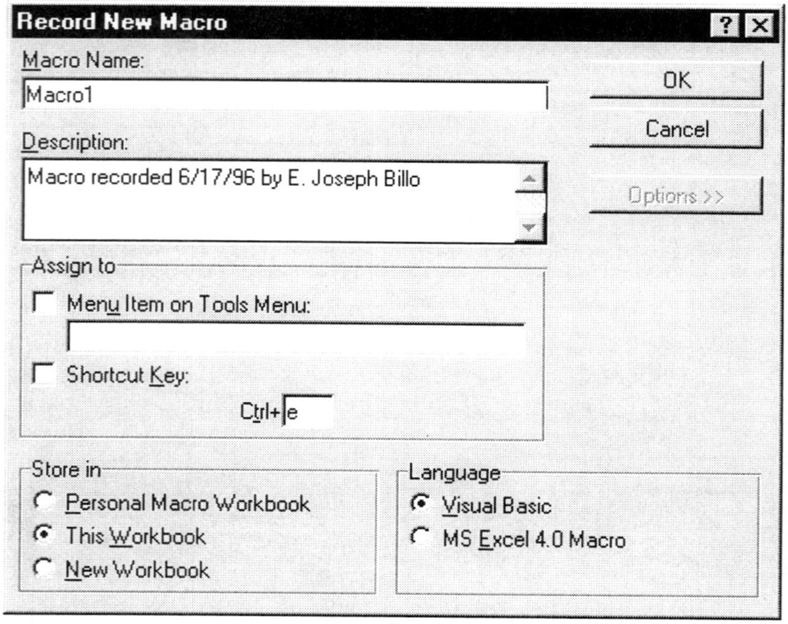

Figure 5-2. The Excel 5.0 Record New Macro dialog box.

To create a macro to do all this automatically, using the Recorder, carry out the following steps. (You must have a open worksheet to operate on.) From the **Tools** menu, choose **Record New Macro....** Then, as indicated in Figure 5-2, you will be asked where you want to store the macro in the Personal Macro Workbook (the Global Macro Sheet if you're using Excel 4.0) or in a macro sheet.

If the macro is stored on the Personal Macro Workbook, it will be available whenever you are using Excel. If it's stored on a separate macro sheet, then the macro sheet must be opened before the macro can be used. Since you'll probably want to use the *FullPage* macro often, choose "Store Macro in Personal Macro Workbook". Then perform the following operations: (i) Choose **Page Setup...** from the **File** menu. (ii) Enter zero in the left, right, top and bottom margin boxes. (iii) Delete the information in the Header box, then press the OK button. (iv) Delete the information in the Footer box, then press the OK button. (v) Complete the operation by pressing the OK button.

When done, press the Macro Stop button (Excel 5.0) or choose **Stop Recorder** from the **Macro** menu (Excel 4.0).

To examine the macro, choose **Unhide** from the **Window** menu (assuming that the Personal Macro Workbook is hidden). It will be in column A if no other global macros have been recorded.

The macro in Figure 5-3 consists of only two macro statements, in cells A2 and A3. The first six arguments in the PAGE.SETUP function are, respectively, header text, footer text, and the left, right, top and bottom margin settings. The other arguments are the default ones for a new worksheet, e.g., don't print row and column headings, print cell gridlines. (See the description of the PAGE.SETUP function in Chapter 9.) The eleventh argument, *orientation*, determines the orientation of the document on the page (1 = Portrait, 2 = Landscape). To make the macro a general one, delete the seventh through tenth arguments (FALSE,TRUE,FALSE,FALSE in Figure 5-3) while leaving the commas as placeholders, and delete all arguments after the eleventh. Arguments not specified will be unchanged from their current values.

To complete the process, select cell A1. Choose **Define Name** from the **Insert** menu (Excel 5.0) or from the **Formula** menu (Excel 4.0) to activate the Define Name dialog box.

If possible, Excel provides a name in the Name box, using the text in the selected cell or from the cell immediately above or to the left of the selected cell if it contains other than text. Here Excel has assigned the name Record1_a, the text in the selected cell, to the macro. (A name can't contain a parenthesis, so Excel used the underscore character instead.) The cell reference $A$1 appears in the Refers To box. You should give the macro a more descriptive name than Record1_a. Name it *FullPagePortrait*. (Names can't contain spaces.) Later you

	Å
1	Record1 (a)
2	=PAGE.SETUP("","",0,0,0,0,FALSE,TRUE,FALSE,FALSE,1,,100,1,1,FALSE)
3	=RETURN()

Figure 5-3. A recorded macro.

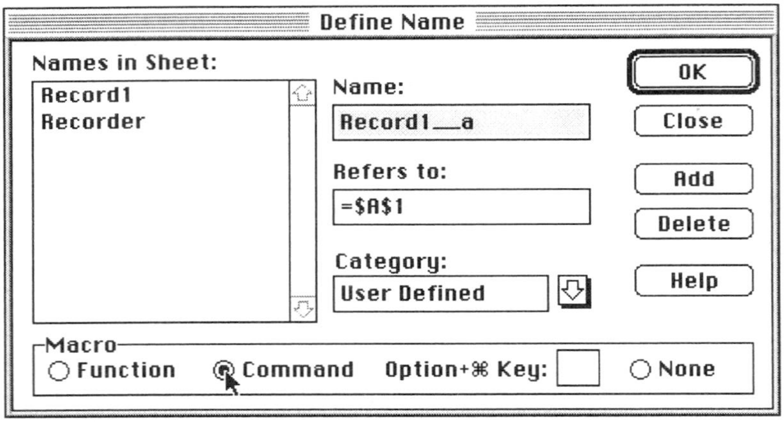

Figure 5-4. The Excel 4.0 for the Macintosh Define Name dialog box.

can create a similar one called *FullPageLandscape*. Then press OK.

To use the macro, choose **Run** from the **Macro** menu. In the Run Macro dialog box (Figure 5-5) you'll see a list of all macros in open macro sheets or in the Global Macro Sheet. Choose the desired macro from the list and then press the OK button, or double-click on the macro name. The margins of the active sheet will be set to zero and the header and footer will be eliminated. Page break lines will be dispayed on the sheet.

As a second example of a command macro created by using the Recorder, let's imagine that you want to take data from non-adjacent areas of one or more spreadsheets and print the data as a report. A simple way to do this is to copy the separate areas to a new worksheet and then print the new sheet. If this is to be performed repeatedly in exactly the same way (for example, to produce a monthly summary report), then creating a macro to automate the process could be worthwhile.

In the following example you copy student quiz grades from a class list, modifying the data and sorting it before printing. In the original worksheet *ClassList*, the student names are in column A, the student ID numbers are in

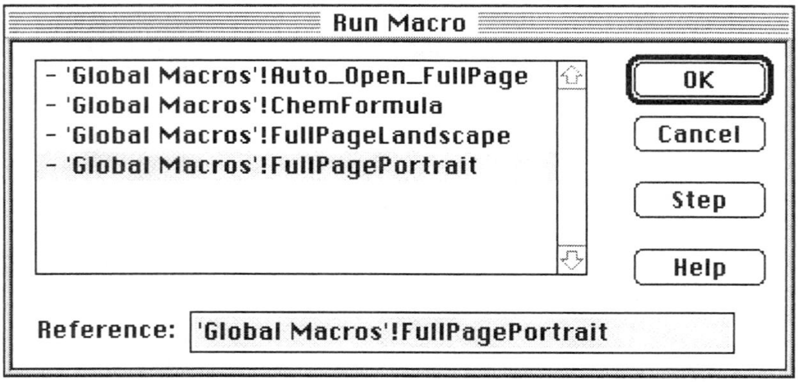

Figure 5-5. The Excel 4.0 for the Macintosh Run Macro dialog box.

column G and current class grades are in column O. You want to prepare a weekly report with the latest student grades, listed only by ID number, to preserve anonymity. The 4-digit ID numbers will be in column A and the class grades in column B of the new worksheet. Before printing, you must sort the list so that the 4-digit identifiers are in ascending order. To do this manually, you carry out the following sequence of menu commands, keystroke operations and mouse operations: (i) select and **Copy** the column of ID# data in the *ClassList* worksheet, (ii) open a new worksheet and **Paste** the data in column A, (iii) return to the *ClassList* worksheet, (iv) select and **Copy** the column of current class grades, (v) activate the new worksheet and **Paste** the data in column B, (vi) select the array of data (rows 1 and 2 contain headings and were not selected) and use the **Sort** command from the **Data** menu to sort in ascending order of ID#.

To use the Recorder to create a macro to carry out the preceding operations, simply do the following: with the *ClassList* worksheet active, choose **Record...** from the **Macro** menu. In the Record Macro dialog box, choose Macro Sheet (see the discussion of macro sheet vs. global macro sheet later in this chapter). Then carry out the sequence of operations listed above. When complete, choose **Stop Recorder** from the **Macro** menu. Now activate the macro sheet entitled Macro1. (Excel numbers macro sheets as they are created, beginning with Macro1.) The macro sheet should look something like that shown in column A of Figure 5-6. (The comments in column B have been added.)

As you will see by examining the macro functions that appear in column A, the Recorder uses the *R1C1 method* for cell references, where R and C indicate the row and column of the selected cell. Thus, for example, in line 2 of the macro listing, C7 refers to column 7, i.e., column G; in line 7 of the macro listing, R3C2 refers to cell B3. Either A1-style references or R1C1-style references can be used in macros; both styles can be used in the same macro. Switching from one method of cell referencing to the other can be a little confusing since, in the A1 method, the column is given first, rather than the row.

	A	B
1	Record1 (a)	
2	=SELECT("C7")	Select ID# column (column G) in Class List worksheet.
3	=COPY()	Copy it.
4	=NEW(1)	Open a new worksheet, using the File menu command.
5	=SELECT("C1")	Select column A in the new worksheet.
6	=PASTE()	Paste the data into the new worksheet.
7	=ACTIVATE("94F Class List")	Use Window menu command to go back to Class List.
8	=SELECT("C15")	Select column O in Class List worksheet.
9	=COPY()	Copy the selected data.
10	=ACTIVATE("Worksheet1")	Use Window to go back to new worksheet.
11	=SELECT("C2")	Select column B as destination...
12	=PASTE()	Paste the data into the new worksheet.
13	=SELECT("R3C1:R150C2")	Select the array of data (A3:B150).
14	=SORT(1,"R3C1",1)	Sort the data.
15	=RETURN()	(appended automatically)

Figure 5-6. A command macro using the Recorder.

The macro functions in cells A2 to A14 of Figure 5-6, such as COPY, NEW, PASTE, are *command-equivalent* macro functions; that is, they correspond to Excel's menu commands. If the menu command has an ellipsis (for example, **New**...) then a dialog box appears when the command is chosen from the menu. The corresponding macro function requires arguments, generally listed in the same order as they appear in the dialog box. For example, the **New** dialog box lists choices in the order: Worksheet, Chart, Macro Sheet, ... The syntax of the NEW macro function is NEW(*type_number*), where 1 = Worksheet, etc.

Finally, use **Define Name** from the **Formula** menu to give a name to the first cell of the macro , press the Command button, then press OK.

As recorded, this macro is not completely general. Cells A7, A10 and A13 need to be modified so that data can be copied from any class list (not just 94F Class List), data can be pasted into any new sheet (not just Worksheet1), and a data array of any size can be selected for sorting (not just rows 3 - 150). Later in this chapter you'll see how to do this.

ABSOLUTE OR RELATIVE REFERENCES WITH THE RECORDER

As you've seen, the *R1C1 method* can be used for cell references in macros. R1C1-style references are always given as text, that is, they are always placed within quotes. Like A1-style references, an R1C1-style reference can be either an *absolute* or a *relative* reference. An *absolute reference* is a reference such as R1C7 or R3C1:R150C2. A *relative reference* is a reference such as R[1]C[5], that indicates the cell one row down and 5 columns to the right, relative to the current selection. Other examples of relative references are RC (the current selection), R[1]C (one row down from the current selection) and RC[-2] (two columns to the left of the current selection).

The **Macro** menu provides the option of recording a macro using either absolute or relative references. When you choose the **Macro** menu, either **Absolute Record** or **Relative Record** will be displayed in the menu. You can toggle back and forth between the two options — if **Relative** is displayed, then **Absolute** is in effect, and vice versa. The default option is **Absolute**.

SAVING MACROS IN A MACRO SHEET
OR IN THE GLOBAL MACRO SHEET

When you choose **Record New Macro...** from the **Tools** menu (Excel 5.0) or **Record...** from the **Macro** menu (Excel 4.0), you are asked whether you want to store the macro in the Personal Macro Workbook (Excel 5.0) or the Global Macro Sheet (Excel 4.0), or in a new macro sheet. The Personal Macro Workbook or Global Macro Sheet is opened every time you open Excel. It is useful for commonly used macros, such as macros associated with custom tools on one of Excel's toolbars (see Chapter 9). Macros that are being developed or that are used only occasionally should be saved on macro sheets. Related macros should be saved on a single sheet; they may make use of common code in the form of subroutines. Specialized macro sheets can be saved in a workbook containing the worksheets that require the macro sheet; that way the macro sheet will be opened each time the workbook is opened.

It's a good idea to use only columns A, B and C in a macro sheet. If you place macros side by side, there's an excellent chance that you'll mistakenly delete lines at a later date as you modify one of the macros.

Summary of Steps to Create a Command Macro

1. From the **File** menu, choose **New**, choose **Macro Sheet**, then press OK.

2. Type in the macro, using Excel's worksheet and macro functions. The Recorder can be used to create fragments of code, which are then copied and pasted into the macro sheet.

3. Select the first statement of the macro, which should be a descriptive name. From the **Formula** menu, choose **Define Name**. Modify the name as necessary.

4. Press the Command option button. You may assign a shortcut key at this time. Then press OK.

5. Save the macro sheet. (Any document name is OK, but note that the macro name appearing in the Macro Run dialog box will be SheetName!MacroName.)

CREATING ADVANCED MACROS

Macros can do much more than the simple keyboard tasks available to the Recorder. With Excel 4.0 Macro Language you can create macros that obtain values from other worksheets, request information from the user, incorporate branching and looping functions, and much more.

GETTING VALUES FROM A WORKSHEET

Your macro may need to get information from other documents and transfer that information to a new worksheet. For example, the macro may prepare a report by copying information from one or more other sheets and pasting it into a new worksheet before printing it.

Information can be obtained from another sheet (the source sheet) simply by using an external reference to that sheet. If the location of the cell from which the information is to be obtained won't change, then simply type an equal sign in the cell in the macro sheet where you want the information to appear. Then activate the source sheet and click on the desired cell. An external reference to the source sheet will be entered in the macro sheet. Then click the check box in the formula bar to complete the formula. For example, if the date that a worksheet *Lab.Inventory* was revised is saved in cell G1 of the worksheet, then the macro statement in Figure 5-7 will return the date value, provided the source sheet is the active document. The exclamation mark preceding the cell reference indicates that it is an external reference (a reference to a cell in the active sheet, not in the macro sheet).

58	Last.date	=!\$G\$1

Figure 5-7. External reference to the active worksheet.

58	Last.date	=Lab.Inventory!\$G\$1

Figure 5-8. External reference to a worksheet.

The statement in Figure 5-8 will obtain the date even if the document is not the active document.

If the document is open, the external reference is as shown in Figure 5-8. If the document is closed, the reference automatically becomes the full path description, i.e.,

='Macintosh HD:Departmental Files:Lab.Inventory'!\$G\$1

The macro function **SELECTION** refers to the currently selected cell in the active worksheet. The syntax of the function is simply SELECTION(). SELECTION can be a single cell or a range. Use **SELECTION** in a macro to operate on a cell selected by the user. Other functions, e.g., **OFFSET**, can use SELECTION as a point of reference.

EXAMPLES

=IF(SELECTION()="",RETURN())

=OFFSET(SELECTION(),x,) returns the value (actually the reference) of the cell x rows down from the current selection. The location of the active cell is not changed.

The macro function **SELECT** moves the active cell to a new location. The syntax is **SELECT**(*reference*). Use **SELECT** with either A1-style or R1C1-style reference to select a cell on a worksheet

EXAMPLES

=SELECT(!A1) selects cell A1 of the active worksheet.

=SELECT("R[1]C") selects the cell one row down from the current selection.

=SELECT("R"&row&"C"&col) selects the cell on the active worksheet with Row = row and Column = col.

Many macro functions take a reference argument. For example, CUT() operates on the current selection in the active worksheet, but CUT(*from_reference*, *to_reference*) will operate on two separate worksheets, neither of which needs to be activated.

To create a completely general macro that operates on open worksheets, you will need to obtain the sheetname of the source document and the destination document. For example, if you open a new worksheet to create the destination document, it will not always have the name Worksheet1. Use GET.DOCUMENT(*info_type_number*) to obtain information about the active document. *Info_type_number* specifies the type of information to be returned;

7	source.sheet	=GET.DOCUMENT(1)
8		=NEW(1)
9	dest.sheet	=GET.DOCUMENT(1)
10		=ACTIVATE(source.sheet)

Figure 5-9. Using GET.DOCUMENT to create code that is not sheetname-specific.

info_type_number = 1 returns the name of the active document. The macro statements in Figure 5-9 save the name of the active document as *source.sheet*, open a new worksheet and save its name as *dest.sheet*, then activate the original source document.

The three macro fragments in Figures 5-10 through 5-12 illustrate methods of copying from one sheet and pasting to another sheet. In the first example (Figure 5-10) the source sheet is open and the name of the destination sheet is known. Column B of the active sheet is copied, then pasted into column C of the sheet *Chem.Inventory.Quarterly.Report*.

In the second example GET.DOCUMENT(1) is used to obtain the names of the source and destination documents. As Figure 5-11 indicates, each document is activated as needed, in order to **Copy** or **Paste**.

The third example (Figure 5-12) determines the size of the reference to be copied by counting the number of non-blank entries in column B of the source sheet. It uses R1C1-style references to select the range of data to be copied in column 2 and then pasted in column 3. Neither document needs to be the active document (although both have to be open, of course.)

	A	B
1		Copies from active sheet to a named sheet
2		(specific sheet names used)
3		=COPY(!$B:$B)
4		=PASTE(Chem.Inventory.Quarterly.Report!$C:$C)

Figure 5-10. External reference to a specific worksheet.

6		Copies from active sheet to a new sheet.
7		(general sheet names used)
8	source.sheet	=GET.DOCUMENT(1)
9		=COPY(!$B:$B)
10		=NEW(1)
11	dest.sheet	=GET.DOCUMENT(1)
12		=PASTE(!$C:$C)
13		=ACTIVATE(source.sheet)

Figure 5-11. External reference to a worksheet using generalized sheet names.

15		Copies from active sheet to a new sheet.
16		(general sheet names used)
17	source.sheet	=GET.DOCUMENT(1)
18	N	=COUNTA(!$B:$B)
19		=NEW(1)
20	dest.sheet	=GET.DOCUMENT(1)
21		=COPY(""&source.sheet&"'!R1C2:R"&N&"C2")
22		=PASTE(""&dest.sheet&"'!R1C3:R"&N&"C3")

Figure 5-12. Obtaining the size of the range to be copied and pasted.

WORKING WITH VALUES WITHIN THE MACRO SHEET

The SET.VALUE function is used in a macro to assign a value to a cell in the macro sheet. The syntax is SET.VALUE(*reference, value*). *Reference* can be a single cell or a range. *Value* can be a single value or an array. If *reference* is a range and *value* is a single value, then *value* is entered in all the cells of the range.

SET.VALUE is used to store intermediate values of a calculation, of an increment counter, or of an index or pointer, etc., during macro execution.

EXAMPLES

=SET.VALUE($C$1,1E-9) enters a value in cell $C$1.

=SET.VALUE(dest,"") enters a null string in the cell dest.

=SET.VALUE(pointer,pointer+1) increments the value stored in the cell pointer.

SET.VALUE assigns a value to a cell without affecting the formula contained in the cell (see Figure 5-13).

SENDING VALUES TO A WORKSHEET

The FORMULA function is used to enter information from a macro sheet into cells on another sheet. The syntax is FORMULA(*formula_text*, *reference*). *Formula_text* can be a value (a number, text, a named variable or an expression) or a formula. The optional *reference* argument specifies where the text is to be entered. If *reference* is omitted, *formula_text* is entered in the active cell.

	A	B
4		**FullPageLandscape**
5	orientation	=2
6		=GOTO(B9)
7		**FullPagePortrait**
8		=SET.VALUE(orientation,1)

Figure 5-13. SET.VALUE assigns a value but does not affect the formula.

EXAMPLES

=FORMULA(0.1023) enters the number in the active cell of the worksheet.

=FORMULA("Gas cylinder usage for the month of "&*month*) enters the text, with the value of *month*, in the active cell.

=FORMULA(X_1+incr,"R"&r_+x&"C"&c_) enters the value of the expression X_1+incr from the macro sheet in the cell of the active sheet with Row = r_+x and Column = c_.

=FORMULA(SUM(B25:F25),Chem.Inventory.Quarterly.Report!$F$17) enters the sum from the macro sheet into the specified cell of the worksheet Chem.Inventory.Quarterly.Report.

To enter a formula in a cell of a worksheet, enclose the formula, complete with equal sign, in quotes. References in *formula_text* must be R1C1-style. Text arguments in formulas must be surrounded by double quotation marks.

EXAMPLES

=FORMULA("=R[-1]C+"&$C$3) enters the formula =A3+0.5 in active cell A4 of the worksheet if the macro sheet contains the value 0.5 in cell C3.

=FORMULA("=""greater than ""&R1C3",!A1) enters the formula ="greater than "&$C$1 in cell A1 of the active sheet; if cell C1 contains the value 100, cell A1 displays the value "greater than 100".

BRANCHING

There are three ways to branch: by using a GOTO statement, by using an IF statement, or by using a subroutine.

The GOTO function is an unconditional branch. It uses the syntax

=GOTO(***reference***)

Reference can be a cell reference or a name within the macro sheet, or an external reference to another macro sheet, which must be open.

This method of branching can easily lead to macros with confusing logic. Such "spaghetti code" should be avoided if possible.

The IF statement, when used in a macro, has two possible syntax forms, the simple IF statement (Syntax 1), whose syntax is almost identical to the IF used in worksheets and the block IF statement (Syntax 2).

Syntax 1: =IF(***logical_test, action_if_true,*** *action_if_false*)

In a macro, *action_if_true* and   *action_if_false* are often GOTO statements.

EXAMPLE

=IF(LEFT(formatcode,1)="S", GOTO(S.Convert),RETURN())

Syntax 2 (or block IF statement): =IF(*logical_test*)

The block IF statement (Figure 5-14) is followed by a series of macro instructions to be carried out if *logical_test* is true, and is terminated by END.IF. Use IF... END.IF when you have more than one statement to execute if *logical_test* is true.

EXAMPLE

	B
14	=IF(name="")
15	=BEEP()
16	=RETURN()
17	=END.IF()

Figure 5-14. Example of the use of a block IF statement.

A block IF can be followed by ELSE or ELSE.IF. ELSE.IF can be followed by another ELSE.IF. END.IF must always be used to terminate a block IF. An example is shown in Figure 5-15.

The block IF format is recommended except in the case of simple branching or looping logic, where the IF(*logical_test*, GOTO...) function is acceptable. If multiple actions must be performed, block IF statements result in more structured programming. An alternative approach is to use subroutines (see "Using Subroutines" later in this chapter).

LOOPING

There are four ways to loop: using a FOR...NEXT loop, using a FOR.CELL...NEXT loop, using a WHILE...NEXT loop or using an endless GOTO loop. Each type has certain advantages and limitations.

USING FOR...NEXT

The FOR...NEXT loop is repeated for a definite, known number of iterations. The loop is initiated by the FOR statement with the following syntax: FOR(*counter_text*, *start_number*, *end_number*, *step_number*). *Counter_text* is a text name that is used to count the number of times the loop has been executed. *Start_number* is the initial value assigned to *counter_text*.

19	=IF(logical_test1)
20	(macro instructions to be carried out if logical_test_1 is true)
21	=ELSE.IF(logical_test2)
22	(macro instructions to be carried out if logical_test_2 is true)
23	=ELSE()
24	(macro instructions to be carried out if neither is true)
25	=END.IF()

Figure 5-15. IF...ELSE.IF...ELSE...END.IF structure.

End_number is the value of *counter_text* after which the loop is terminated. *Step_number* is the value by which *counter_text* is incremented. *Counter_text, start_number* and *end_number* must be supplied, but *step_number* is an optional parameter; if it is omitted, it is assumed to be 1.

Following the FOR function are the formulas that are to be evaluated each time the loop is carried out. The NEXT() function delineates the end of the loop. Figure 5-16 illustrates the structure of a FOR...NEXT loop.

EXAMPLE

30	=FOR("x", 1, 100)
31	(macro instructions to be carried out on each cycle of the loop)
32	=NEXT()

Figure 5-16. A FOR...NEXT loop.

A simple macro to enter the names of the months in a worksheet is shown in Figure 5-17.

In cell B3 the OFFSET function is used to calculate a reference to a cell relative to another cell. Here the referenced cell is *x* rows down and zero columns to the right from the base reference, cell $C$1 in the macro sheet. The loop counter in cell B2 goes from 0 to 11 to correspond with the OFFSET from cell C1. The macro function FORMULA(*formula_text*) inserts the text in the active cell of the worksheet (much more about FORMULA later in this chapter). In cell B4 the active cell (the selection) is moved one column to the right. Then the process is repeated.

Figure 5-18 shows an alternative approach involving the macro function FORMULA(*formula_text* , *reference*). *Reference* specifies where *formula_text* is to be entered, relative to the current selection. Using concatenation of text and variables, the R1C1-style reference changes with each cycle of the loop, from RC[0] to RC[11].

	C	D
1	**loop macro1**	January
2	=FOR("x",0,11)	February
3	=FORMULA(OFFSET($D$1,x,))	March
4	=SELECT("RC[1]")	April
5	=NEXT()	May
6	=RETURN()	June
7		July
8		August
9		September
10		October
11		November
12		December

Figure 5-17. Example of a FOR...NEXT loop.

	E	F
1	loop macro2	January
2	=FOR("x",0,11)	February
3	=FORMULA(OFFSET($F$1,x,),"RC["&x&"]")	March
4	=NEXT()	April
5	=RETURN()	May
6		June
7		July
8		August
9		September
10		October
11		November
12		December

Figure 5-18. Another example of a FOR...NEXT loop.

There is a big difference between these two approaches: the first moves the selection, the second doesn't. Using the latter method, you can write to cells that are not on the screen without changing the selection. Using the former approach, each time the selection moves off-screen, the screen is updated to keep the selected cell visible. This can be annoying if screen updating occurs repeatedly.

Remember that counter text is a text variable and must be placed in quotes.

USING FOR.CELL...NEXT

The FOR.CELL...NEXT loop is useful for operating on a range of cells. The syntax is FOR.CELL(*cell_text*, *range_reference*, *skip_blanks_logical*). *Cell_text* is the name used to identify each cell in turn. *Range_reference* is the range of cells to be operated on; if omitted, the current selection is used. If *skip_blanks_logical* is TRUE or 1, blank cells in the selected range are skipped. Cells are operated on from left to right in a row, then downwards to the next row; see Figure 5-19.

USING WHILE...NEXT

The WHILE...NEXT loop is an open-ended loop, terminated by a logical test. The syntax is WHILE(*logical_test*). As long as *logical_test* is TRUE, the loop will be carried out. If *logical_test* is FALSE, the loop is not carried out. As in the FOR...NEXT loop, the loop is delimited by the NEXT() function. The WHILE...NEXT loop (Figure 5-20) is useful for reading successive values from a table and

34	=FOR.CELL("z",,1)
35	=SELECT(z)
36	(macro instructions to be carried out on each cell
37	in the selected range, skipping blank cells)
38	=NEXT()

Figure 5-19. The FOR.CELL...NEXT loop.

40	=WHILE(SELECTION()<>"")
41	(Perform some action)
42	=SELECT("R[1]C")
43	=NEXT()
44	=RETURN()

Figure 5-20. The WHILE...NEXT loop.

46	begin	(Perform some action)
47		=SELECT("R[1]C")
48		=IF(SELECTION()<>"",GOTO(begin),HALT())

Figure 5-21. A GOTO endless loop.

performing an action until the end of the table is reached, as indicated by reading a null value.

AN ENDLESS LOOP BY USING GOTO

The GOTO endless loop terminated by a logical test essentially performs the same function as the WHILE...NEXT loop. The logical test can be placed anywhere in the loop. This type of loop (Figure 5-21) is seldom preferred.

USING SUBROUTINES

Command macros can utilize subroutines in the same way as other programming languages. Most subroutine macros will be located on the same sheet as the calling macro. To create a subroutine macro, choose **Define Name** from the **Formula** menu, type a name in the Name box, press the None option button, then press OK. The name cannot include spaces; use the underline or period instead.

To call the subroutine macro, use the expression =subroutine_name(). Alternatively you can use the expression =address(), e.g., =$B$21(). Arguments can be passed to a subroutine in the same way as in a function macro (see next chapter).

The use of subroutines to break a complicated task up into smaller, more manageable parts is strongly recommended.

SUMMARY OF TYPES OF CELL REFERENCE IN EXCEL 4.0 MACROS

In a command macro, cell references can have a variety of forms, and each form is used for a specific purpose. Cell references can be in either A1-style or R1C1-style, and they can be either absolute or relative. They can refer to the macro sheet (the reference alone is given), to the active worksheet (the reference is preceded by !) or to another worksheet (the reference is preceded by *SheetName!*). Thus there are 12 possible reference types. Not all 12 reference

Table 5-1. Reference Styles for Macro Sheets

To ...	Use the following reference style(s)	Example
... make a reference to a cell in the macro sheet.	$A$1*	=GOTO($B$51) (transfer macro execution to cell B51)
... make a reference to a cell in the active worksheet.	!$A$1	=FORMULA("Sample Number",!$E$1) (enter the text "Sample Number" in cell E1 of the active worksheet)
or	"!R1C1"	=SELECT("R1C2:R3C14") (select cells B1:N3 in the active worksheet)
... make a relative reference to a cell in the active worksheet.	"R[x]C[y]"	=SELECT("R[-1]C") (select the cell one row to the left of the active cell in the active worksheet.)
... make a reference to a cell in a named sheet.**	Name!$A$1	=Budget!$D$20 (returns the value contained in cell D20 of the sheet Budget).

*   Either the A1 or $A$1 forms of the reference may be used.
**   The sheet does not need to be open.

forms are useful in macro sheets: the most commonly used ones are shown in the Table 5-1, with examples.

Generally A1-style references are more convenient and understandable. When R1C1-style references are used, they must appear as text (that is, in quotes). Some functions require references to be in R1C1 style. Because R1C1-style references are given as text, you can use the & operator to concatenate text and variables in order to define a reference. An example of this approach was shown in Figure 5-18.

INTERACTIVE MACROS

Macros can be interactive: they can pause and issue a message to the user or request data. Messages are provided by means of the ALERT function; data is requested by means of the INPUT function.

USING ALERT BOXES

The ALERT function displays an Alert box on the screen. The syntax is ALERT(*message_text, type_number*). There are three types of Alert box: Type 1 displays a message with OK and Cancel buttons. The ALERT function returns TRUE if the OK button is pressed and FALSE if the Cancel button is pressed.

Figure 5-22. Types 1, 2 and 3 Macintosh Alert box icons.

Types 2 and 3 display a message and an OK button only; they differ only in the type of icon displayed (See Figure 5-22). A type 2 Alert box is suitable for providing information; a type 3 Alert box provides a warning. An optional Help button can also be displayed in each type of box.

EXAMPLE

The formula

=ALERT("Increment must be greater than zero. You entered "&incr&". Please re-enter.",2)

produces the box shown in Figure 5-23, if the named reference incr contains the value –10.

GETTING USER INPUT

There are two convenient ways to get user input: by using the INPUT function, or by means of the ?-form of certain commands. (User input can also be obtained by means of a custom dialog box. Creating custom dialog boxes is not discussed in this book; see *Complete Guide to Microsoft Excel Macros* for guidance.)

The INPUT function displays a dialog box for user input. The dialog box can have a title in the title bar, and a message within the box. The dialog box has an edit box into which the data is entered. A default value for the input can appear in the box.

The syntax of the INPUT command is INPUT(*message_text*, *type_number*, *title_text*, *default_value*, *x_position*, *y_position*, *help_reference*). *Type_number* is a number specifying the type of data to be entered. *Title_text* is the title to be displayed in the title bar; if *title_text* is omitted, the title will be "Input". *Default_value* is the default value to be displayed in the edit box. *X-position*, *y-position* and *help_reference* are described in the Excel reference manuals or in the On-Line Help.

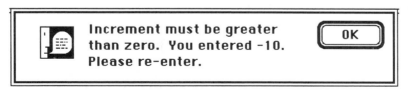

Figure 5-23. An Alert box message.

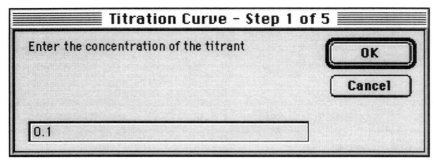

Figure 5-24. An Input box.

The statement (all in one line, of course)

=INPUT("Enter the concentration of the titrant",1,"Titration Curve - Step 1 of 5",0.1)

produces the dialog box shown in Figure 5-24.

If a value is entered and the OK button pressed, the cell containing the input statement contains the value entered. If the Cancel button is pressed, the cell contains **FALSE**.

THE ?-FORM OF A COMMAND

Every menu command has a command-equivalent macro function. If the menu command has an ellipsis, indicating that a dialog box will appear when the command is chosen from the menu, then the corresponding macro function has a ?-form that will produce the same dialog box. Thus for example NEW(1) creates a new worksheet, but NEW?() opens the New dialog box.

AUTOEXEC MACROS

An AutoExec macro is a macro that runs automatically whenever a certain event occurs: when a certain document is opened or closed, or when a window is activated or deactivated.

To create an AutoExec macro, simply give it a name that begins with Auto_Open, Auto_Close, Auto_Activate or Auto_Deactivate. Usually you'll want to press the "None" button in the Define Name dialog box, so that the macro doesn't appear in the list of available macros in the Macro Run dialog box.

EXAMPLE. Figure 5-25 shows macro code at the beginning of a macro sheet that automatically hides the macro sheet when it is opened.

	A	B
1		**Auto_Open_Macro**
2		=HIDE()
3		=RETURN()
4		(additional macros)

Figure 5-25. A simple Auto_Open macro.

Excel Tip.   *To open or close an AutoExec macro sheet without running the macro, press SHIFT while you* **Open** *or* **Close** *the sheet.*

DEBUGGING A MACRO

During the development of a macro, it is almost certain that errors will be revealed when the macro is run. When an error is encountered during execution, the Macro Error dialog box (Figure 5-26) will be displayed. The only information provided to you will be the cell in which an error caused the macro to halt. If the macro sheet is not hidden, the Goto button will be enabled. By pressing it you will be able to go directly to the offending cell; otherwise press the Halt button.

To locate errors in macro code, you can examine the value contained in a cell or the current value of a variable or expression. As well, you can examine the step-by-step execution of the macro.

EXAMINING VALUES OF CELLS OR VARIABLES

If an error is encountered, Excel displays the Macro Error dialog box, which indicates the cell in which the error occurred. You can use the Goto button to jump quickly to that cell. To examine the expression in the cell, choose **Display**... from the **Options** menu and uncheck the Formulas check box, to show values rather than formulas. An error value displayed in a cell, and the error type, may be useful in tracking down the error (see "Error Values and Their Meanings" in Chapter 3).

Excel Tip. *To view cell values rather than formulas, use COMMAND+ (Macintosh) or CONTROL+SHIFT+ ~ (Windows). Repeat the keystroke series to toggle back to formulas.*

To view the value of a named variable or an expression, select it in the formula bar, then press COMMAND+= to display its value. Click on the X-box in the formula bar or press the Undo key to revert to the formula.

EXAMINING A MACRO STEP BY STEP

As another aid in locating the source of error in a command macro, you can use the Step option in the Run Macro dialog box (see Figure 5-5). This permits you to examine each formula and its result as it is calculated (Figure 5-27). Use the Evaluate button to observe the calculation of a complicated formula one function or operator at a time.

Figure 5-26. Macro Error dialog box.

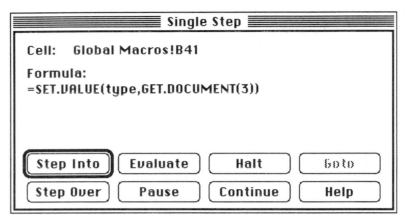

Figure 5-27. The Single Step dialog box

Alternatively you can insert a STEP() expression into the macro. When you **Run** the macro, it will switch to Step mode when it reaches the cell containing the STEP function.

> *Excel Tip.* Use *STEP* in an *IF* statement to examine a macro step by step only when a certain condition occurs.

> *Excel Tip.* Delete the equal sign in a macro statement to temporarily prevent that statement from executing.

A MACRO TO PERFORM A ROW-BY-ROW SORT

Excel's **Sort** command in the **Data** menu can sort a two-dimensional array either by rows or by columns. For example, a class list of student names (e.g., in column A), ID numbers (column B) and grades (in columns C through N) could be sorted by ID number.

Occasionally it is desirable to be able to sort a two-dimensional array of data both horizontally *and* vertically. In the example above, the class list contains quiz grades for 12 quizzes in columns C through N. If it is desired to drop the two lowest grades for each student, then a convenient way to do this would be to sort each student's grades in descending order and then sum the first 10 columns for each student. This requires a row-by-row sort. For a small class the most efficient way would be to manually select each student's row of 12 grades and sort them, but for large classes this becomes tedious. The following macro carries out such a "row-by-row" sort automatically. (It is assumed that the initial sort by rows, to put the class list in alphabetical order, has already been done.) The macro is written to sort individual rows of columnar data in descending order.

The data array to be sorted is selected in the same fashion as for the normal Excel **Sort** command, by clicking and dragging with the mouse. Then the *Row-by-Row Sort* macro is run by using the **Run** command in the **Macro** menu.

	A	B
1	**Row-by-Row Sort**	
2	=ECHO(0)	Turn off screen update.
3	=CALCULATION(3)	Turn off recalculation.
4	=COLUMN(SELECTION())	Get col# of leftmost column.
5	=COLUMNS(SELECTION())	Get # of columns in selection.
6	=ROW(SELECTION())	Ditto for rows.
7	=ROWS(SELECTION())	
8	=FOR("x",A6,A6+A7-1)	
9	=SELECT("R"&x&"C"&A4&":R"&x&"C"&A4+A5-1)	Select the range of cells.
10	=SORT(2,"RC",2)	Sort it.
11	=NEXT()	
12	=CALCULATION(1)	Turn on recalculation.
13	=ECHO(1)	Turn screen update back on.
14	=RETURN()	

Figure 5-28. *Row-by-Row Sort* macro.

The macro listing is shown in Figure 5-28. This macro illustrates the use of R1C1-style references and concatenation of text and variables to select cells in a worksheet. Because of its relative simplicity, named variables were not used.

THE MACRO. To speed up execution, cell A3 turns off automatic recalculation. Otherwise, class averages, etc., would be recalculated after every sort, slowing down the process significantly. (A small additional increase in speed is obtained by using ECHO(0) to turn off screen update.) Next, the position and size of the selected array of cells are saved: cell A4 returns the column number of the leftmost column of the selection, A5 the number of columns in the selection, A6 the row number of the uppermost row of the selection and A7 the number of rows in the selection. For example, if the array selected is C3:N140, then row# = 3, column# = 3, #columns = 12, #rows = 138.

In cell A9 the row of cells to be sorted is selected: R(x)C(column#):R(x)C(column# + #columns - 1). In the example above, the initial selection would be R3C3:R3C14, or C3:N3.

Cell A10 performs the sort; the syntax is SORT(*sort_by_flag*, *sort_key*, *order_flag*). *Sort_by_flag* is a number specifying whether to sort by rows (*sort_by_flag* = 1) or columns (*sort_by_flag* = 2). Here each row is being sorted by columns, so *sort_by_flag* = 2. *Sort_key* identifies which row to sort by when sorting columns, or which column to sort by when sorting rows. Here the *sort_key* used is "C"&A4. Actually, where a single row is being sorted, any of the following expressions for *sort_key* are suitable: "RC"&A4, "RC", "R", or even blank, i.e., SORT(2,,2). *Order_flag* is a number specifying whether to sort in ascending order (*order_flag* =1) or descending order (*order_flag* =2).

The FOR...NEXT loop increments the row number from the initial value in A6 to the final value (A6+A7-1), whereupon automatic recalculation and screen update are turned back on and return.

USEFUL REFERENCE

Complete Guide to Microsoft Excel Macros, second edition, Chris Kinata and Charles W. Kyd, Microsoft Press, Redmond, WA, 1993.

6

CREATING CUSTOM FUNCTIONS IN EXCEL 4.0 MACRO LANGUAGE

Like Excel's built-in functions, a function macro returns a value to the calling worksheet. Most function macros use one or more arguments from the worksheet. In this chapter you'll learn how to create simple custom function macros and examine the logic of a more complicated one.

THE STRUCTURE OF A CUSTOM FUNCTION

The structure of a custom function macro is similar to that of a command macro — it consists of a series of statements executed downwards, ending in a RETURN statement. In a function macro (Figure 6-1), the data passed to the macro are specified by a series of ARGUMENT functions. These must precede the macro statements that perform the calculations. The macro is terminated by a RETURN statement, which specifies the value to be passed back to the worksheet. The optional RESULT function may be used to specify the data type returned to the worksheet; it must precede any other statements, including the ARGUMENT statements.

The order of the ARGUMENT statements determines the order in which the arguments must be provided in the function; the name used in an ARGUMENT function specifies the argument name that appears as a placeholder argument in the function when **Paste Function** or CONTROL+A (see "A Shortcut to a Function" in Chapter 3) are used. Up to 29 arguments can be passed.

Data_type_number specifies the type of argument that will be accepted by a custom function. The data type numbers for arguments are: Number, 1; Text, 2; Logical, 4; Reference, 8; Error, 16; Array, 64.

	A
1	**MacroName**
2	=RESULT(type_number)
3	=ARGUMENT(name_text,data_type_number)
4	:
5	Macro Statements
6	:
7	=RETURN(value)

Figure 6-1. The structure of a function macro.

113

To permit an ARGUMENT function to accept more than one data type, enter the sum of the numbers for the different data types in *data_type_number*. For example, if *data_type_number* is 3, either number or text will be accepted. If an argument of the wrong data type is passed, the function returns #VALUE!, but first Excel tries to convert the argument to the specified type. If *data_type_number* is not specified, the default value is 7 (numbers, text or logical values are allowed).

The ARGUMENT function can also take an optional *reference* argument, specifying where on the macro sheet the argument value will be saved. The full syntax is ARGUMENT(*name_text*, *data_type_number*, *reference*). See Appendix D for details.

OPTIONAL ARGUMENTS IN FUNCTION MACROS

A custom function can have optional arguments, that is, ones that may be omitted. Optional ARGUMENT functions should follow any required ARGUMENT functions.

You will need to handle the presence or absence of optional arguments in a macro, for example by using the ISNA function or by providing default values.

USING ARRAYS AS ARGUMENTS (INPUT) OR RESULT (OUTPUT)

If you want the function to accept an array as an input, you must specify the argument as an array using *data_type_number* = 64. If you specify an argument value as an array, it's most often advisable to use the optional *reference* argument of the ARGUMENT function, so that the array will be saved on the macro sheet. Be sure to specify a range large enough to hold the largest array that may be entered, otherwise some of the data will be lost. You can use the INDEX function to obtain the value of an individual array element, and SET.VALUE to change individual values. If the optional reference argument isn't used, you can still use INDEX but you will not be able to change individual values within the array.

If your custom function returns an array result, you must use the RESULT function with *data_type_number* = 64.

USING A REFERENCE AS AN ARGUMENT

If an argument is specified as a reference variable, e.g., =ARGUMENT("known_ys",8), the variable can be treated as an array variable. For example, =AVERAGE(known_ys) returns the average of the values in the reference. If the actual reference is required, it can be extracted using the ROW, ROWS, COLUMN and COLUMNS functions.

BEWARE OF COMMAND-EQUIVALENT MACRO FUNCTIONS

Command-equivalent macro functions can't be used in function macros. Since a custom function macro returns a calculated value or array of values to a single worksheet cell, you usually won't try to include functions like OPEN or PASTE in your function macros. But, for example, you might try to use the macro function CLEAR to erase an area of the macro sheet that you intended to use as a "scratchpad" area. The statement containing the CLEAR function, a command-equivalent macro function, will simply not be executed when the custom function is calculated.

Unfortunately Excel does not prevent you from using command-equivalent functions in a function macro. They appear in the Paste Function list box and can be pasted into a macro that has been designated a function macro.

CREATING A SIMPLE FUNCTION MACRO

As an example of a simple function macro, we'll create a macro to convert temperatures in Fahrenheit into Celsius. The argument passed to the function macro is a number, the Fahrenheit temperature *degF*. The value returned by the function is *degC*. The equation for the conversion is $degC = (degF - 32) \times 5/9$. Figure 6-2 illustrates the macro. Note that while *degC* is a named variable and must be defined using **Define Name** from the **Formula** menu, the function argument *degF*, which must appear within quotes in the ARGUMENT function, is defined automatically by the ARGUMENT function. Figures 3 and 4 illustrate alternate versions of the macro.

	A	B
1		degF.to.degC
2		=ARGUMENT("degF",1)
3		=5*(degF-32)/9
4		=RETURN($B$3)

Figure 6-2. A simple function macro.

	A	B
1		degF.to.degC
2		=ARGUMENT("degF",1)
3	degC	=5*(degF-32)/9
4		=RETURN(degC)

Figure 6-3. A different version of a simple function macro.

	A	B
1		degF.to.degC
2		=ARGUMENT("degF",1)
3		=RETURN(5*(degF-32)/9)

Figure 6-4. Another version of a simple function macro.

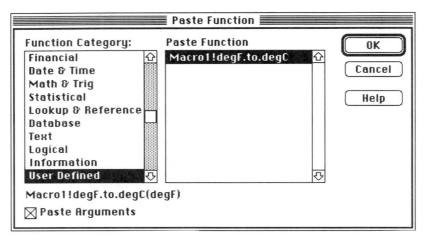

Figure 6-5. The Paste Function dialog box.

After typing the macro on a macro sheet, select the first statement of the macro (here, cell B1) and choose **Define Name** from the **Formula** menu. Excel proposes a name for the macro: *degF.to.degC*, the text in the selected cell. Press the Function button in the Define Name dialog box ("User Defined" should appear in the Category list box), then press the OK button.

USING A FUNCTION MACRO

To use the macro, select the worksheet cell where you want to enter the function. The macro sheet must be open. Choose **Paste Function** from the **Formula** menu to display the Paste Function dialog box, as in Figure 6-5. Scroll through the Function Category list and select the User Defined category. The *degF.to.degC* macro name will appear in the Paste Function list box. When you select a function, the function name and syntax appear in the bottom left corner of the dialog box. The Paste Arguments box should be checked. Now press the OK button. The function will be pasted into the cell, with the argument *degF* selected (highlighted), as illustrated in Figure 6-6.

Type the cell reference of the argument, or click on the cell containing the argument to enter the reference, as illustrated in Figure 6-7.

	A	B
1	°F	°C
2	32	=Macro1!degF.to.degC(degF)

Figure 6-6. Step 1 in entering a custom function.

	A	B
1	°F	°C
2	32	=Macro1!degF.to.degC(A2)

Figure 6-7. Step 2 in entering a custom function.

	A	B
1	**°F**	**°C**
2	32	0
3	75	23.88888889
4	212	100
5	-40	-40

Figure 6-8. Results returned by the *degF.to.degC* custom function.

When you click the check box in the formula bar, the value returned by the function is displayed. Figure 6-8 illustrates some values returned by the *degF.to.degC* custom function.

NAMING FUNCTION MACROS

The strategy for naming a custom function is a little different from that for a command macro. The name of a command macro should be long enough to indicate the macro's use. A function macro name, on the other hand, should be as short as possible, since it will appear in worksheet cells just like any other worksheet function. A function with a long name, especially if used repeatedly in the same cell, will make for long statements that are difficult to read or print.

Remember that the name of the function macro as it will appear when pasted into a worksheet cell will be *SheetName!FunctionName*, which will make the name even longer. Give the macro sheet a short descriptive name and the function macro a short descriptive name. For example, a custom function macro to calculate R^2, the correlation coefficient, might be given the name R^2 and saved along with other statistical functions on a macro sheet named STAT. The function name then would appear as STAT!R^2 when pasted into a worksheet cell.

SAVING FUNCTION MACROS
IN THE GLOBAL MACRO SHEET

It makes sense to save often-used function macros in the Global Macro Sheet (or the Personal Macro Workbook in Excel 5.0), but you may have developed, tested and debugged a function macro on a separate macro sheet. If you want to transfer it to the Global Macro Sheet, remember to **Cut** instead of **Copy** from the macro sheet and then **Paste** to the Global Macro Sheet, so that addresses will transfer correctly. If you want to **Copy** and **Paste**, first **Unhide** the Global Macro Sheet and determine the beginning reference of the location where the function macro will be located. Return to the macro sheet and **Copy** and **Paste** the macro to the same location on the macro sheet. Now **Copy** the macro from its new location on the macro sheet and **Paste** it to the same location in the Global Macro Sheet. All named references will be transferred correctly. However, you'll still have to use **Define Name** from the **Formula** menu to enter the macro name and define it as a function macro.

Summary of Steps to Create a Function Macro

1. From the **File** menu, choose **New**, choose **Macro Sheet**, then OK.

2. Type in the macro, using Excel's worksheet and macro functions.

3. Select the first statement of the macro, which should be a descriptive name. From the **Formula** menu, choose **Define Name**. Modify the name as necessary. Then press OK.

4. Click the Function option button. You can't assign a shortcut key. Then press OK.

5. Save the macro sheet. (Any name is OK, but note that the macro name which will appear in the Paste Function dialog box will be *SheetName!MacroName*.)

A CUSTOM FUNCTION MACRO TO CALCULATE MOLECULAR WEIGHTS

As an example of a more advanced custom function macro, let's examine MOL.WT, a macro that calculates the molecular weight of a compound from its chemical formula. This function is useful for laboratory inventory lists, for creating tables to be imported into a word-processing document, or for spreadsheet calculations involving moles.

Some examples of formulas that the macro can handle are NaCl, $C_{12}H_{22}O_{11}$, $Ca(NO_3)_2$ and $BaCl_2 \cdot 2H_2O$. Although Excel 4.0 doesn't provide subscript numbers, you learned how to implement them in Chapter 1, using the **Font** command in the **Format** menu. The macro is able to handle formulas either with subscripts or with regular numbers. Since there is no convenient way in Excel 5.0 to provide the centered dot ($\cdot$) in formulas, an asterisk or space is used instead, i.e., $BaCl_2\ 2H_2O$. The centered dot is available by using a custom font.

The syntax of the custom function is MOL.WT(***formula***, *decimals*). *Formula* is a chemical formula, as text, or a reference to a cell containing a formula. *Decimals* is the number of decimal places displayed in the value that is returned by the function. *Decimals* is an optional argument; if omitted, the number of decimal places displayed is determined according to the rules governing significant figures, by the smallest number of decimal places in the atomic weights used to calculate the value. Some examples of the results returned by MOL.WT are shown in Figures 6-9 and 6-10.

	A	B	C
79	Compound	Formula	Mol. Wt.
80	Sodium chloride	NaCl	58.4425
81	Calcium nitrate	Ca(NO3)2	164.088
82	Barium chloride dihydrate	BaCl2*2H2O	244.263
83	Acetone	(CH3)2CO	58.080

Figure 6-9. Molecular weights calculated using the MOL.WT custom function; formulas were entered using the Geneva font, i.e., without subscripts.

A	B	C
84 Compound	Formula	Mol. Wt.
85 Sucrose	$C_{12}H_{22}O_{11}$	342.301
86 Triethylamine	$(CH_3CH_2)_3N$	101.192
87 Phosphomolybdic acid	$H_3PMo_{12}O_{40} \cdot 30H_2O$	2365.71
88 Potassium aluminum sulfate	$Al_2(SO_4)_3 \cdot K_2SO_4 \cdot 24H_2O$	948.782

Figure 6-10. Molecular weights calculated using the MOL.WT custom function; formulas were entered using a custom font.

The macro parses the chemical formula into its component atomic symbols, numbers and parentheses, looks up the atomic weights of the symbols in a table, multiplies atomic weights by the appropriate coefficients and calculates the sum. To parse the formula, the macro steps through the formula one character at a time, examining each character in turn. Parsing a chemical formula is a bit complicated; we can be sure that a symbol or number has been determined unambiguously only when the next symbol or number is encountered. For example, C could be carbon or the first letter of calcium, cerium, etc. Only when the subsequent character H in the formula $CH_3CH_2NH_2$ is read is it established that C is carbon.

The macro consists of a main program and a subroutine. The main program scans and parses the formula and sets the appropriate status flags (tasks 1 through 7 in the following list).

1. Each character in the formula is examined to see whether it is an uppercase letter, a lowercase letter, a number, a parenthesis, or a space or centered dot. Two flags (*cflag* and *parenflag*) are used to indicate that a particular formula element has been encountered. *Cflag* = 1 indicates that a space or centered dot was encountered, and that a multiplicative coefficient may follow. *Parenflag* = 1 indicates that an opening parenthesis was encountered, *parenflag* = 2 indicates that a closing parenthesis was encountered.

2. If the character is an uppercase letter, it is stored as the variable *symbol*.

3. If the character is a lowercase letter, it is appended to the uppercase letter in *symbol*. A non-null value in *symbol* serves as the flag to indicate, in later numerical calculations, that a symbol is pending.

4. If the character is a number, it is stored as the variable *num*. Subsequent numbers are appended to the first. A non-null value in *num* serves as the flag to indicate that a number is pending.

5. If the character is an opening parenthesis, the parenthesis flag *parenflag* is set to 1.

6. If the character is a space or centered dot, the multiplicative coefficient flag *cflag* is set to 1.

7. When a subsequent uppercase letter or a parenthesis is encountered, the reading of the preceding symbol, number or symbol-number pair is complete. For example, when the "C" in NaCl is read, this indicates that *symbol* contains "Na". When the "H" in $C_{12}H_{22}O_{11}$ is read, *symbol* contains "C" and *num* contains 12.

The subroutine *Eval* performs the numerical calculations, based on the status flags (tasks 8 through 12).

8. If a symbol is pending, the atomic weight of the element is obtained by table lookup and added to the formula weight sum stored as the variable *MW*. If both a symbol and a number are pending, the atomic weight is multiplied by the number and added to the formula weight sum stored as *MW*.

9. If *parenflag* = 1, the above values are stored as the variable *pFW* instead.

10. If *cflag* = 1, *number* is transferred to *coeff*. All calculations in 8 and 9 above are multiplied by *coeff*, with a default value of 1.

11. If *parenflag* = 2, the values in *pFW* are transferred to *MW*.

12. The smallest number of decimal places in the atomic weights used is determined as the calculation proceeds and is stored as *numdec*.

THE MACRO. The listing of the main program is shown in Figure 6-11. Cells B4 and B5 establish the syntax of the macro, namely MOL.WT(***formula***,*decimals*). *Decimals* is the number of decimals in the returned result. In cell B6, if no formula has been passed, the function returns the error message n/a.

Cells B7 through B16 initialize the various flags and values. The loop to examine each character begins in cell B17. In B18 a character is obtained using the MID function. If the character is an uppercase letter, *Eval* is first used to handle anything that is pending (see task 7). Then *char* is transferred to *symbol*. In B27, if the character is a lowercase letter, and there is no preceding uppercase letter, then n/a is returned. Otherwise *char* is appended to *symbol*. If the character is a subscript number, with symbol CODE in the range 162-193, then in B33 the MATCH function is used to find the corresponding character code for the number; and this character is stored in *char*. If the character is a number or the decimal point, it is saved as *number* or appended to *number* if *number* already contained a value. In B40-B53, if the character is an opening parenthesis, a closing parenthesis, or a space or centered dot, *Eval* is called to process what is pending; then the flags are set as appropriate.

When all characters have been examined, *Eval* is called a final time to process whatever is pending, after which the result is returned. If the formula weight is zero or if a symbol error was encountered (e.g., "NuCl" rather than "NaCl" was passed), n/a is returned. If the formula weight is non-zero, then the number of decimal places in the result is set using the FIXED function. If *decimals* was not specified, the result is set using *num.dec*. If *decimals* was specified, then the result is set using the smaller of *decimals* and *num.dec*.

The *Eval* subroutine (Figure 6-12) performs the molecular weight calculations. In B62-B64, if a symbol is pending, the EXACT function is used to see if the symbol matches one in the table (otherwise VLOOKUP returns the closest element symbol that is less than or equal to the lookup value). If a match is found, the atomic weight is obtained by using the VLOOKUP function; if not, then *error.flag* is set to 1 in B69. In B65 the number of decimal places in the atomic

Table 6-1. Calculations Performed by the Eval Subroutine

parenflag	symbol	number	Calculation*	Example
0	0	0	No action taken	
0	1	0	Add AW to MW	NX**
0	1	1	Add number x AW to MW	H_2X
1	0	0	No action taken	()
1	1	0	Add AW to pFW	(...CX
1	1	1	Add number x AW to pFW	(...H_3X
2	0	0	Add pFW to MW	...)X
2	0	1	Add number x pFW to MW	...)2X

* AW = atomic weight; MW = molecular weight; pFW = formula weight of portion of formula within parentheses.

** X represents a character that causes evaluation of pending formula elements, i.e., those immediately preceding X. X can be an upper-case letter, a left or right parenthesis or a space. End-of-string also causes evaluation.

weight is calculated. In B66, if this is the first evaluation of decimal places (*temp.dec* has not been calculated previously), then *num.dec*, the number of decimal places in the result, is set equal to *temp.dec*. Otherwise, *num.dec* is set equal to the smaller of *temp.dec* and *num.dec*.

The statements in B72-B93 are a series of nested IF...ELSE.IF statements. The appropriate calculation is performed, based on the values of *parenflag, symbol* and *number*. The possible values of *parenflag, symbol* and *number* and the corresponding calculations are shown in Table 6-1. The possible values are also shown in column A of the subroutine, to the left of the IF statements.

In B94-B102 multiplicative coefficients are handled (e.g., formulas such as $(C_2H_5)_3N \cdot HCl$ or $BaCl_2 \cdot 2H_2O$). If a space or centered dot had previously been encountered, then *cflag* is set to 1. In B95-B96, if a number is pending, then *coeff* is set equal to *number* and *number* is cleared. Otherwise *coeff* is set equal to 1.

Finally, in cells B103 and B104, *symbol* and *number* are cleared before the subroutine returns to the calling program.

	A	B
1		**MOL.WT Macro**
2		Syntax : MOL.WT(formula,#decimal places).
3		Number of decimal places is optional.
4		=ARGUMENT("formula")
5		=ARGUMENT("decimals")
6		=IF(formula=0,RETURN("n/a"))
7	parenflag	=0
8	cflag	=0
9	error.flag	=0
10	coeff	=1
11	symbol	=""
12	number	=""
13	AW	=0
14	FW	=0
15	pFW	=0
16	num.dec	=0
17		=FOR("x",1,LEN(formula))
18	char	=MID(formula,x,1)
19		● *Upper-case letters*
20		=IF(AND(CODE(char)<=90,CODE(char)>=65))
21		=Eval()
22		=SET.VALUE(symbol,char)
23		=GOTO(loop)
24		=END.IF()
25		● *Lower-case letters*
26		=IF(AND(CODE(char)<=122,CODE(char)>=97))
27		=IF(symbol="",RETURN("n/a"))
28		=SET.VALUE(symbol,symbol&char)
29		=GOTO(loop)
30		=END.IF()
31		● *Numbers*
32		=IF(AND(CODE(char)<=193,CODE(char)>=162))
33		=SET.VALUE(char,CHAR(MATCH(CODE(char),subscript.codes,0)-1+48))
34		=END.IF()
35		=IF(OR(AND(CODE(char)<=57,CODE(char)>=48),CODE(char)=46))
36		=IF(number="",SET.VALUE(number,char),SET.VALUE(number,number&char))
37		=GOTO(loop)
38		=END.IF()
39		● *Parentheses*
40		=IF(char="(")
41		=Eval()
42		=SET.VALUE(parenflag,1)
43		=GOTO(loop)
44		=ELSE.IF(char=")")
45		=Eval()
46		=SET.VALUE(parenflag,2)
47		=GOTO(loop)
48		=END.IF()

Figure 6-11. The MOL.WT custom function macro.

49		● *(space or centered dot)*
50		=IF(OR(char=" ", CODE(char)=198))
51		=Eval()
52		=SET.VALUE(cflag,1)
53		=END.IF()
54	loop	=NEXT()
55		=Eval()
56		=IF(fw=0,RETURN("n/a"))
57		=IF(ISNA(decimals),RETURN(FIXED(fw,num.dec,1)))
58		=IF(num.dec<decimals,RETURN(FIXED(fw,num.dec,1)))
59		=RETURN(FIXED(fw,decimals,1))
60		

Figure 6-11. (continued) The MOL.WT custom function macro.

Since the number of decimal places in the returned result is determined within the macro by the FIXED function, you can't use number formatting to change the number of decimal places. You must use the optional argument *decimals*. If you want to eliminate the requirement to enter a value for *decimals* each time you use the MOL.WT function, you can delete line B6 from the macro (better still, simply delete the equal sign). Later, if you need to reduce the number of places displayed, **Copy** and **Paste Special (Values),** then **Format** the number.

	A	B
61		**Eval**
62		=IF(symbol<>"")
63		=IF(EXACT(symbol,VLOOKUP(symbol,$A$107:$B$189,1)))
64		=SET.VALUE(AW,VLOOKUP(symbol,$A$107:$B$189,2))
65	temp.dec	=LEN(MID(AW,FIND(".",AW),9))-1
66		=IF(num.dec=0,SET.VALUE(num.dec,temp.dec))
67		=IF(num.dec>temp.dec,SET.VALUE(num.dec,temp.dec))
68		=ELSE()
69		=SET.VALUE(error.flag,1)
70		=END.IF()
71		=END.IF()
72	P S N	*=IF(parenflag=0)*
73	0 1 0	=IF(AND(symbol<>"",number=""))
74		=SET.VALUE(fw,fw+AW*coeff)
75	0 1 1	=ELSE.IF(AND(symbol<>"",number<>""))
76		=SET.VALUE(fw,fw+VALUE(number)*AW*coeff)
77		=END.IF()
78		*=ELSE.IF(parenflag=1)*
79	1 1 0	=IF(AND(symbol<>"",number=""))
80		=SET.VALUE(pfw,pfw+AW*coeff)
81	1 1 1	=ELSE.IF(AND(symbol<>"",number<>""))
82		=SET.VALUE(pfw,pfw+VALUE(number)*AW*coeff)
83		=END.IF()
84		*=ELSE.IF(parenflag=2)*
85		=SET.VALUE(parenflag,0)
86	2 0 0	=IF(AND(symbol="",number=""))
87		=SET.VALUE(fw,fw+pfw)
88		=SET.VALUE(pfw,0)
89	2 0 1	=ELSE.IF(AND(symbol="",number<>""))
90		=SET.VALUE(fw,fw+VALUE(number)*pfw)
91		=SET.VALUE(pfw,0)
92		=END.IF()
93		=END.IF()
94		**=IF(cflag<>0)**
95		=IF(number<>"")
96		=SET.VALUE(coeff,VALUE(number))
97		=SET.VALUE(number,"")
98		=ELSE()
99		=SET.VALUE(coeff,1)
100		=END.IF()
101		=SET.VALUE(cflag,0)
102		=END.IF()
103		=SET.VALUE(symbol,"")
104		=SET.VALUE(number,"")
105		=RETURN()

Figure 6-12. The Eval subroutine.

CREATING ADD-IN FUNCTION MACROS

Saving custom functions as Add-In macros is by far the most convenient method to use them. Here are some of the advantages:

- Add-in custom functions are listed in the Paste Function list box without the macro sheetname preceding the name of the function, making them virtually indistinguishable from Excel's built-in functions (see Figure 6-13). They are listed both in the User-Defined category and alphabetically in the All category.

- An add-in macro sheet is hidden and can't be displayed by using the **Unhide** command.

- If the add-in macro sheet is placed in the Excel Startup folder in the Preferences folder, the add-ins will be available every time you start Excel.

To save a macro sheet as an add-in macro sheet, choose **Save As** from the **File** menu. In Excel for the Macintosh, press the Options button, choose Add-In as the File Format and press the OK button, then press the **Save** button. In Excel for Windows, choose Add-In as the Save File As type, then press OK. In Excel for Windows, Add-In macros are automatically given the filename extension .XLA.

Your custom functions will now be indistinguishable from Excel's built-in functions. Command macros can also be saved as add-ins.

To **Open** an add-in macro sheet for viewing or editing, choose **Open** from the **File** menu. Choose the name of the add-in macro sheet from the list box. Hold down the SHIFT key while pressing the OK button (Windows) or the Open button (Macintosh).

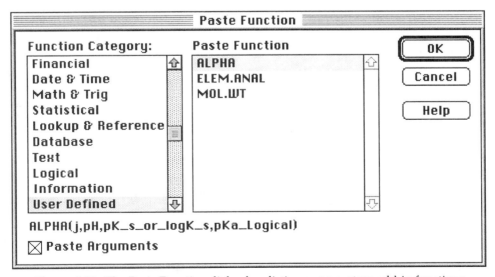

Figure 6-13. The Paste Function dialog box listing some custom add-in functions.

ADVANTAGES AND DISADVANTAGES
OF FUNCTION MACROS

Some function macros perform calculations that could be performed by using a worksheet formula or formulas, while in other cases implementing the function on the worksheet is impossible: when looping or branching is required, for example.

The main advantage of using a function macro instead of worksheet formulas is that errors caused by entering the formulas are minimized. The main disadvantage is that custom functions are slow. For example, in a sheet that calculates the four alpha values (see Chapter 18) for phosphoric acid in 0.1 pH increments from pH 0 to 14 (calculation of 564 cells), there was almost a tenfold increase in time compared to direct calculation. Using the alpha function required 18 seconds to recalculate the sheet, while using worksheet functions to perform the calculations required 2 seconds.

The alternative to function macros? — A library of formulas that can be copied into cells.

7

EXCEL 5.0 MACRO LANGUAGE:
VISUAL BASIC FOR APPLICATIONS

In this chapter you'll get an overview of macro programming using Microsoft's Visual Basic Programming System, Applications Edition; Excel's implementation of Visual Basic is referred to hereafter as Visual Basic for Applications, or VBA.

VBA has a wide range of commands, functions and methods that can be used to create custom applications. This chapter concentrates almost exclusively on the commands and functions for performing numeric calculations. If you are familiar with programming in other dialects of BASIC, or in FORTRAN, or in MS Excel 4.0 Macro Language, many of the programming techniques described in this chapter will be familiar.

VISUAL BASIC PROCEDURES AND MODULES

VBA macros are usually referred to as *procedures*. They are written or recorded on a module sheet. There are two types of VBA procedure: **Sub** procedures, which correspond to command macros in Excel 4.0 Macro Language, and **Function** procedures, which correspond to custom function macros.

RECORDING OR TYPING A VBA MACRO

To create a macro, you can either record it using the Recorder (see "Using the Recorder" later in this chapter) or type the code on a module sheet. To insert a module sheet in a workbook, choose **Macro** from the **Insert** menu, then **Module** from the submenu.

When you type VBA code in a module, it's good programming practice to use TAB to indent related lines for easier reading. As you type the VBA code, Excel checks each line for errors. Any line that contains one or more errors will be displayed in red, or another color if previously specified. Variables usually appear in black. Other colors are also used: comments (see later) are usually green and VBA keywords (**Function**, **GoSub**, etc.) usually appear in blue.

You can enter numbers in E format; they will automatically be converted to floating point. You can't enter numbers as percent; the percent symbol has another meaning in VBA (See "VBA Data Formats" later in this chapter).

If you type a long VBA expression, it will not wrap to the next line, but will simply disappear off the screen. You will have to insert a line-continuation character (a space followed by the underscore character) to cause a line break in a line of VBA code, as in the following example:

ReturnValu **= InputBox**("Enter validation code number", _
"Validation of this copy of SOLVER.STATS")

The line-continuation character can't be used within a string.

Several VBA statements, separated by colons, can be combined in one line.

COMPONENTS OF VISUAL BASIC STATEMENTS

VBA macro code consists of *statements*. Statements are constructed using VBA commands, functions, operators, variables and objects. (VBA Help refers to keywords such as **Beep**, **Do** or **Exit** as statements, but here they'll be referred to as commands, and we'll use "statement" in a general way to refer to a line of VBA code.) Commands, operators, variables and functions in VBA are similar to their analogs in other computer languages, while objects are somewhat different. Let's look at the familiar parts first.

Table 7-1. Some Useful VBA Commands

Beep	Makes a "beep" sound.
Dim	Declares an array and allocates storage for it.
Do...Loop	Delineates a block of statements to be repeated.
Else	Optional part of **If...Then** structure.
ElseIf	Optional part of **If...Then** structure.
End	Terminates a procedure.
End If	Terminates block of statements begun by **If**.
Exit	Exits a **Do...**, **For...**, **Function...** or **Sub...** structure.
For Each...Next	Delineates a block of statements to be repeated.
For...Next	Delineates a block of statements to be repeated.
Function	Marks the beginning of a **Function** procedure.
GoSub...Return	Delineates a subroutine.
GoTo	Unconditional branch.
If...Then...Else	Delineates a block of conditional statements.
On...GoSub	Branch to one of several specified subroutines.
On...GoTo	Branch to one of several specified lines.
Select Case	Executes one of several blocks of statements.
Set	Assigns an object reference to a variable or property.
Stop	Stops execution.
Sub	Marks the beginning of a **Sub** procedure.
Until	Optional part of **Do...Loop** structure.
While	Optional part of **Do...Loop** structure.
With...EndWith	Delineates a block of statements to be executed on a single object.

SOME USEFUL VBA COMMANDS

Commands in VBA are similar to commands in BASIC or FORTRAN. Table 7-1 lists some of the most useful VBA commands.

OPERATORS

VBA operators include the arithmetic operators (+, -, *, /, ^), the text concatenation operator (&), the comparison operators (=, <, >, <=, >=, <>) and the Boolean or logical operators (**And, Or, Xor, Not**), which are discussed later.

Excel Tip.   *Be sure to leave a space on either side of the concatenation operator; otherwise it will be mistaken for a type-declaration character. See "VBA Data Types" later in this chapter.*

VARIABLES AND ARGUMENTS

Variables are the names you create to indicate the storage locations of values or references. Arguments are variables that are passed from a worksheet to a procedure, for example.

The value of a variable is determined by an *assignment statement*. An assignment statement assigns the result of an expression to a variable or object; the form of an assignment statement is:

> *variable = expression*

Unlike Excel 4.0 Macro Language, you don't have to Assign Names to variables or arguments. They are automatically assigned as you type the VBA code in a module. There are just a few rules for naming variables or arguments:

- You can't use any of the VBA reserved words, such as Function, Range or Value.
- The first character must be a letter.
- A name cannot contain a space or a period.
- The type-declaration characters (%, $, #, !, &) cannot be embedded in a name.

You can use either upper- or lowercase letters. If you change the case of a variable name in the line you are currently typing (e.g., from *variable1* to *Variable1*), VBA will change all instances of the old form of the variable to the new form.

You should make variable names as descriptive as possible, but avoid overly long names since you'll have to type them. You can use the underscore character to indicate a space between words, as in e.g., formula_string. Don't use a period to indicate a space , since VBA reserves the period character for use with objects. The most popular form for variable names uses upper- and lowercase letters, as in e.g., FormulaString.

Table 7-2. Some Useful VBA Functions

Abs	Returns the absolute value of a number.
Asc	Returns the character code of a character.
Chr	Returns the character corresponding to a code.
Exp	Returns e raised to a power.
Fix	Returns the integer part of a number (truncates).
Int	Returns the integer part of a number (rounds down).
IsArray	Returns **True** if the variable is an array.
IsNull	Returns **True** if the expression is null, i.e., contains no valid data.
IsNumeric	Returns **True** if the expression can be evaluated to a number.
LBound	Returns the lower limit of an array dimension.
LCase	Converts a string into lowercase letters.
Left	Returns the leftmost characters of a string.
Len	Returns the length (number of characters) in a string.
Log	Returns the natural (base-*e*) logarithm of a number.
Mid	Returns a specified number of characters from a string.
Right	Returns the rightmost characters of a string.
RTrim	Returns a string without trailing spaces.
Sqr	Returns the square root of a number
Str	Converts a number to a string
UBound	Returns the upper limit of an array dimension.
UCase	Converts a string into uppercase letters.

SOME USEFUL FUNCTIONS

The list of functions available in VBA is similar to, but not exactly the same as the functions that are available in Excel itself. There are 108 VBA functions listed in VBA Help. Table 7-2 lists some of the more useful ones for numerical calculations.

> *Excel Tip.* To use one of Excel's built-in worksheet functions in VBA, append the function name to the Application object. For example, the VBA **Log** function returns the natural (base-*e*) logarithm of a number, but **Application.Log**(number) returns the common (base-10) logarithm.

OBJECTS, PROPERTIES AND METHODS

VBA is an *object-oriented* programming language. *Objects* in Microsoft Excel are the familiar components of Excel, such as a Worksheet, a Chart, a Toolbar or a Range. Objects have *properties* and *methods* associated with them.

OBJECTS

A complete list of objects in Microsoft Excel is listed in Excel's On-line Help. You can also use the Object Browser to see the complete list of objects. To display the Object Browser dialog box, choose **Object Browser** from the **View** menu.

In this chapter, the names of objects are capitalized to identify them as objects. There is a hierarchy of objects: a Workbook object can contain a Worksheet object, which in turn can contain a Range object, and so on. Figure 7-1 shows an abbreviated hierarchical list of the most useful objects.

You can also refer to *collections* of objects: groups of objects of the same kind. A collection has the plural form of the object's name, e.g., **Worksheets**. **Worksheets** refers to all worksheets in a particular workbook. To reference a particular worksheet, you use the reference **Worksheets**(NameText), e.g., **Worksheets**("Spectrum1").

To distinguish a particular collection of worksheets in a particular workbook, use the period operator to connect object names. For example, to refer to the Worksheet *Spectrum1* in the Workbook *Deconvolution*, use the reference **Workbooks**("Deconvolution").**Worksheets**("Spectrum1"). To refer to a particular cell within the worksheet, use

Workbooks("Deconvolution").**Worksheets**("Spectrum1").**Range**("E5")

If *Deconvolution* is the only open workbook, and Spectrum1 is the active sheet, you can omit references to them and simply use the reference **Range**("E5").

SOME USEFUL OBJECTS

The objects that you will probably use most often in the beginning are the Workbook object, the Worksheet object and the Range object. (For reasons that will become clear in a while, there is no Cell object.) Later you may use the Dialog, MenuBar, ToolBar, or Chart objects and the objects they in turn contain.

Application	MenuBar	Menu	MenuItem	
	ToolBar	ToolbarButton		
	Workbook	Chart	Axis	AxisTitle
				GridLines
				TickLabels
			etc.	
		Worksheet	PageSetup	
			Range	Areas
				etc.

Figure 7-1. Partial list of objects, arranged in hierarchical order.

"OBJECTS" THAT ARE REALLY PROPERTIES

The Application object has properties which you can treat just as though they were objects: the ActiveWindow, ActiveWorkbook, ActiveSheet, ActiveCell, Selection and ThisWorkbook properties. Thus the reference Application.ActiveCell refers to the active cell in the active workbook in the application. Since there is only one Application object, you can omit the reference to Application and simply use ActiveCell.

YOU CAN DEFINE YOUR OWN OBJECTS

VBA allows you to equate a variable to an object, but the variable does not automatically become an object. If you then attempt to use the variable in an expression that requires an object, you'll get an "Object required" error message. The **Set** command lets you define a variable or property as an object. The macro code in Figure 7-2 combines two separate references into a single multiple reference and then changes the background color of the range of cells to blue.

SOME USEFUL PROPERTIES

Objects have *properties* and *methods* associated with them. Using VBA you can examine an object's properties, or change the condition of an object by changing the value of one or more of its properties. You can perform tasks by utilizing the methods that an object can perform.

There is a large and confusing number of properties: VBA Help lists a total of 486 property names. The list of properties belonging to the Range object alone contains more than 40 entries. Some of the most useful properties of the Range object are listed in Table 7-3.

USING PROPERTIES

In a VBA macro, many times you'll need to determine the value or state of an object's property, or change it. There are two kinds of property: read-only properties, and read-write properties. Properties can have values that are either numeric, string or logical.

Table 7-3. Some Useful VBA Properties

Column	Returns a number corresponding to the first column in the range.
ColumnWidth	Returns or sets the width of all columns in the range.
Count	Returns the number of items in the range.
Font	Returns or sets the Font of the range.
Formula	Returns or sets the formula.
Name	Returns or sets the name of the range.
NumberFormat	Returns or sets the format code for the range.
Row	Returns a number corresponding to the first row in the range.
RowHeight	Returns or sets the height of all rows in the range.
Text	Returns or sets the text displayed by the cell.
Value	Returns or sets the contents of the cell or range.

```
Set Area1 = Range(Cells(1, 5), Cells(2, 5))
Set Area2 = Range(Cells(3, 5), Cells(5, 5))
Set BigArea = Union(Area1, Area2)
BigArea.Interior.ColorIndex = 8
```

Figure 7-2. Using **Set** to define a variable as an object.

To return an object's property, use the following syntax:

VariableName = ObjectName.PropertyName

To set an object's property (only if it is a read-write property, of course), use the following syntax:

ObjectName.PropertyName = expression

For example, to change the background color of a range of cells to blue, use the statement

Range("A5:E5").Interior.ColorIndex = 8

METHODS

Many objects also have methods. A method is an operation that can be performed on an object, or on a property of an object. In the example immediately above, **Range** is an object, **Interior** is a property and **ColorIndex** is a method.

SOME USEFUL METHODS

VBA Help lists 267 methods. Many of them correspond to familiar menu commands. For example, **Copy**, **Cut**, **Clear** and **Sort** are some methods that can be performed on a range of cells. Some VBA methods that you may find useful are listed in Table 7-4.

USING THE RECORDER

The Recorder can be used to record macros in VBA as well as in Excel 4.0 Macro Language. In fact, VBA is the default recorder language in Excel 5.0.

Just as we did in chapter 5, we'll record the *FullPage* macro, which sets Top, Bottom, Left and Right margins to zero and eliminates the default Header and Footer text. First, choose **Record Macro** from the **Tools** menu, then **Record New**

Table 7-4. Some Useful VBA Methods

Activate	Activate an object (sheet, etc.)
Clear	Clears an entire range.
Close	Closes an object
Copy	Copies an object to a specified range or to the Clipboard
Cut	Cuts an object to a specified range or to the Clipboard
FillDown	Copies the cell(s) in the top row into the rest of the range.
Select	Selects an object.

Macro... from the submenu. Press the Options button, then press the Store in This Workbook and Visual Basic language buttons, followed by OK. The Macro Stop button will appear, indicating that a macro is being recorded. Now choose **Page Setup...** from the **File** menu, choose the Margins tab and set Top, Bottom, Left and Right margins to zero; choose the Header/Footer tab and choose (none) for both the Header and Footer text, then press OK. Finally, press the Macro Stop button. You'll now find a new macro module inserted at the end of the

```vba
'
' Macro1 Macro
' Macro recorded 2/16/96 by E. Joseph Billo
'
'
Sub Macro3()
   With ActiveSheet.PageSetup
      .PrintTitleRows = ""
      .PrintTitleColumns = ""
   End With
   ActiveSheet.PageSetup.PrintArea = ""
   With ActiveSheet.PageSetup
      .LeftHeader = ""
      .CenterHeader = ""
      .RightHeader = ""
      .LeftFooter = ""
      .CenterFooter = ""
      .RightFooter = ""
      .LeftMargin = Application.InchesToPoints(0)
      .RightMargin = Application.InchesToPoints(0)
      .TopMargin = Application.InchesToPoints(0)
      .BottomMargin = Application.InchesToPoints(0)
      .HeaderMargin = Application.InchesToPoints(0.5)
      .FooterMargin = Application.InchesToPoints(0.5)
      .PrintHeadings = False
      .PrintGridlines = True
      .PrintNotes = False
      .PrintQuality = -4
      .CenterHorizontally = False
      .CenterVertically = False
      .Orientation = xlPortrait
      .Draft = False
      .FirstPageNumber = xlAutomatic
      .Order = xlDownThenOver
      .BlackAndWhite = False
      .Zoom = 100
   End With
End Sub
```

Figure 7-3. Page Setup macro recorded in VBA.

A
1 Macro1
2 =SET.PRINT.TITLES("","")
3 =SET.PRINT.AREA("")
4 =PAGE.SETUP("","",0,0,0,0,FALSE,TRUE,FALSE,FALSE,1,,100,#N/A,1,FALSE,-4,0.5,0.5,FALSE,FALSE)
5 =RETURN()

Figure 7-4. Page Setup macro recorded in Excel 4.0 Macro Language.

workbook, after the last sheet. VBA automatically provides the header information, a MacroN name, the date and the name of the person recording (actually the name of the person to whom the copy of Excel is registered). The macro should look like the example shown in Figure 7.3.

Compare this with the Excel 4.0 Macro Language macro in Figure 7-4.

Although the arguments do not appear in exactly the same order, and there aren't the same number of arguments, you can see that the VBA procedure is essentially identical.

CREATING ADVANCED MACROS IN VBA

VBA is designed primarily to create custom applications, using custom dialog boxes, toolbars, etc. There are literally hundreds of objects, properties, methods, etc. This chapter focuses on the tools you need to create macros to automate chemical worksheet calculations: how to transfer values from a sheet to a VBA module, how to perform calculations within a VBA module, how to perform logical branching and iterative looping, and how to send values back from a VBA module to a worksheet.

Excel Tip. If you're an experienced MS Excel 4.0 Macro Language programmer, and are beginning to learn VBA, you can find the VBA equivalent of Excel 4.0 Macro Language functions by using the VBA On-line Help. From the Help Contents screen, choose Reference Information, then Microsoft Excel Macro Functions Contents, then Visual Basic Equivalents for Macro Functions and Commands. To see the VBA equivalent of an Excel 4.0 Macro Language function, press the button corresponding to the first letter of the function, then scroll through the list. (Don't use the Search button: that will take you to the Excel 4.0 Macro Language information screens.) You'll see VBA expressions and examples.

A REFERENCE TO A CELL OR RANGE OF CELLS

You can reference a particular cell or range within a worksheet in a number of ways: by selecting a cell or range, by reference to the **ActiveCell**, by simply referring to a cell reference in square brackets, or by using the **Range** method or the **Cells** method.

The following references all refer to cell B3. The syntax of **Cells** is **Cells**(*RowIndex*, *ColumnIndex*):

[b3]

Range("B3")

Cells(3,2)

```
Range("D1:D20").Select
Selection.Copy
Sheets("Sheet15").Select
Range("A1").Select
ActiveSheet.Paste
```

Figure 7-5. VBA code fragment by the Recorder.

```
Range("D1:D20").Copy (Sheets("Sheet15").Range("A1"))
```

Figure 7-6. A more efficient way to accomplish the same thing, without selecting cells.

As you will see in several examples later in this chapter, the **Cells** method is very versatile, since *RowIndex* and *ColumnIndex* can be calculated easily by means of expressions. Contrast this with the rather cumbersome concatenation of text and variables required to construct a reference in R1C1-style in Excel 4.0 Macro Language (see Figure 5-28, for example).

Don't use **Select** unless you actually need to select cells in a worksheet. For example, to copy a range of cells from one worksheet to another, you could use the statements shown in Figure 7-5, and in fact this is exactly the code you would generate using the Recorder. But you can do the same thing much more efficiently, and without switching from one worksheet to another, by using the code shown in Figure 7-6.

REFERENCES USING THE UNION OR INTERSECT METHOD

VBA can create references using methods that are the equivalents of the union operator or intersection operator described in Chapter 3. The **Union** method creates a reference that includes multiple selections, e.g., A1,B5 or G3:L3,G5:L5. The syntax of the **Union** method is **Union**(*range1*, *range2*). The **Intersect** method creates a reference that is common to two references, e.g., F4:F6 E5:I5. The syntax of the **Intersect** method is **Intersect**(*range1*, *range2*). Both *range1* and *range2* must be range objects.

GETTING VALUES FROM A WORKSHEET

The following examples illustrate how to get the contents of a single cell in a worksheet and assign it to a variable in VBA.

variable1 = **ActiveCell**

variable2 = **Worksheets**("Sheet1").**Range**("A9")

variable3 = [c1]

variable4 = **Cells**(StartRow+x,StartCol)

SENDING VALUES TO A WORKSHEET

The **Value** method is used to enter values from the macro sheet into cells in a worksheet.

Cells(1, 2).**Value** = variable1

Range("E1").**Value** = "Jan.-Mar."

Range("area").**Value** = variable2

The .**Value** can often be omitted, as illustrated in the following examples.

[a1] = Address1

Range("F1:F10") = 5

Worksheets("Sheet1").**Range**("A1") = variable2

The corresponding **Formula** method is used to enter a formula in a cell.

Cells(1, 3).**Formula** = "=sum(F1:F10)"

VISUAL BASIC ARRAYS

Arrays are easier to work with in VBA than in Excel 4.0 Macro Language. Normally you'll give an array a name, such as Sample. You can then specify any element in the array by using an index number: Sample(1), Sample(7), etc.

The **Dim** statement is used to declare the size of an array. Unless directed otherwise, VBA arrays begin with a index of 0. The statement

Dim Sample(10)

establishes array storage for 11 elements, Sample(0) through Sample(10). If you are more comfortable with arrays that begin with the index number 1, or if you aren't going to use the zeroth element of the array and want to conserve storage, use the statement

Dim Sample (1 to 10)

Excel Tip.   *To declare all arrays in a module as arrays with index beginning at 1, place the following VBA statement at the beginning of the module:*
Option Base 1.

Arrays can be multidimensional. Excel permits arrays with up to 60 dimensions; one-, two- and three-dimensional arrays are common. To create an array with dimensions 2 x 500, use the statement

Dim Spectrum (2,500)

Since multidimensional arrays such as the one above can use up significant amounts of memory, it's important to define the data type of the variable. (See "VBA Data Types" later in this chapter.) The complete syntax of the **Dim** statement is

Dim *VariableName*(*Lower* **To** *Upper*) **As** *Type*

Type can be **Integer**, **Single**, **Double**, **Variant**, etc. See the complete list of data types in Table 7-5.

Several variables can be dimensioned in a single **Dim** statement, but there must be a separate **As *Type*** for each variable.

It's considered good programming practice to put the **Dim** statements at the beginning of the procedure.

If you don't know the size of an array, you can dimension it using the **Dim** command using empty parentheses, then use the **ReDim** command later.

ARRAYS AS ARGUMENTS IN FUNCTIONS

If a range argument is passed in a function macro, the range automatically becomes an array in the VBA procedure. A one-row or one-column reference becomes a one-dimensional array; a rectangular range becomes a two-dimensional array of dimensions *array (rows, columns)*.

VBA DATA TYPES

VBA uses a range of different data types. Table 7-5 lists the built-in data types.

Unless you declare a variable's type, VBA will use the **Variant** type. You can save memory space if your procedure deals only with integers, for example, by declaring the variable type.

You can also declare a variable's type by appending a type-declaration character to the variable name, a technique from older versions of BASIC. The

Table 7-5. VBA's Built-In Data Types

Data Type	Storage Required	Range of Values
Logical	2 bytes	**True** or **False.**
Integer	2 bytes	-32,768 to 32,767.
Long integer	4 bytes	-2,147,483,648 to 2,147,483,647.
Single precision	4 bytes	-3.402823E38 to -1.401298E-45 for negative values; 1.401298E-45 to 3.402823E38 for positive values.
Double precision	8 bytes	-1.79769313486232E308 to -4.94065645841247E-324 for negative values; 4.94065645841247E-324 to 1.79769313486232E308 for positive values.
Currency	8 bytes	-922,337,203,685,477.5808 to 922,337,203,685,477.5807.
Date	8 bytes	
Object	4 bytes	Any Object reference.
String	1 byte/ character	
Variant	16 bytes + 1 byte/character	Any numeric value up to the range of a Double or any character text.

type-declaration characters include % for integer variables and $ for string variables.

THE VARIANT DATA TYPE

The **Variant** data type is the default data type in VBA. Like Excel itself, the **Variant** data type handles and interconverts between many different kinds of data — integer, floating point, string, etc. The **Variant** data type automatically chooses the most compact representation. But if your procedure deals with only one kind of data, it will be more efficient and usually faster to declare the variables as, for example, **Integer** .

STRING DATA TYPES

Strings can be stored either as *variable-length strings* (the default data type) or as *fixed-length strings*. To declare a string variable as fixed-length, use the statement in a **Dim** statement

> **String** * *length*

For example, the following statement sets aside storage for a two-dimensional array of names and addresses, containing fixed-length strings of 32 characters:

> **Dim** AddressList(4, 500) **As String** * 32

If a string of length less than 32 characters is assigned to the array AddressList, trailing spaces are added to fill out the string length. If a string of more than 32 characters is assigned to AddressList, the string will be truncated.

THE BOOLEAN (LOGICAL) DATA TYPE

Logical variables can only have the values **True** or **False**. The keywords **True** and **False** are often implied; thus the expressions

> **If** (j > N) = **True Then** ALPHA = "n/a"

and

> **If** (j > N) **Then** ALPHA = "n/a"

are equivalent.

You can use other data types as Boolean variables. When a variable is used in a logical expression, zero is converted to **False** while any non-zero value is converted to **True**. Thus the expression

> **If** j **Then** *expression*

tests for a non-zero value of the variable j.

When Boolean variables are converted to other data types, **False** becomes zero but **True** is converted to –1.

DECLARING VARIABLES OR ARGUMENTS IN ADVANCE

VBA uses the **Variant** data type as the default data type for variables and arguments. The **Variant** data type permits Excel to switch between floating-point, integer or string variables as required.

You can force a particular variable or argument to take a specified data type. For variables, use the **Dim** statement, e.g.,

> **Dim** ChemFormula **As String**

You don't need to specify dimensions for an array.

SPECIFYING THE DATA TYPE OF AN ARGUMENT

You can specify the data type of an argument passed to a **Function** procedure by using the **As** keyword in the **Function** statement. For example, if the **Function** procedure MolWt has two arguments, formula (a string) and decimals (an integer), then the statement

> **Function** MolWt (formula **As String**, decimals **As Integer**)

declares the type of each variable. If an argument of an incorrect type is supplied to the function, a #VALUE! error message will be displayed.

SPECIFYING THE DATA TYPE RETURNED BY A FUNCTION PROCEDURE

You can also specify the data type of the return value. If none is specified, the **Variant** data type will be returned. In the example of the preceding section, MolWt returns a floating-point result. The **Variant** data type is satisfactory; however, if you wanted to specify double precision floating-point, use an additional **As Type** expression in the statement, e.g.,

> **Function** MolWt (formula **As String**, decimals **As Integer**) **As Double**

PROGRAM CONTROL

If you are familiar with Excel 4.0 Macro Language, or computer languages such as BASIC or FORTRAN, you will find most of the material in this section quite familiar.

DECISION-MAKING (BRANCHING)

VBA supports **If...Then** and **If...Then...Else** or **ElseIf** structures, very similar to Excel 4.0 Macro Language. In addition, VBA provides the **Select Case** decision structure, similar to the ON *value* GOTO statement in BASIC.

As in Excel 4.0 Macro Language, the **If...Then** statement can be on a single line:

> **If** (x = j) **Then** numerator = 10 ^ (logbeta - pH * x)

or it can be followed by multiple statement lines as in Figure 7-7, similar to the Block If statement in Excel 4.0 Macro Language.

```
If (pKa_logical = False) Then
    logbeta = logbeta + pKs_or_logKs(x)
    denom = denom + 10 ^ (logbeta - pH * x)
    etc.
Endif
```

Figure 7-7. Example of VBA **If...Endif** structure.

If...Then...Else or **ElseIf** structures are also possible. For example:

If LogicalExpression **Then** statement **Else** statement

or, as illustrated in Figure 7-8, you can employ several logical expressions.

```
If LogicalExpression1 Then
    statements
ElseIf LogicalExpression2
    statements
ElseIf LogicalExpression3
    statements
    etc.
Endif
```

Figure 7-8. The VBA **If...ElseIf...Endif** structure.

The **Select Case** statement provides an efficient alternative to the series of **ElseIf** conditionN statements shown above when conditionN is a single expression that can take various values. The syntax of the **Select Case** statement is illustrated in Figure 7-9.

TestExpression is evaluated and used to direct program flow to the appropriate **Case**. ExpressionListN can be a single value, a list of values separated by commas or a range of values. The optional **Case Else** statement is executed if TestExpression doesn't match any of the values in any of ExpressionListN.

The example shown in Figure 7-10 illustrates the use of **Select Case** to calculate the pK_a value of a polyprotic acid. Since data at or near the equivalence

```
Select Case TestExpression
    Case ExpressionList1
    statements
    Case ExpressionList2
    statements
    Case ExpressionList3
    statements
    Case Else
    statements
End Select
```

Figure 7-9. The VBA **Select Case** structure.

```
NBar = (ZP * CR + CA + COH - CH - CNa) / CR

Select Case NBar
Case 3.2 To 3.8
    pK = pH + Application.Log((NBar - 3) / (4 - NBar))
Case 2.2 To 2.8
    pK = pH + Application.Log((NBar - 2) / (3 - NBar))
Case 1.2 To 1.8
    pK = pH + Application.Log((NBar - 1) / (2 - NBar))
Case 0.2 To 0.8
    pK = pH + Application.Log((NBar) / (1 - NBar))
Case Else
    pK = ""
End Select
    ⋮
    ⋮
End Sub
```

Figure 7-10. An example of the **Select Case** structure.

points cause large calculation errors, the pK_a is calculated only for n-bar values in the range 0.2-0.8, 1.2-1.8, 2.2-2.8 or 3.2-3.8. The expression used to calculate the pK_a from the n-bar parameter depends on the number of protons bound, i.e., on the value of n-bar. The **Select Case** statement is used to direct program flow to the appropriate expression.

Note that a range of values is indicated by using the **To** keyword.

LOGICAL OPERATORS

You are already familiar with the **And** and **Or** operators, but VBA provides in addition the **Xor** (exclusive or) operator. The operators have the following syntax:

expression1 **And** expression2	**True** if both expressions are **True**.
expression1 **Or** expression2	**True** if either expression is **True**.
expression1 **Xor** expression2	**True** if one expression is **True**, the other **False**.

The expressions above must evaluate to **True** or **False**; that is, they must be logical expressions. The logical operators are almost always used in combination with **If** statements.

More than one **And** or **Or** can be combined in a single statement, as illustrated by the following example.

> **If** Char = " " **Or** Char = "*" **Or** Char = "," **Or** Char = "(" **Or** Char = "/" **Then**...

evaluates to **True** if any one of the logical expressions is **True**.

Parentheses are often necessary in order to control the logic of the expression. For example, each of the expressions

>   **If** (expression1 **And** expression2) **Or** expression3 **Then**...

>   **If** expression1 **And** (expression2 **Or** expression3) **Then**...

has eight different possible combinations of expression1, expression2 and expression3; two of them give different outcomes depending on which expression is used.

LOOPING

The loop structures in VBA are similar to those available in Excel 4.0 Macro Language.

FOR...NEXT LOOPS

The syntax of the **For...Next** loop is given in Figure 7-11.

```
For Counter = Start To End Step Increment
    statements
Next Counter
```

Figure 7-11. The VBA **For...Next** structure.

Both **Step** Increment in the **For** statement, and Counter following the **Next** are optional. If Increment is omitted, it is set equal to 1. Increment can be negative.

FOR EACH...NEXT LOOPS

The **For Each...Next** loop structure is similar to the **For...Next** loop structure, except that it executes the statements within the loop for each element of an array, or each object within a group of objects. Figure 7-12 illustrates the syntax of the statement.

```
For Each Element In Group
    statements
Next Element
```

Figure 7-12. The VBA **For Each...In... Next** structure.

DO WHILE... LOOP

The **Do While...Loop** is similar to the WHILE...NEXT loop in Excel 4.0 Macro Language. The syntax, shown in Figure 7-13, is best used when you don't know beforehand how many times the loop will need to be executed.

```
Do While LogicalExpression
    statements
Loop
```

Figure 7-13. The VBA **Do While...Loop** structure.

```
Do
     statements
Loop While LogicalExpression
```

Figure 7-14. Alternate form of **Do While...Loop** structure.

An alternate format of this type of loop places **While** *LogicalExpression* at the end of the loop, as exemplified in Figure 7-14.

Note that this form of the **Do While** structure executes the loop at least once.

EXITING FROM A LOOP OR FROM A PROCEDURE

Often you use a loop structure to search through an array or collection of objects, looking for a certain value or property. Once you find a match, you don't need to cycle through the rest of the loops. You can exit from the loop using the **Exit For** (from a **For...Next** loop or **For Each...Next** loop) or **Exit Do** (from a **Do While**... loop). The **Exit** statement will normally be located within an **If** statement. For example,

If CellContents.Value <= 0 **Then Exit For**

Use the **Exit Sub** or **Exit Function** to exit from a procedure. Again, the **Exit** statement will normally be located within an **If** statement.

Exit statements can appear as many times as needed within a procedure.

SUBROUTINES

Although all **Sub** procedures are subroutines, by "subroutine" we mean a sub-program that is called by another VBA program. It's good programming practice to break a complicated task up into simpler tasks, and write subroutines to do each task. The separate subroutines are called by a main program.

There are several ways to execute a subroutine within a main program. The two most common are by using the **Call** command, or by simply using the name of the subroutine. These are illustrated in Figure 7-15. MainProgram calls subroutines Task1 and Task2, each of which requires arguments.

The two methods use different syntax if the subroutine requires arguments. If the **Call** command is used, the arguments must be enclosed in parentheses. If only the subroutine name is used, the parentheses must be omitted. Note that the variable names of the arguments in the calling statement and in the subroutine do not have to be the same.

If a subroutine argument is an array, use the arrayname with empty parentheses to pass the whole array.

You can call any subroutine in the same workbook using a statement such as the one in Figure 7-15. To call a subroutine in a different workbook, you must establish a reference to that workbook. Choose **References** from the **Tools** menu. The Excel 5.0 References dialog box (Figure 7-16) initially displays all open workbooks that contain VBA modules, plus two essential object libraries, Visual

```
Sub MainProgram()
etc.
Call Task1(argument1,argument2)
etc
Task2 argument3,argument4
etc
End Sub

Sub Task1(ArgName1,ArgName2)
etc
End Sub

Sub Task2(ArgName3,ArgName4)
etc
End Sub
```

Figure 7-15. A main program and two subroutines. Note the different syntax of the two subroutine calls.

Basic for Applications and Microsoft Excel 5.0 Object Library. To add another workbook to the Available References list, you don't need to open the workbook; just press the Browse button and then use the Browse dialog box to select the desired workbook. It will be added to the list of references.

SCOPING A SUBROUTINE

A subroutine can be Public or Private. Public subroutines can be called by any subroutine in any module. The default for any **Sub** procedure is **Public**. A **Private** subroutine can be called only by other subroutines in the same module. To declare the subroutine Task3 as a private subroutine, use the statement

> **Private Sub** Task3()

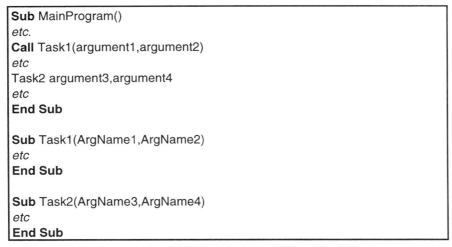

Figure 7-16. Excel 5.0 References dialog box.

CREATING A SIMPLE USER-DEFINED FUNCTION IN VBA

A user-defined or custom function is a **Function** procedure in VBA. A **Function** procedure cannot change the worksheet environment, e.g., change font style, make a cell Bold.

The structure of a **Function** procedure is shown in Figure 7-17.

The procedure begins with the VBA keyword **Function**, followed by the name of the function and the argument list in parentheses. Arguments are separated by commas. If there are no arguments, *FunctionName* must still be followed by parentheses; if you omit them, VBA will automatically supply them. The function name and the arguments automatically become stored variables. The procedure must contain at least one statement of the form *FunctionName = expression*. This defines the value that is returned by the procedure. The procedure is terminated by **End Function.**

To create a custom function, activate a VBA module in an open workbook or **Insert** a new Module. Type the VBA code, beginning with **Function** *FunctionName(Argument1, ...)*. You can have several procedures on one module sheet.

Type the VBA code. Once you've finished typing, and the macro code is error-free, you're done. Unlike Excel 4.0 Macro Language, you don't have to use **Define Name**, or declare a macro as a function or a command macro. As soon as the sheet is done, it's a macro. In fact, as soon as you type the statement **Function** FunctionName, the function appears in the Insert Function list box.

CREATING A SIMPLE CUSTOM FUNCTION

We'll create a simple custom function to convert temperatures in degrees Celsius to degrees Fahrenheit. In Figure 7-18, DegC is the argument passed by the function from the worksheet to the module.

Notice the similarity between this custom function macro and the Excel 4.0 Macro Language macro in Figure 6-2.

```
Function FunctionName(Argument1, ...)
    VBA statements
    FunctionName = expression
End Function
```

Figure 7-17. Structure of a user-defined function.

```
Function DegCtoDegF(DegC)
   DegCtoDegF = 1.8 * DegC + 32
End Function
```

Figure 7-18. Celsius to Fahrenheit custom function.

USING A VBA CUSTOM FUNCTION

When used in a worksheet, a VBA custom function is indistinguishable from an Excel 4.0 Macro Language custom function. You enter the function in a worksheet cell by typing it, by using **Function...** from the **Insert** menu or by using the Function Wizard.

ASSIGNING A CUSTOM FUNCTION
TO A FUNCTION WIZARD CATEGORY

Your custom functions will automatically be listed in the User Defined category in the Function Wizard dialog box. However, you can assign a function to any of the built-in categories (Math & Trig, Statistical, etc.) if you wish. To do this, use the Object Browser.

VBA SUB PROCEDURES

An Excel 4.0 Macro Language command macro is a **Sub** procedure in VBA. The structure of a **Sub** procedure is shown in Figure 7-19.

The procedure begins with the VBA keyword **Sub**, followed by the name of the function and the argument list. Arguments are separated by commas. If there are no arguments, *ProcedureName* must still be followed by parentheses; if you omit them, VBA will automatically supply them. The procedure name and the arguments automatically become stored variables. The procedure is terminated by **End Sub**. **Sub** procedures can change the worksheet environment.

CREATING A SIMPLE SUB PROCEDURE
TO FORMAT TEXT AS A CHEMICAL FORMULA

As an example, we'll create the *ChemFormula* command macro to format text as a chemical formula. Each numeric character in a text string will be made a subscripted character , as described in Chapter 1.

When developing a complex macro, it's often helpful to do so in steps of increasing difficulty. Here we begin by creating a simple **Sub** procedure, VBAChemFormula1, that examines each character in turn and formats any numeric characters as subscripts. To do this we'll obtain the text string as a variable, then use a loop to examine each character in turn. The complete expression to subscript specified characters in a text string in the active cell uses the Characters(start, length) method to specify the characters:

Range.**Characters**(start, length).**Font.Subscript = True**

```
Sub ProcedureName(Argument1, ...)
    VBA statements
End Sub
```

Figure 7-19. Structure of a **Sub** procedure.

```
Sub VBAChemFormula1()

For x = 1 To Len(ActiveCell)
    Char = Mid(ActiveCell, x, 1)
If Char >= "0" And Char <= "9" Then ActiveCell.Characters(x,
1).Font.Subscript = True
Next
End Sub
```

Figure 7-20. A simple **Sub** procedure to format text as a chemical formula.

The complete macro is shown in Figure 7-20. This macro is clumsy, since it formats each number character separately; a formula such as $C_{12}H_{22}O_{11}$ would require six text-formatting operations. A more efficient approach is to determine the *position* and *count* of the numeric substrings within the string, and format the text only once for each substring (three times for $C_{12}H_{22}O_{11}$). The macro is shown in Figure 7-21.

Again, we use the **Mid** function to examine each character. **SubFlag** is used to indicate when we are within a numeric substring. When the first numeric character is encountered, **SubFlag** is set to **True** and **SubPosition** is set equal to the string pointer x used in the **Mid** function. The **SubCount** variable is initialized to zero and then incremented by 1. If the next character examined is also numeric, then, since **SubFlag** is **True**, only   **SubCount** is incremented.

```
Sub VBAChemFormula2()
SubFlag = False
SubCount = 0
SubPosition = 1
For x = 1 To Len(ActiveCell)
    Char = Mid(ActiveCell, x, 1)
If (Char >= "0" And Char <= "9") Then
    If SubFlag = False Then
            SubFlag = True
            SubPosition = x
            SubCount = 0
    End If
SubCount = SubCount + 1
Else
    If SubFlag = True Then ActiveCell.Characters(SubPosition,
SubCount).Font.Subscript = True
    SubFlag = False
    SubCount = 0
End If
Next x
If SubFlag = True Then ActiveCell.Characters(SubPosition,
SubCount).Font.Subscript = True
End Sub
```

Figure 7-21. A more sophisticated **Sub** procedure to format text as a chemical formula.

If the character is non-numeric, and **SubFlag** is **True**, then the immediately preceding character(s) must have been numeric, and the characters specified by **SubPosition** and **SubCount** are therefore subscripted, **SubFlag** is set to **False** and **SubCount** is set to zero.

INTERACTIVE MACROS

VBA provides dialog boxes for display of messages or for input. VBA is more versatile than Excel 4.0 Macro Language in the implementation of these boxes.

MSGBOX

The VBA equivalent of the Excel 4.0 Macro Language Alert dialog box function is the **MsgBox** function. **MsgBox** dialog boxes are essentially identical in appearance to Excel 4.0 Macro Language Alert boxes — there are the same three message icons, for example — but with VBA there is much greater versatility in the availability and display of buttons and in the return values.

The syntax of **MsgBox** is

> **MsgBox** *prompt_text, buttons, title_text, helpfile, context*

or

> **MsgBox** (*prompt_text, buttons, title_text, helpfile, context*)

where *prompt_text* is the message displayed within the box, *buttons* specifies the buttons to be displayed and *title_text* is the title to be displayed in the Title Bar of the box. For information about *helpfile* and *context*, refer to *Microsoft Excel Visual Basic Reference*.

For example, the VBA expression

> **MsgBox** "You entered " & incr & "." & **Chr**(13) & **Chr**(13) & _
>
> "That value is too large." & **Chr**(13) & **Chr**(13) & "Please try again.", 64

produces the message box shown in Figure 7-22.

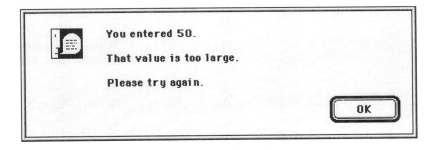

Figure 7-22. Excel 5.0 for the Macintosh **Msgbox** display.

The value of *buttons* determines the type of message icon and the number and type of response buttons; it also determines which button is the default button. The possible values are listed in Table 7-6. The values of *buttons* are built-in constants — for example, the value 64 for *buttons* can be replaced by the variable name vbInformation. (For a complete list of names of these built-in constants, refer to *Microsoft Excel Visual Basic Reference*.) The values 0-5 specify the number and type of buttons, values 16-64 specify the type of message icon, and values 0, 256, 512 specify which button is the default button. You can add together one number from each group to form a value for *buttons* . For example, to specify a dialog box with a Warning Query icon, with Yes, No and Cancel buttons, and with the No button as default, the values 32 + 3 + 256 = 291. The same result can be obtained by using the expression buttons = vbInformation + vbYesNoCancel + vbDefaultButton2 and using the variable buttons in the **MsgBox** function.

MSGBOX RETURN VALUES

MsgBox returns a value that indicates which button was pressed.

INPUTBOX

The VBA equivalent of the Excel 4.0 Macro Language INPUT dialog box function is **InputBox**. **InputBox** dialog boxes are essentially identical in appearance to Excel 4.0 Macro Language INPUT boxes. There are both an **InputBox** Function and an **InputBox** method.

Table 7-6. Values for the **buttons** Parameter of **MsgBox**

buttons	Description
0	Display OK button only.
1	Display OK and Cancel buttons.
2	Display Abort, Retry, and Ignore buttons.
3	Display Yes, No, and Cancel buttons.
4	Display Yes and No buttons.
5	Display Retry and Cancel buttons.
16	Display Critical Message icon.
32	Display Warning Query icon.
48	Display Warning Message icon.
64	Display Information Message icon.
0	First button is default.
256	Second button is default.
512	Third button is default.

The syntax of the **InputBox** Function is

InputBox*(prompt_text, title_text, default, x_position, y_position, helpfile, context)*

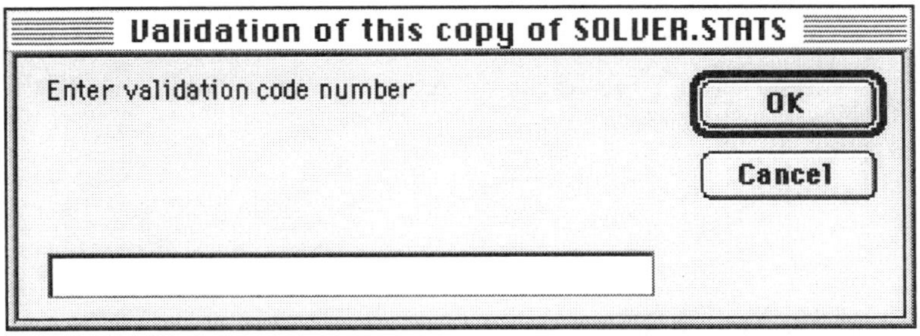

Figure 7-23. Excel 5.0 for the Macintosh InputBox display.

where *prompt_text* and *title_text* are as in **MsgBox**. *Default* is the expression displayed in the input box, as a string. The horizontal distance of the left edge of the box from the left edge of the screen, and the vertical distance of the top edge from the top of the screen are specified by *x_position* and *y_position*, respectively. For information about *helpfile* and *context*, refer to *Microsoft Excel Visual Basic Reference*.

If the user presses the OK button or the RETURN key, the **InputBox** function returns as a value whatever is in the text box. If the Cancel button is pressed, the function returns a null string. The following example produces the input box shown in Figure 7-23.

ReturnValu = **InputBox**("Enter validation code number", _

"Validation of this copy of SOLVER.STATS")

The syntax of the **InputBox** method is

Object.InputBox*(prompt_text, title_text, default, x_position, y_position, helpfile, context, type_num)*

The differences between the **InputBox** function and the **InputBox** method are the following: (i) *default* can be any data type, (ii) the additional argument *type_num* specifies the data type of the return value. The values of *type_num* and the corresponding data types are listed in Table 7-7. Values of *type_num* can be added together. For example, to specify an input dialog box that would accept number or string values as input, use the value $1 + 2 = 3$ for *type_num*.

Table 7-7. InputBox Data Type Values

type_num	Data type
0	Formula
1	Number
2	String
4	Logical
8	Reference (as a Range object)
16	Error value
64	Array

TESTING AND DEBUGGING

Unlike Excel 4.0 Macro Language, which can only halt and issue the message "Macro error at...", VBA provides a large number (over 50) of informative error messages when an error occurs during execution. These are called run-time errors. For example:

Subscript out of range	Attempted to access an element of an array outside its specified dimensions.
Property or method not found	Object does not have the specified property or method.
Argument not optional	A required argument was not provided.

These error messages are for the most part self-explanatory. For a complete list of error messages and their meanings, go to VBA Help, choose Reference Information, then Trappable Errors.

TRACING EXECUTION

When your program produces an error during execution, or executes but doesn't produce the correct answer, it is often helpful to examine the values of variables during execution. To do this you use the Debug window. A convenient way to display the Debug window is by adding a *breakpoint* in your VBA code. A breakpoint halts execution of VBA code and displays the Debug window. At this point you can examine the values of variables or step through the VBA code. There are several ways to enter break mode:

- Place the cursor in the line of code where you want to set a breakpoint. Press the Toggle Breakpoint button 🖐 on the Visual Basic toolbar. The line will become (and remain) highlighted.

- Insert a **Stop** statement in the VBA code.

- Enter a break expression in the Add Watch dialog box (see "Examining the Values of Variables" later in this chapter).

When you run the macro, the code will execute until the breakpoint is reached, at which point execution will stop and the Debug window will be displayed. You can now step through the code one statement at a time or examine the values of selected variables.

To remove a breakpoint, place the cursor on the highlighted line and press the Toggle Breakpoint button, or delete a **Stop** statement.

Excel Tip. Make sure that the VBA toolbar is displayed before you run the macro. You'll need some of the VBA toolbuttons.

STEPPING THROUGH CODE

Once execution has been halted by a breakpoint, you can step through code by using the Step Into button 📶 on the Visual Basic toolbar. As you step through the macro, the next statement to be executed is indicated by a rectangular outline, as shown in Figure 7-24.

EXAMINING THE VALUES OF VARIABLES

You can also display the values of selected variables in the upper Watch Pane of the Debug window. There are at least two ways to select variables or expressions to be displayed:

- While the VBA module is active, highlight the variable or expression that you want to be displayed in the Watch Pane, and then press the Watch button 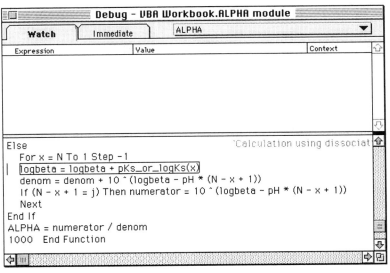 on the Visual Basic toolbar.

- Choose **Add Watch** from the **Tools** menu and type the variable name or expression in the Expression text box.

*Excel Tip. If you're already in break mode, select the desired variable or expression in the lower half of the Debug window, then choose **Add Watch**. The variable name will be displayed in the Expression text box.*

```
┌──────────── Debug – VBA Workbook.ALPHA module ────────────┐
│ ┌──────┐ ┌──────────┐ ┌ ALPHA                       ▼ ┐  │
│ │ Watch │ │Immediate │                                   │
│ ├──────────────────┬────────────┬─────────────────┐      │
│ │ Expression       │ Value      │ Context          │↑    │
│ │                  │            │                  │     │
│ │                  │            │                  │     │
│ │                  │            │                  │     │
│ │                  │            │                  │↓    │
│ ├──────────────────┴────────────┴─────────────────┤↑    │
│ │ Else                       'Calculation using dissociat│
│ │   For x = N To 1 Step -1                          │     │
│ │ │ logbeta = logbeta + pKs_or_logKs(x)             │     │
│ │   denom = denom + 10 ^ (logbeta – pH * (N – x + 1))    │
│ │   If (N – x + 1 = j) Then numerator = 10 ^ (logbeta – pH * (N – x + 1))│
│ │   Next                                            │     │
│ │ End If                                            │     │
│ │ ALPHA = numerator / denom                         │     │
│ │ 1000  End Function                                │↓    │
│ └───────────────────────────────────────────────────┘    │
└──────────────────────────────────────────────────────────┘
```

Figure 7-24. Excel 5.0 for the Macintosh Debug window.

Now when execution is halted by a breakpoint, the Watch Pane will display a list of expressions and their current values (Figure 7-25); as you step through the VBA code, the values will be updated.

To remove variables or expressions from the Watch Pane, choose **Edit Watch** from the **Tools** menu. You will be able to delete unwanted watch expressions.

Excel Tip.   If your keyboard has a DEL key, you can delete a watch expression by selecting the variable or expression to be deleted from the list of expressions in the Watch Pane, then press the DEL key.

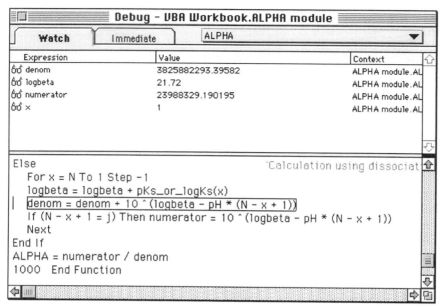

Figure 7-25. Watch expressions in the Watch Pane of Excel 5.0 for the Macintosh Debug window.

USING CONDITIONAL WATCH EXPRESSIONS

A conditional Watch expression causes VBA to enter break mode only when a variable reaches a certain value or changes in value, or when an expression evaluates to **True**. The following examples illustrate some useful conditional watch expressions:

    x = 100

    numerator>denominator

    FormulaString =""

To establish conditional watch expressions, choose **Add Watch** from the **Tools** menu and type the expression in the Expression text box, then press the appropriate Watch Type button. There are three possibilities, which are indicated by different icons in the Watch Pane:

    Watch expression (current value is displayed in the Watch Pane when VBA enters break mode)

    Conditional break expression (break occurs when expression is True)

    Conditional break expression (break occurs when expression changes)

Excel Tip.   A long Watch expression may extend into the Value column. You can adjust the widths of the Expression, Value and Context columns by placing the mouse pointer on the separator bar to the left of the Value or Context headers; the pointer will change to the ✥ pointer shape and you can drag the separator bar to adjust the column width.

USEFUL REFERENCES

Microsoft Excel, Version 5.0:Visual BASIC User's Guide, Microsoft Press, Redmond, WA, 1993.

Microsoft Excel Visual Basic Step by Step, Microsoft Press, Redmond, WA, 1994.

Microsoft Excel/Visual Basic Reference, Second Edition, Microsoft Press, Redmond, WA, 1995.

John Walkenbach, *Excel 5 for Windows Power Programming Techniques*, IDG Books Worldwide, San Mateo, CA, 1994.

8

CREATING CUSTOM MENUS AND MENU BARS

In this chapter you'll learn how to add a new command to a menu, how to add a new menu to a menu bar and how to create a whole new menu bar with separate menus and commands. In this way you can add new capabilities to Excel or even create a complete custom application.

MODIFYING MENUS OR MENU BARS

You may find that you use a particular macro so often that it would be more convenient to have it on one of Excel's drop-down menus, rather than having to actuate it by means of the **Run** command from the **Macro** menu. Running a command macro by means of an Excel menu command makes the custom command accessible to Excel users who are not familiar with, or not comfortable with, the use of macros.

In Excel 5.0 you can use the Menu Editor to modify menus or menu bars. In Excel 4.0 you must write a macro in Excel 4.0 Macro Language to modify menus or menu bars.

THE EXCEL 5.0 MENU EDITOR

By means of the Menu Editor, you can add commands to any of Excel's menus, add menus to existing menu bars, or add menu bars. You can also delete any of Excel's built-in commands, menus or menu bars (they can be restored later if you wish).

Here's how to add a new command to an existing menu. Of course, the custom command that you're going to install will be executed by means of a macro that you've written (either a command macro written in Excel 4.0 Macro Language or a **Sub** procedure written in VBA). Make sure the macro document is open before you begin.

To display the Menu Editor dialog box (Figure 8-1), press the Menu Editor button ⊞ on the Visual Basic toolbar, or choose the **Menu Editor...** command from the **Tools** menu (you'll see this command only when you have a VBA module as the active sheet).

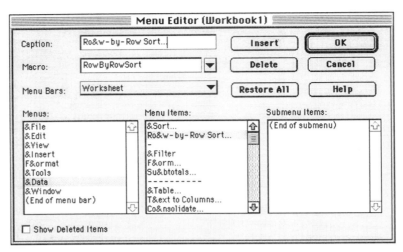

Figure 8-1. The Excel 5.0 Menu Editor dialog box.

From the Menus list box, select the menu to which you want to add the new command. The commands in that menu will be displayed in the Menu Items list box. Select the command above which you want to insert the new command, and then press the Insert button. The cursor will be activated in the Caption text input box; type the name of the new command. To underline a letter in the command name, to indicate the key to be pressed to choose the command, precede that letter in the name by the & (ampersand) symbol. To insert a separator bar, enter a single hyphen in place of a menu command.

Next, press the drop-down list button in the Macro input box. Select the macro to be assigned to the command, then press the OK button. Figure 8-2 shows the command **Row-by-Row Sort...** inserted below the **Sort...** command in the **Data** menu, with a separator bar.

You can add commands in submenus, or new menus or menu bars, in the same way.

The new menu command that you added by using the Menu Editor is saved with the workbook that was open when you created the custom menu and is only available when that workbook is open. To have a custom menu available each time you start Excel, write an Auto_Open macro to customize the menu and place it in the Global Macro Sheet (Excel 4.0) or the Personal Macro Workbook (Excel 5.0). The macro can be written either in Excel 4.0 Macro Language or in VBA. The following sections describe how to customize Excel's menus using Excel 4.0 Macro Language. For details on performing the same tasks using VBA, see, for example, *Microsoft Excel Visual Basic Step by Step*.

ADDING A COMMAND TO A MENU (EXCEL 4.0)

To add a custom command to one of Excel 4.0's built-in menus, you must write a short macro to install the command. You can include the macro statements to install the command as part of the macro that executes the new command.

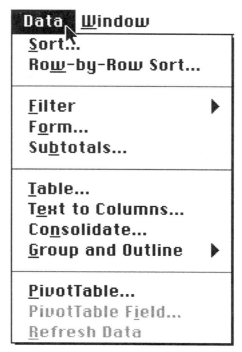

Figure 8-2. A menu command added to the **Data** menu by using the Excel 5.0 Menu Editor.

After a macro has been written and debugged, use ADD.COMMAND to add it as a custom menu command to a menu on a built-in or custom menu bar (Figure 8-3) ADD.COMMAND permits you to locate the new menu command in any one of Excel's existing menus. The syntax for the ADD.COMMAND function is ADD.COMMAND(***bar_number***, ***menu***, ***command_table_reference***, ***position***). *Bar_number* is the menu bar ID number. (To see a table of bar ID numbers for built-in menu bars, use Excel's On-Line Help. Choose "Reference Information"; then choose "Microsoft Excel Macro Functions Information" and search under ADD.COMMAND.) The ID number of the active menu bar can be obtained by using the GET.BAR function described in the paragraph following. *Menu* can be either the number of the menu or the name of the menu as text. For bar numbers 1 through 6, menus are numbered starting with 1 for the leftmost menu. Bar numbers 7, 8 and 9 are shortcut menus. *Command_table_reference* (Figure 8-4) is an array or reference on the macro sheet that describes the new command. It must be at least two cells wide. The first cell contains the command name as it will appear in the menu. The second cell indicates the macro that corresponds to the command, in the form of either a name or an R1C1-style

	A	B
14	active.bar	=GET.BAR()
15		=GET.BAR(active.bar,"Formula","Solver...")
16		=ADD.COMMAND(active.bar,"Formula",$A$25:$D$25,B15+1)

Figure 8-3. Adding a new menu command to the Excel 4.0 **Formula** menu.

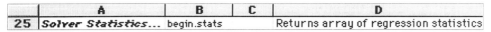

Figure 8-4. A command table specifying the new menu command.

reference as text. The optional third, fourth and fifth columns contain, respectively, the shortcut key, status bar message and custom help topic. *Position* specifies the placement of the new command. It can be either a number indicating the position of the command, or the name of an existing command, as text, above which the new command will be added. If position is omitted, the new command will be added to the bottom of the menu.

EXAMPLES

=ADD.COMMAND(1,1,$A$20:$B$20) adds the command specified in cells A20:B20 to the bottom of the **File** menu.

=ADD.COMMAND(1,"Data",$A$20:$B$20,"Sort...") adds the command specified in cells A20:B20 to the **Data** menu, immediately following the **Sort...** command.

Excel Tip. Wherever possible, use menu and command names rather than position as arguments. The position of a particular menu or command may change if new menus or commands have been added. Using names ensures that the desired menu or command is selected.

GET.BAR(*bar_number, menu, command*) returns the ID number of the active menu bar, the position of a specified menu on a menu bar or the position of a specified command in a menu. All of the arguments are optional. If *bar_number, menu* and *command* are all specified, GET.BAR returns the position of the command in the specified menu; if *bar_number* and *menu* are specified, GET.BAR returns the position of the menu on the menu bar; if GET.BAR() is used, the ID number of the menu bar is returned. *Menu* can be the number of the menu or the name of the menu as text. For bar numbers 1 through 6, menus are numbered starting with 1 for the leftmost menu. Bar numbers 7, 8 and 9 are shortcut menus. *Command* can be the number of the command or the name of the command as text. If the name of the command is given, the position of the command in the menu is returned. Commands are numbered from 1 from the top of the menu; separator lines are included in the count. If the command name includes an ellipsis (...), then the ellipsis must be included in the command name. If command is given as a number, the name of the command at that position in the menu is returned. Bar numbers 7, 8 and 9 (shortcut menus) require special numbering of the commands within the menu.

If the menu or command name or position specified does not exist, then #N/A! is returned. This can be used to determine whether a particular command is already located on a menu.

EXAMPLES

=GET.BAR(1,"File","Print...") returns 13, the position of the **Print** command in the **File** menu (separator bars are included in the count). GET.BAR(1,1, "Print...") is equivalent.

=GET.BAR(1,"Format",1) returns the name of the first command in the **Format** menu.

=GET.BAR(1,"Data",) returns 5, the position of the **Data** menu on the menu bar.

=GET.BAR() returns the ID number of the active menu bar.

ADDING A MENU COMMAND BY MEANS OF AN AUTO_OPEN MACRO

An **Auto_Open** macro that installs its command in an appropriate menu is an ideal way to customize a menu. The user need only **Open** the macro sheet; the macro command is then available in the familiar Excel working environment.

Figure 8-5 shows a portion of a macro to add a command "Row-by-Row Sort" to the **Data** menu, immediately below the **Sort...** command. GET.BAR(1,"Data","Row-by-Row Sort") returns #N/A! if the command is not present in the **Data** menu, enabling us to check whether the command has already been installed. The statement in cell A3 returns the position of the **Sort...** command in the menu, so that the new command can be inserted *below* the **Sort...** command by the statement in cell A4. The statement

ADD.COMMAND(1,"Data",$A$19:$D$19,"Sort...")

installs the new command immediately *above* the **Sort...** command.

ADDING A MENU TO A MENU BAR (EXCEL 4.0)

To add a menu to an existing menu bar, use the ADD.MENU macro function. The   syntax   of   this   command   is   ADD.MENU(**bar_number,** **menu_table_reference,**   *position*) To create a new menu in an existing menu bar, you set up a *menu table* in the macro sheet, and use ADD.MENU to insert the new menu into the desired menu bar. For *bar_number*, see the preceding section.

The menu table consists of a *command table*, five columns wide, containing command names, command macro names, shortcut keys, status bar messages and Help topics. The first cell in the first row contains the *menu name*. Subsequent rows contain the command table information for each menu command (see "Adding a Command to a Menu" earlier in this chapter). If a command name cell contains a hyphen, then a *separator line* will appear in that

	A	B
1	Auto_Open_Sort_Macro	EJB 3/11/94
2	=IF(ISERROR(GET.BAR(1,"Data","Row-by-Row Sort")),,RETURN())	
3	=GET.BAR(1,"Data","Sort...")	
4	=ADD.COMMAND(1,"Data",$A$7:$D$7,A3+1)	
5	=HIDE()	
6	=RETURN()	
7	Row-by-Row Sort	DoSort
8	DoSort	

Figure 8-5. An Auto_Open macro to install a new menu command.

menu position. If *position* is not specified, the new menu is added to the right of the existing menus. If the active menu bar is the worksheet menu bar (bar ID# = 1), with seven built-in menus, there is room for one additional menu.

CREATING A NEW MENU BAR (EXCEL 4.0)

If you have written a number of related macros, you may wish to create a custom menu bar, with several custom menus, each containing custom commands. To add a new menu bar, use the ADD.BAR function. This creates a new, empty menu bar and returns its bar ID#, but does not display it. Use ADD.MENU (see "Adding a Menu to a Menu Bar (Excel 4.0)" earlier in this chapter) to add menus to the menu bar. To display the menu bar, use the SHOW.BAR macro function. A maximum of 15 custom menu bars may be defined.

Figure 8-6 shows a portion of a macro to create a custom menu bar for calculating and plotting titration curves, with four drop-down menus, **File**, **Initial Conditions**, **Select Weak Acid/Base** and **Select Plot Type**. The menu table for the **File** menu is shown in Figure 8-7.

Menu commands from any of Excel's built-in menus can be included in menus on a custom menu bar. Use the appropriate macro function to duplicate the command as it appears in the built-in menu. The Recorder can be useful in obtaining the macro functions for a particular menu command. Commands such as **Close** or **Save** often correspond to a single Excel macro function. If the command employs a dialog box (i.e., the command name includes an ellipsis) then there will exist a ? form of the Excel function. For example, to duplicate the

	A	B	C	D
	Menu	Menu List	Macro Name	CMD Key
2		(hyphen=separator bar)		
3	=ADD.BAR()	File		
4	=ADD.MENU(A3,B3:D14)	New Plot	New	N
5	=ADD.MENU(A3,B16:D18)	Open...	Open	O
6	=ADD.MENU(A3,B21:D30)	Close	Close	W
7	=ADD.MENU(A3,B32:D38)	-		
8	=SHOW.BAR(A3)	Save	Save	S
9		Save As...	Save_As	
10		-		
11		Print Preview	Print_Preview	
12		Print...	Print	P
13		-		
14		Exit to Excel	Exit1	Q
15				
16		Initial Conditions		
17		Select Titrant	Macro1	
18		Strong Acid/Base	Macro2	
19				

Figure 8-6. Portion of a macro sheet that installs a new menu bar with four menus.

Save As... menu command, use the SAVE.AS? form of the Excel function, which will produce the appropriate dialog box. (The SAVE.AS form of the function could be used if the filename is already available.)

```
┌─────────────────────────────────────┐
│                  E                  │
├─────────────────────────────────────┤
│           "File" Macros             │
│                                     │
│ New                                 │
│ =NEW(2)                             │
│ =RETURN()                           │
│ Open                                │
│ =OPEN?()                            │
│ =RETURN()                           │
│ Close                               │
│ =FILE.CLOSE()                       │
│ =RETURN()                           │
│ Save                                │
│ =SAVE()                             │
│ =RETURN()                           │
│ Save_As                             │
│ =SAVE.AS?()                         │
│ =RETURN()                           │
│ Print_Preview                       │
│ =PRINT.PREVIEW()                    │
│ =RETURN()                           │
│ Print                               │
│ =PRINT?()                           │
│ =RETURN()                           │
│ Exit1                               │
│ =SHOW.BAR()                         │
│ =DELETE.BAR(A3)                     │
│ =RETURN()                           │
└─────────────────────────────────────┘
```

Figure 8-7. A macro fragment that implements the **File** menu commands.

9

CREATING CUSTOM TOOLS
AND TOOLBARS

In this chapter you'll learn how to customize Excel's built-in toolbars, and how to create new toolbuttons to simplify some of the operations that you perform often.

CUSTOMIZING TOOLBARS

Many of the tools on Excel's toolbars are not useful for scientific calculations. You can remove the tools you don't use, giving a less cluttered workspace and providing space for other, more useful ones.

MOVING AND CHANGING THE SHAPE OF TOOLBARS

Excel's toolbars are usually located along the edges of the screen, most commonly along the top edge. Excel initially displays the Standard toolbar, which provides tools for common Excel actions. There are a number of other toolbars provided with Microsoft Excel, including the Formatting, Utility, Chart, Drawing and Macro toolbars. You can also create custom toolbars, containing the tools you use most.

A toolbar can be moved from its position at the top of the screen and placed anywhere on the screen. Simply place the mouse pointer anywhere on the toolbar between the toolbuttons, and drag it to the new location. A toolbar that is moved from a position along the edge of the screen is called a *floating toolbar*. It appears in its own window, with a title bar, a Close box and a Size box. If you change the height or width of a floating toolbar using the size box, the tools are automatically adjusted to fit the new shape. If a floating toolbar (Figure 9-1) is dragged near the edge of the screen, Excel places it in a *toolbar dock*. There are four toolbar docks, one along each of the edges of the Excel application window.

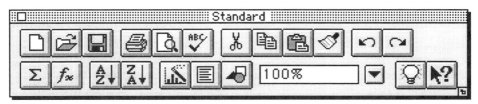

Figure 9-1. A floating toolbar.

Excel Tip . *A toolbar that contains a drop-down list box, such as the Style box, cannot be placed in a vertical position in the left or right toolbar docks.*

If a toolbar does not fill the whole window (from left to right or vertically, depending on its orientation) the toolbar dock is visible as the extra blank space at either side of the toolbar, separated from the toolbar by the toolbar border.

Excel Tip. *Double-click on the title bar of a floating toolbar or in a blank area of the toolbar to return the toolbar to its previous docked position. Double-click in a blank area of a docked toolbar to return it to its previous floating position.*

ACTIVATING OTHER TOOLBARS

You can remove a toolbar from the Excel window by dragging it from the toolbar dock and then pressing the Close box. Restore toolbars by using **Toolbars...** from the **View** menu (Excel 5.0) or from the **Options** menu (Excel 4.0). You can have several toolbars in the Excel window at once. Select the desired toolbar(s) from the Toolbars list box (Figure 9-2) and press the OK button.

ADDING OR REMOVING TOOLBUTTONS FROM TOOLBARS

Removing seldom-used toolbuttons from a toolbar provides a less-cluttered working environment and makes room for other, more useful tools. To delete a tool from a toolbar, follow the procedure in the box on the following page.

Excel Tip. *Instead of choosing **Toolbars...** from the **View** menu (Excel 5.0) or the **Options** menu (Excel 4.0), you can double-click on the toolbar dock background to display the Toolbars dialog box.*

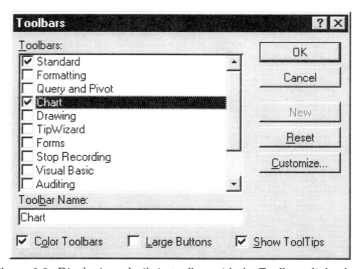

Figure 9-2. Displaying a built-in toolbar with the Toolbars dialog box.

To Delete a Tool from a Toolbar

1. Choose **Toolbars...** from the **View** menu (Excel 5.0) or the **Options** menu (Excel 4.0).

2. Drag the tool off the toolbar. When you click on it, the tool outline is highlighted. Place the tool anywhere on the desktop (the dialog box is a good place). The toolbutton will disappear.

Excel Tip. Press the right mouse button (Windows) or COMMAND+ OPTION (Macintosh) and click in a blank area of the toolbar to display the Toolbar Shortcut menu.

There are some useful formatting tools for scientific use that are not found on the default Standard toolbar. For example, the "Add one decimal place to number format" and "Remove one decimal place from number format" tools are useful. These are located on the Formatting toolbar. They can easily be moved to the Standard toolbar by means of the procedure in the box below.

To Add a Built-in Tool to a Toolbar

1. Choose **Toolbars...** from the **View** menu (Excel 5.0) or the **Options** menu (Excel 4.0).

2. Press the Customize button.

3. Select the desired tool from the various displays. Drag the tool to the desired place on the toolbar. Spaces between tools may be added or removed by using the procedure in the following box.

The built-in Microsoft Excel toolbars can be restored by using the Reset button on the Toolbars dialog box. Individual tools can be restored by following the procedure in the box above. Custom toolbars cannot be Reset.

Tools can be organized into logical groups by inserting spaces between tools or grouping them together, by using the procedure in the box below.

To Insert or Remove Space Between Tools on a Toolbar

1. Choose **Toolbars...** from the **View** menu (Excel 5.0) or the **Options** menu (Excel 4.0).

2. Press the Customize button.

3. To insert a space between two adjoining tools, drag the left-hand tool to the left so that it overlaps the next tool slightly.

4. To remove a space between two adjoining tools, drag the right-hand tool to the right so that it overlaps the next tool slightly.

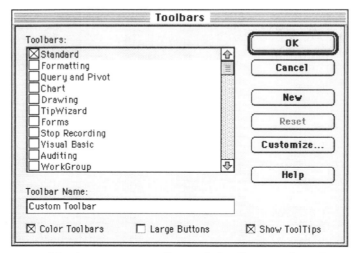

Figure 9-3. Naming a custom toolbar with the Show Toolbars dialog box.

CREATING A NEW TOOLBAR

There are two ways to create a custom toolbar. One way is to modify an existing toolbar (such as the Standard toolbar). The other way is to choose **Toolbars**... from the **View** menu (Excel 5.0) or the **Options** menu (Excel 4.0), type the name of the new toolbar in the Toolbar Name box, as in Figure 9-3, and press the Add button. A small floating toolbar will appear, along with the Customize dialog box (Figure 9-4). You can then proceed to add built-in tools to the toolbar, as described earlier, or custom tools, as described later, in the section "Creating Custom Tools".

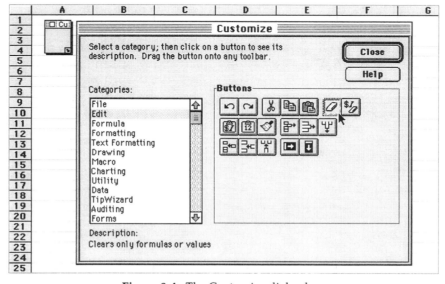

Figure 9-4. The Customize dialog box.

A TOOLBAR FOR CHEMISTS

The Standard toolbar (Figure 9-5) contains many tools that are not useful for scientific calculating. I've removed many of the tools and replaced them with some of Excel's built-in tools from other toolbars. In addition, I've created a few new tools that are useful for chemists.

Figure 9-5. Excel 5.0 for the Macintosh Standard toolbar.

Figure 9-6. A toolbar for chemists.

The toolbar shown in Figure 9-6 is a modified version of the Excel 5.0 Standard toolbar. A number of the tools provided on the default Standard toolbar have been removed, including the New, Open, Save and Print tools, the Cut, Copy and Paste tools, the Undo, Repeat, Sort Ascending and Sort Descending tools, and the TipWizard and Help tools. The Bold, Italic, Left Align, Center Align, Right Align, Center Across Columns, AutoSum and ChartWizard tools were retained. The Increase Font Size, Decrease Font Size, Add One Decimal Place, Remove One Decimal Place, Light Shading, Clear Formulas or Values, Paste Formats, Paste Values and Text Box tools were added from Excel's other menus. In addition, the custom tools FullPagePortrait, FullPageLandscape, ChemFormula and Toggle Between Floating Point and Scientific were added, as described in the sections that follow.

CREATING CUSTOM TOOLS

Some macros that you'll create will be written for a very specific purpose, such as to prepare a specialized report. The macro sheet will be opened only when you want to assemble the report. Other macros automate tasks that you perform often, and you'll want to have them available whenever you're using Excel. These macros should be saved on the Global Macro Sheet or in the Personal Macro Workbook. To make a command macro even easier to use, you can add a custom toolbutton to a toolbar and assign the macro to it. The three macros described in this section — the *NumberFormatConvert* macro, the *FullPage* macros and the *ChemFormula* macro — are ones that are particularly convenient to use when they are assigned to a button. The *NumberFormatConvert* macro toggles a number between floating-point and scientific format. The *FullPage* macros set margins to zero and delete the header and footer, to provide maximum space on a page for printing the spreadsheet. The *ChemFormula* macro converts text in a cell to a chemical formula, e.g., H2SO4 becomes H_2SO_4.

When a macro is completely debugged, you can assign it to a toolbar button.

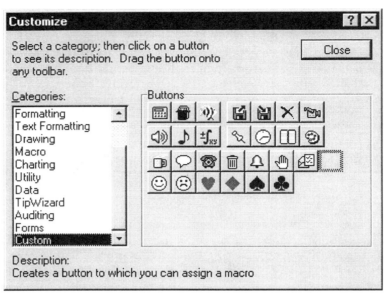

Figure 9-7. Using the Customize dialog box.

The Customize dialog box (Figure 9-7) contains buttons that can be assigned to a macro by following the procedure in the following box. Usually you'll select the blank tool and create a toolface for it later. When a custom tool is moved to a toolbar, Excel opens the Assign Macro dialog box (Figure 9-8) and displays the macros that are on the Global Macro Sheet or the Personal Macro Workbook, or on any open macro sheet.

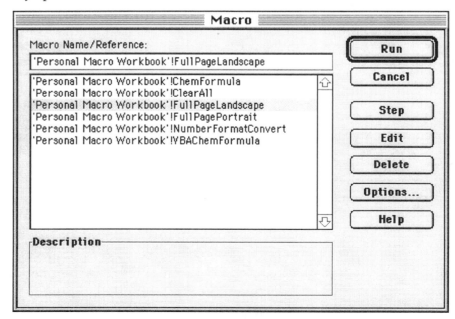

Figure 9-8. Assigning a macro to a toolbutton.

> **To Add a Custom Tool to a Toolbar and Assign a Macro to It**
>
> 1. Choose **Toolbars...** from the **View** menu (Excel 5.0) or the **Options** menu (Excel 4.0).
> 2. Press the Customize button.
> 3. Scroll through the Categories box and choose "Custom".
> 4. Select the blank tool (you'll create a face for it later) and drag it to the desired location on the toolbar.
> 5. Excel displays the Assign Macro dialog box. Select the name of the command macro to be assigned to the tool (best located in the global macro sheet). Click the OK button.

Excel Tip. When you place a custom tool on a toolbar, you don't have to assign a macro to it right away. Later, if you click a custom tool that doesn't have a macro assigned to it, the Assign to Tool dialog box will appear.

ADDING THE *NUMBER FORMAT CONVERT* MACRO COMMAND AS A TOOLBUTTON

In scientific spreadsheet computing, it's common to convert numbers from floating-point format to scientific, and vice versa. Although Excel provides tools in the Formatting toolbar to format numbers with Currency, Percent or Comma formats, there isn't a tool for Scientific format. The macro listing in Figure 9-9 toggles a number between Floating Point or General formats and Scientific format.

THE MACRO. In cell B4, if the selected cell does not contain a number, no action is taken. In cell B5, the CELL(*infotype*) function returns information about the selected cell. If *infotype* = format, CELL returns G if the number format is general, F4 if the format is 0.0000, S2 if the format is 0.00E+00, P2 if the format is 10%, etc. In cells B6 to B12, the IF... ELSE.IF... END.IF logic is used. If the

	A	B
1		**NumberFormatConvert**
2		Toggles between Number format (General or Floating Point)
3		and Scientific format
4		=IF(ISNUMBER(SELECTION()),,RETURN())
5	fmtcode	=CELL("format")
6		=IF(OR(fmtcode="G",LEFT(fmtcode,1)="F",LEFT(fmtcode,1)="P"))
7		=FORMAT.NUMBER("0.00E+00")
8		=RETURN()
9		=ELSE.IF(LEFT(fmtcode,1)="S")
10		=FORMAT.NUMBER("general")
11		=RETURN()
12		=END.IF()
13		=RETURN()

Figure 9-9. Simple number-formatting macro to assign to a toolbutton.

number format is G, F or P, the formatting is changed to the built-in Scientific format, 0.00E+00. If the number format is S, the number format is changed to General. If the number format is any other type, no action is taken.

Writing a macro that performs Floating Point/Scientific number formatting and returns a number with the same number of significant figures as in the original number is much more difficult. You may wish to try to write one (hint: the CELL("format') worksheet function will be useful).

ADDING THE *FULL PAGE* MACRO COMMANDS AS TOOLBUTTONS

The FullPage macros maximize the space on a page that is available for printing a worksheet, macro sheet or chart, by eliminating margins, header and footer. To do this by using menu commands requires choosing **Page Setup** from the **File** menu, setting Left Margin, Right Margin, Top Margin and Bottom Margin to zero, choosing Header and deleting the header text, then choosing Footer and deleting the footer text. The FullPage macro was written to do this at the click of a tool; the listing is shown in Figure 9-10.

THE MACRO. Upon entry to either of the macro entry points (shown in bold), a value (1 = Portrait, 2 = Landscape) is assigned to the variable orientation. Because the syntax of the PAGE.SETUP function is slightly different for charts and either worksheets or macro sheets, the value of type is obtained by the GET.DOCUMENT function. If there is no sheet open, the GET.DOCUMENT function returns #N/A!, whereupon the macro exits without doing anything. Otherwise, the margins are then set to zero and the header and footer to null.

Two custom toolbuttons were created, using Excel 5.0's Button Editor, and positioned on the left side of the toolbar, with the Portrait button in position 1 and the Landscape button in position 2. The macros FullPagePortrait and FullPageLandscape were assigned to them, as described earlier.

	A	B	C
1		**FullPageLandscape**	
2	orientation	=2	2=Landscape
3		=GOTO(B6)	
4		**FullPagePortrait**	
5		=SET.VALUE(orientation,1)	1=Portrait
6	type	=GET.DOCUMENT(3)	1=Sheet or macro,2=chart
7		=IF(ISNA(type))	#N/A if no window open.
8		=BEEP()	... so beep
9		=RETURN()	... and return.
10		=END.IF()	
11		=IF(type=2)	If it's a chart...
12		=PAGE.SETUP("","",0,0,0,0,3,,orientation)	Use syntax for charts
13		=RETURN()	
14		=END.IF()	
15		=PAGE.SETUP("","",0,0,0,0,,,,,orientation)	Otherwise, use syntax for sheets
16		=RETURN()	

Figure 9-10. FullPage Landscape and Portrait macros to assign to toolbar buttons.

ADDING THE *CHEMFORMULA* MACRO COMMAND
AS A TOOLBUTTON

In Chapter 1 we saw that subscripts and superscripts can be produced in Excel 4.0 by using a custom font designed for that purpose. The subscripts in chemical formulas were typed by using Option + (number), after choosing the custom font from **Font** in the **Format** menu. The following macro, which converts all numbers in a selected cell containing text into subscript numbers. Thus text can be typed as H2SO4; then, selecting the cell and clicking the ChemFormula button converts the contents of the cell to H_2SO_4. The ChemFormula macro is listed in Figure 9-11.

THE MACRO. Cell B7 transfers the cell contents to the macro sheet. Cell B8 is the destination at which the converted string will be assembled. In cell B9 the contents of the cell are examined; if the cell contains null or non-text, no action is taken. Then, in cells B11 to B17, we loop through the characters of the source string using the MID function, examining each character in turn. If the character is not a digit (greater than "9" or less than "0"), the character is concatenated in the destination string. If the character is a number, we replace it with the corresponding subscript character. The digits 0 - 9 correspond to the character codes 48 - 57, but because of the way the Apple keyboard is mapped, the OPTION + number codes are 188, 193, 170, 163, 162, 176, 164, 166, 165, 187. Thus it's necessary to use a LOOKUP function to obtain the code of the subscript number. Here the CHOOSE function is used. Since the index argument CHOOSE uses to select the value to be returned must be 1 or greater, we use CHOOSE(digit+1, Value1, Value2, ...). After the new string has been assembled, cell B18 changes the font to the custom font by using FORMAT.FONT and the string is entered into the selected cell by using the FORMULA function.

A custom toolbutton was created, using Excel 5.0's Button Editor, and the ChemFormula macro was assigned to it.

	A	B
1		**ChemFormula**
2		Converts a chemical formula in standard Geneva font, e.g. H2SO4,
3		to one in GenevaChem font w/ subscripted numbers
4		(All numbers are converted into subscripts)
5		Operates on a single cell.
6		
7	source	=SELECTION()
8	dest	
9		=IF(OR(SELECTION()="",ISNONTEXT(SELECTION())),RETURN())
10		=SET.VALUE(dest,"")
11		=FOR("x",1,LEN(source))
12		=IF(OR(MID(source,x,1)>"9",MID(source,x,1)<"0"))
13		=SET.VALUE(dest,dest&MID(source,x,1))
14		=ELSE()
15		=SET.VALUE(dest,dest&CHAR(CHOOSE(MID(source,x,1)+1,188,193,170,163,162,176,164,166,165,187)))
16		=END.IF()
17		=NEXT()
18		=FORMAT.FONT("GenevaChem")
19		=FORMULA(dest)
20		=RETURN()

Figure 9-11. Simple form of the ChemFormula macro to assign to a toolbar button.

TO SAVE A CUSTOM TOOLBAR
OR TO TRANSFER A CUSTOM TOOLBAR TO ANOTHER COMPUTER

Each time you quit Excel, the latest toolbar is saved in a file named Excel Toolbars in the Preferences folder in the System folder (Macintosh), or in EXCEL.XLB (Windows). Make a copy of the file (quit Excel before copying if you have made changes in this session).

Excel Tip   *To save a custom toolbar, just rename the Excel Toolbars file. Excel will create a new Excel Toolbars file the next time you quit Excel.*

Excel Tip. *If you mistakenly press the Close button on the Global Macro Sheet, the sheet will be put away in the directory where EXCEL.INI is located (Windows) or in the Preferences folder (Macintosh). You won't be able to use the **Unhide** command in the **Window** menu to unhide it. If you have installed a custom tool that is assigned to a macro on the Global Macro Sheet, just click the tool to open the Global Macro Sheet again.*

CUSTOMIZING A TOOLFACE

With Excel 5.0, you can edit an existing image or create a new one using the Button Editor. First, choose **Toolbars**... from the **View** menu. Click on Customize... . Select the tool image you wish to edit while holding down the COMMAND+OPTION buttons to display the Toolbar shortcut menu. Choose Edit Button Image from the shortcut menu to display the Button Editor (Figure 9-12).

The button image will be displayed in the Picture area, 16 pixels wide by 15 pixels high. You can add or remove pixels from the image. Click on any pixel in the Picture area to add a pixel of a selected color. Click a second time if you want to remove that color. Make the background color "Erase". This will provide a gray background identical to the rest of the button.

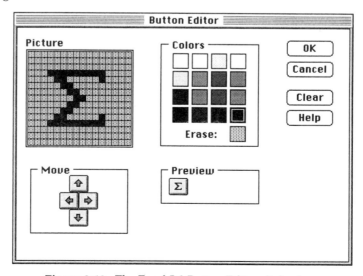

Figure 9-12. The Excel 5.0 Button Editor dialog box.

The Move buttons allow you to shift the image somewhat, either up, down or to the left or right, but only if the background is "Erase". If you want to create a new picture, press the Clear button before you begin. You will start with a complete "Erase" background. The Preview window shows the appearance of the toolface.

To edit the large toolbuttons, choose **Toolbars...** from the **View** menu and check the Large Buttons checkbox. Then display the button image to be edited, as described on the previous page. The pixel area displayed is 24 pixels wide by 23 high.

With Excel 4.0, you will have to use a separate drawing program to edit a tool face or create a new one. The image should be 16 pixels wide by 15 pixels high. As an alternative you can label a tool face with a text label (see box below). Type the button text in MS Word, copy it, switch to Excel and paste the image on the toolbutton. For readability, the text should be no more than two characters, or three characters using condensed spacing. Ten-point Geneva seems to be suitable, and Bold often improves readability. A little experimenting with font and size will probably be necessary. For example, the Excel 4.0 *ChemFormula*

button was labeled with the text "H_2O" to produce the toolbutton image $\boxed{H_2O}$.

To Paste Text or a Graphic on a Toolface

1. Copy the text or graphic to be pasted. Then switch to Excel.

2. Choose **Toolbars...** from the **View** menu (Excel 5.0) or the **Options** menu (Excel 4.0).

3. Press the Customize button.

4. Click on the toolbutton to be customized. It will become highlighted.

5. Choose **Paste Tool Face** (Excel 5.0) or **Paste Button Image** (Excel 4.0) from the **Edit** menu.

HOW TO ADD A TOOLTIP TO A CUSTOM BUTTON

If you create a custom button by dragging one of the Custom buttons onto a toolbar, the ToolTip message is "Custom". Figure 9-13 shows a custom ToolTip message displayed with the FullPageLandscape toolbutton.

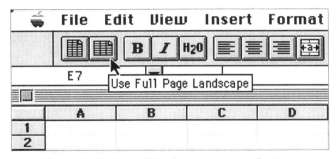

Figure 9-13. A ToolTip for a custom toolbutton.

To add or change ToolTip text, you must use VBA. Instead of writing and running a macro, you can use VBA's Immediate mode. But before you start, you need to know the name of the toolbar where your button is located; get it from **Toolbars...** in the **View** menu. You also have to identify the position number of the custom button for which you want to add a ToolTip. Beginning at the left end of the toolbar, count all the buttons *and* the spaces in between to get the position of the custom button. Then follow the steps outlined in the box below.

To Add a ToolTip to a Custom Toolbutton

1. Open a VBA module sheet by choosing **Macro** from the **Insert** menu, then **Module** from the submenu.

2. Open the Debug window by choosing **Debug Window** from the **View** menu.

3. Type the following VBA statement, then press RETURN.

   Toolbars("ToolbarName").Toolbarbuttons(position number).Name = "TooltipText"

   (you supply whatever is appropriate for the three arguments: "ToolbarName", position number, "TooltipText").

4. Close the Debug window and discard the module sheet. The ToolTip will be saved when you exit from Excel.

HOW TO ADD A STATUS BAR MESSAGE FOR A CUSTOM BUTTON

If you create a custom button, the Status Bar message is "Creates a button to which you can assign a macro". To enter your own Status Bar text, follow the procedure outlined below.

To Update the Status Bar Text

1. Choose **Macro...** from the **Tools** menu.

2. Select the macro for which you want to add or change the Status Bar message text.

3. Press the Options... button to display the Macro Options dialog box.

4. Place the cursor in the Status Bar Text input box and type the desired message, then press OK.

CREATING TOOLS OR TOOLBARS BY MEANS OF A MACRO

There are a number of macro commands for creating and manipulating toolbuttons or toolbars. You can use either Excel 4.0 Macro Language or VBA. Refer to Excel's On-line Help for more information.

10

ADVANCED SPREADSHEET TECHNIQUES

In this chapter you'll learn how to create links between worksheets, so that data from one sheet can be used in another. You'll also learn how to copy worksheets or charts and paste them into other applications, and how to create a button on a spreadsheet.

LINKING WORKSHEETS

By linking worksheets, you can:

- utilize the data from one worksheet in a number of other worksheet applications.
- merge data from several worksheets in a summary sheet.
- simplify a complicated model by breaking it up into manageable portions.

CREATING LINKS BETWEEN WORKSHEETS

A *dependent worksheet* contains one or more references to a *source worksheet*, from which it obtains values. Linked worksheets are linked dynamically; that is, when you make changes in a source worksheet, the changes automatically occur in the dependent sheet.

Links between worksheets are established by means of *external references*. An external reference is simply a reference that includes the filename of the source worksheet enclosed in single quotes and separated from the reference or name by an exclamation point: *'SheetName'!CellReference.*

For example, to enter a value from a source document into a cell of a dependent document, first select the cell where the value is to be entered (let's say cell B2 of the worksheet Summary Sheet) and type an equal sign, then switch to the source document and select the cell containing the binding constant value (let's say it's in cell H12 of the worksheet Expt #XVIII-32). When you press the Enter button, Excel returns you to the dependent document and the formula ='Expt #XVIII-32'!$H$12 is entered in cell   B2 of Summary Sheet.

For a more sophisticated example, consider a grade sheet that contains columns for student names, ID numbers, exam grades, and homework grades.

The homework grades are kept in a separate worksheet Homework Grades 95F; the student names are in column A of this worksheet and the final averaged homework grades are in column Y. In the grade sheet (the dependent document), the column for homework grades contains the formula

=VLOOKUP(name,'Homework Grades 95F'!$A$3:$Y$57,25,0).

which looks up the student's name in column A3:A57 of the source worksheet Homework Grades 95F, then finds that student's homework grade in column Y3:Y57 (offset by 25 columns from column A), and returns the value to a cell in the dependent worksheet.

If the source sheet is not open, Excel provides the complete path of the directory or folder containing the sheet, e.g.,

=VLOOKUP(name,'Macintosh HD:Lecture and Laboratory:CH351 Lecture:All 95F: Homework Grades 95F'!$A$3:$Y$57,25,0)

To create a formula containing an external reference, you can type the reference in the format shown above, but there's a much simpler way to do it: simply activate the source sheet (by clicking on it or by using the **Window** menu), then select the cell or range you want to insert in the formula. Excel will automatically insert the reference using the correct syntax.

If you open a dependent worksheet before you open its source document or documents, you'll get an "Update references to unopened documents?" message. If you press the Yes button, there can be a long delay while the data is updated (depending on the size and complexity of the documents). If you press No, you'll get either whole columns of #VALUE! error values, or even worse, values that are not updated. So it's advisable to open all source documents before beginning to open dependent documents. For the same reason, always save the source worksheet before saving the dependent sheet linked to it.

If you edit a source worksheet, the cell references in the dependent worksheet will be updated provided the dependent worksheet is open. The same is true if you change the name of a source document or move it to a different directory. If you rename or move a source document, you'll get a "Can't find document" message. You'll have to re-establish the links between the documents.

LINKING FROM EXCEL TO MICROSOFT WORD

You can also create a link from Excel to Microsoft Word. This is a dynamic link; when values change in the Excel worksheet, the data will be updated in the MS Word document. Use the procedure in the box on the following page.

The Microsoft Word commands **Paste Link** or **Paste Special...** may not be present in the **Edit** menu. If the command is not present, you can add it to Word's **Edit** menu using **Commands...**.

The price you pay for the convenience of linking documents is that each time you open the Word document, you will get the message box shown in Figure 10-2; thus the linking of data between applications should be practiced with some circumspection.

To Link Data in an Excel Worksheet to an MS Word Document

1. Select the cell, cell range or chart to be linked to Word, and **Copy** it.

2. Switch to Word and select an appropriate location for the data from Excel.

3. Choose **Paste Link** from the **Edit** menu of Word. The Excel data will appear in the Word document.

    or...

4. Choose **Paste Special...** from the **Edit** menu of Word. The Paste Special dialog box will appear (Figure 10-1). You have the option of pasting the data with or without links. To paste the data from a single worksheet cell into a Word document, choose Formatted Text (RTF means Rich Text Format).

LINKING FROM OTHER APPLICATIONS TO EXCEL

Very rarely, you may have occasion to link from another application to Excel. The procedure is essentially identical to that described in the preceding section. Excel's **Paste Link** command is in the **Edit** menu.

If you link some text from Microsoft Word to a cell in an Excel spreadsheet, you will be able to examine the remote reference formula in the formula bar.

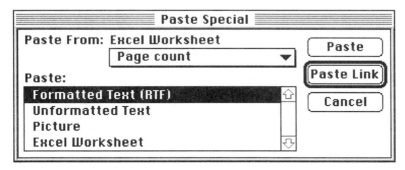

Figure 10-1. The Paste Special dialog box from Microsoft Word.

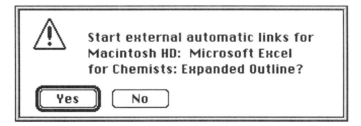

Figure 10-2. MS Word's Establish External Links? message.

COPYING FROM EXCEL TO MICROSOFT WORD

There are three ways to copy from Excel to Word: by using **Copy** and **Paste**, by using the Camera icon on the toolbar or by making a "screen shot". Each produces a different "product". The first two can be linked dynamically to the source worksheet.

USING COPY AND PASTE

If you **Copy** a portion of a spreadsheet and **Paste** it into MS Word, the copied data will appear in the Word document as a table (Figure 10-3). Dotted lines will indicate the table cells on the screen, but these do not appear when the document is printed. Data in individual cells in the table can be selected and edited.

To copy a chart, select the chart by choosing **Select Chart** from the **Chart** menu or by clicking in the chart border. White "handles" appear at the corners and edges of the chart, indicating that the whole chart has been selected. Then choose **Copy** from the **Edit** menu. A marquee will appear around the chart. Activate the Word document, position the cursor where you want the chart to appear in the document, then choose **Paste**. If the chart is an embedded one, select it by clicking on it, then **Copy** it and **Paste** it to the Word document. The appearance of the copied chart will be a little different, depending on whether you copied a chart in a chart sheet or an embedded chart. Most noticeably, the copy of an embedded chart will have a border; you can remove the border from the chart, before copying, by choosing **Object...** from the **Format** menu, then choosing the Patterns tab, then Border = None.

The chart in the Word document will be the same size that it was on the screen. To re-size it, click on it. Black handles will appear. Selecting one of the handles and dragging it crops the image; selecting a handle and dragging while pressing SHIFT re-sizes the image. To re-size the image to a desired size, display the ruler and scroll the page until the image is adjacent to the ruler, then re-size.

It's generally desirable to adjust the size and proportions of a chart in the document window before copying. If a pasted chart has to be reduced significantly in size to fit on a page, the axis labels, text characters and data markers may become unreadably small.

							QUIZ #					
	1	2	3	4	5	6	7	8	9	10	11	12
AGRAWAL, Sunda	5	4	2	5	4	2	5	1	2	3	4	2
ALI, M Saqwat	0	2	0	2	3	2	0					
ALVARADO, Annelise	5	5	5	5	5	4	5	5	4	5	5	5
AMATO, John	2	4	2	2	2	2	4	3	1	1	2	1
ANTOINE, Mark	2	2	3	2	5	0	3	1	1	2	4	3
ATKINSON, Gordon		3	5	3	3	2	5	3		2		3

Figure 10-3. Portion of a worksheet after **Copy** and **Paste**.

USING THE CAMERA TOOLBUTTON

You can use the Camera tool [📷] to copy a portion of an Excel spreadsheet into a Word document. (See Chapter 9 for instructions on how to make this button available.) To copy using the Camera tool, select the cells to be copied, then click on the toolbutton (you'll hear a "shutter" sound). Activate the Word document, position the cursor where you want the spreadsheet to appear, and **Paste**. For Excel 4.0 only, to include row and column headings in the copied spreadsheet, choose **Page Setup...** from the **File** menu and check the Row & Column Headings box. Figure 10-4 shows a typical result.

The copied spreadsheet image is a dynamic object; values in its cells are updated when the source data are changed. Values in the graphic object can't be edited. The image can be re-sized by using the technique described earlier.

MAKING A "SCREEN SHOT" (MACINTOSH)

You can take a picture of the whole Excel screen (a "screen shot"), then use familiar graphic editing tools to select any portion of it. You'll need to have the SimpleText or TeachText utility available.

First, arrange the window as desired. Then press COMMAND+SHIFT+3 (you'll hear the "shutter" sound). A screen shot will be placed in the hard disk, numbered consecutively Picture1, etc. Open PictureN. You are now in SimpleText or TeachText; the cursor will be a crosshair. Click and drag the cursor to select the desired portion of the image, or select the whole image using **Select All** from the **Edit** menu, then **Copy**. Finally, activate the Word document and **Paste**. The graphic object (see e.g., Figure 10-5) can be re-sized as described previously. The data are not dynamically limked to the source worksheet.

	A	B	C	D	E
1	Bemegride Dimerization Studied by NMR				
2	D proton fit				
3			K=	4.02	
4			delta1 =	7.715	
5			delta2 =	9.588	
6					

Figure 10-4. Portion of a worksheet copied using the Camera tool.

	A	B	C	D	E
1	Bemegride Dimerization Studied by NMR				
2	D proton fit				
3			K=	4.02	
4			delta1 =	7.715	
5			delta2 =	9.588	
6					

Figure 10-5. Portion of a worksheet copied as a "screen shot".

MAKING A "SCREEN SHOT" (WINDOWS)

To make a screen shot of the entire screen, press the Print Screen key, or, to make a screen shot of the active window, press ALT+Print Screen. Then activate MS Word and **Paste** the graphic (which was captured on the Clipboard) into your MS Word document. You can move or re-size the graphic in MS Word. To crop a portion of the graphic, you'll have to import the Clipboard contents into a graphics program such as Paintbrush or Photoshop.

PUSHBUTTONS ON SPREADSHEETS

You can create a button on a spreadsheet that, just like a button on a toolbar, will run a command macro. Of course, the macro should be one that performs a task specific to the spreadsheet that it's on.

Here's an example of how a button can simplify and automate a worksheet task. Suppose you use a macro to create and format a report form that is produced on a regular basis. A copy of each version of the report is saved as a separate document with the name *ReportName (Date)*, e.g., 97F by Fac (05/16/96). Instead of manually entering the report name each time (with the current date) in the Save dialog box, the report can be saved with the appropriate name and date by a simple macro that uses the SAVE.AS? form of the SAVE.AS command. This macro is linked to a button that is placed in a suitable location on the worksheet. Then, to save the document, the user needs only to press the Save Document button on the worksheet (see Figure 10-6). The button will not appear on the sheet when it is printed.

The document name is created by the macro from information on the sheet and entered in the text input box of the Save As dialog box (Figure 10-7). The button was created by macro statements on the same macro sheet that created the report form. The macro listing is shown in Figure 10-8.

	A	B	C	D
1	[SAVE DOCUMENT]		SEMESTER 97F BY FACULTY	
2			*(Assignments in italics are changes from last listing.)*	
3			(Where rooms are not listed, those courses have rooms assigned by the registrar.)	
4	BILLO E	CH35101	ANALYTICAL CHEMISTRY	MWF 9
5	BILLO E	CH35301	ANALYTICAL CHEM LAB	M 1-5
6	BILLO E	CH35301	ANALYTICAL CHEM LAB	T 1-5
7	BILLO E	CH35301	ANALYTICAL CHEM LAB	W 1-5
8	BILLO E	CH35501	ANALYT CHEM DISCUSSION	M 4
9	BILLO E	CH35502	ANALYT CHEM DISCUSSION	W 4

Figure 10-6. Portion of a worksheet incorporating a button.

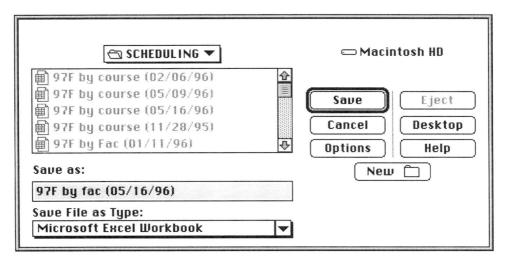

Figure 10-7. The Save dialog box with a document name added by a macro.

102	**Create button on sheet**	
103	=CREATE.OBJECT(7,"R1C1",5,3,"R2C1",115,8)	Position and size of button.
104	=ASSIGN.TO.OBJECT("'SEM. BY FACULTY Macro'!SaveSheet")	Assign macro to button.
105	=TEXT.BOX("SAVE DOCUMENT")	Add button text.
106	=FONT.PROPERTIES("Chicago","Regular",12)	Default button text font & size.
107	=SELECT("R1C5")	Park the cursor.
108	=RETURN()	
109	**SaveSheet**	
110	(saves document when button on sheet pressed)	
111	=SAVE.AS?(LEFT(sourcename,4)&"by fac ("&TEXT(last.date,"MM/DD/YY")&")")	Assign filename.
112	=RETURN()	

Figure 10-8. Macro statements to create a button on a sheet.

First, here's how to create a button manually. To create your own button by means of a macro, you can record the steps using the Recorder and paste them into a macro sheet or module. Follow the procedure to create a button on a spreadsheet that is outlined in the following box.

Once a button has been created and a macro assigned to it, the mouse pointer becomes a hand when the mouse pointer is positioned over the button (see Figure 10-6).

To move, re-size or delete a button to which a macro has been assigned, hold down the CONTROL key while selecting the button.

To Create a Button on a Spreadsheet

1. Display the Drawing toolbar (Excel 5.0) or the Utility toolbar (Excel 4.0).

2. Press the Button button ⬭ . The mouse pointer becomes a crosshair. Use the crosshair pointer to draw a rectangle on the worksheet for the desired location of the button (it can be moved or sized later).

3. The button is a graphic object; it has the name "Button n". Select the button text and type a suitable name for the button (e.g., Save Document).

4. The Assign to Object dialog box is also displayed, so that you can assign a macro to the button. (If you press the Cancel button, you can assign a macro to the button later.)

5. To assign a macro to an already-created button, select the button (black handles appear), then choose **Assign Macro...** from the **Tools** menu (Excel 5.0) or **Assign to Object...** from the **Macro** menu (Excel 4.0).

11

ADVANCED CHARTING TECHNIQUES

Excel does an excellent job of producing a chart from your data automatically. But there are many options available for improving the appearance or usefulness of a chart. As well, this chapter shows several ways to create *combination charts*, and how to plot multiple data series, customize charts and link spreadsheet data to chart text.

GOOD CHARTS VS. BAD CHARTS

Most charts produced by chemists are X-Y charts and contain one or more sets of data points. Excel connects plotted points by straight-line segments; thus, a chart of data that should produce a smooth curve might look like the "bad chart" shown in Figure 11-1.

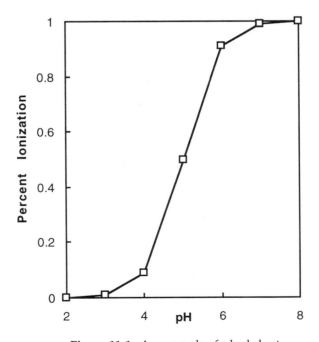

Figure 11-1. An example of a bad chart.

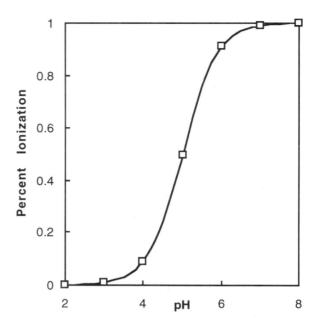

Figure 11-2. Smooth curves are produced by using small increments between data points.

To produce a smooth curve, you need to use small X-increments between points. Figure 11-2 is a good example of the use of small charting increments to produce smooth curves. If you want to minimize the number of calculations, you can limit your use of small plotting increments to regions where the curvature is changing rapidly.

CHARTS WITH MORE THAN ONE DATA SERIES

Excel can plot several data series in the same chart. If the values in the data series are similar in magnitude, then plotting two or more sets of data is not any different from plotting one set. If the numbers are very different (for example, if one set of Y-values is in the range 1-10 and the other set is in the range 0.001-0.010) then you must use an overlay or combination chart (see "Combination Charts" later in this chapter).

PLOTTING TWO DIFFERENT SETS OF Y VALUES IN THE SAME CHART

If more than two columns (or rows) of data are selected for plotting, Excel uses the leftmost row or uppermost column as the independent variable (plotted on the X-axis) and the remaining rows or columns as the dependent variables (plotted on the Y-axis). The fragment of a spreadsheet shown in Figure 11-3 illustrates one column of X data and two columns of Y data to be selected for a chart. If the data series are non-adjacent, hold down the CONTROL key (Excel 4.0 for the Macintosh) or the COMMAND key (Excel 5.0 for the Macintosh) while you select the separated columns of data. (Excel will always use the leftmost row or column for the independent variable data, though.)

	A	B	C
3	X	Y1	Y2
4	0	0	0
5	1	3	5
6	3	9	15
7	5	15	25
8	7	21	35
9	9	27	45
10	11	33	55

Figure 11-3. Spreadsheet layout for two Y-data series.

Excel uses plotting symbols of different colors for each data series (or, if you have a monochrome monitor, different symbols: solid squares for the first data series, open squares for the second data series, etc.). If the chart is printed in black and white, plotting symbols of different colors are replaced by the monochrome symbols shown in Figure 11-4. You can change the plotting symbols or remove them if you customize the chart. If the plotting symbols are removed, Excel distinguishes the different data series by using a solid line, a broken line, etc. Figure 11-5 illustrates the two data series from the spreadsheet of Figure 11-3.

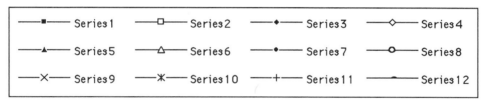

Figure 11-4. Excel's plotting symbols.

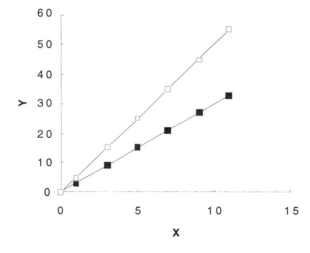

Figure 11-5. Chart with one X- and two Y-data series.

	A	B	C
28	X	Y1	Y2
29	0	0	
30	1	3	
31	3	9	
32	5	15	
33	7	21	
34	9	27	
35	11	33	
36	0		0
37	2		10
38	4		20
39	6		30
40	8		40
41	10		50

Figure 11-6. Spreadsheet layout for two X- and two Y-data series.

PLOTTING TWO DIFFERENT SETS OF X AND Y VALUES IN THE SAME CHART

To plot two sets of X-Y data on a single chart, use the data layout shown in Figure 11-6. Occasionally you'll need to insert a blank row between the two sets of data.

A chart produced from the two data series in the spreadsheet of Figure 11-6 is shown in Figure 11-7.

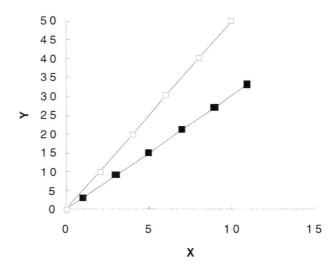

Figure 11-7. Chart with two X- and two Y-data series.

CUSTOMIZING CHARTS

The **Format** menu allows you to customize an Excel chart. You can change the weight and style of lines, the color and shape of plotting symbols, the numerical ranges of axes, etc. The various components of a chart — Chart, Plot Area, Axes, Data Series, Arrows, Text, Gridlines, etc. — are the *chart elements*. To format a chart element, first select the element by clicking on it. The selected element will be indicated by the appearance of "handles". In Excel 4.0 there are five menu choices in the **Format** menu: **Patterns**, **Font**, **Text**, **Scale** and **Legend**. In Excel 5.0 there is a single context-sensitive menu command in the **Format** menu, which appears as **Selected Axis...**, **Selected Data Series...**, etc., depending on which chart element is selected.

A dialog box will appear with the formatting possibilities for the selected chart element. There are different **Patterns** dialog boxes or tabs for Chart, Plot, Series, Axis, Arrow, Gridline, Text, etc., different **Font** dialog boxes or tabs for Chart, Axis, Text, etc., and so on. For example, with a Series selected, Patterns permits you to change or remove the plotting symbol or the line. With an Axis selected (Figure 11-8), you can use Patterns to change the style or weight of the axis, change or remove or add major and minor Tick Marks, and change or remove the Tick Labels.

With an axis selected, the Scale tab (Figure 11-9) enables you to change the scale range, and where the X-axis crosses the Y-axis, for example. The Number tab (not shown) permits you to change the number formatting of the axis labels, by using the same number formats that are available in the **Format Cells...** command. The number formatting option is available in Excel 5.0 only.

Excel Tip (Excel 5.0 only). You can also number-format chart axes by using number-formatting toolbuttons, such as the Increase Decimal [+.0/.00] *or Decrease Decimal* [.00/+.0] *buttons.*

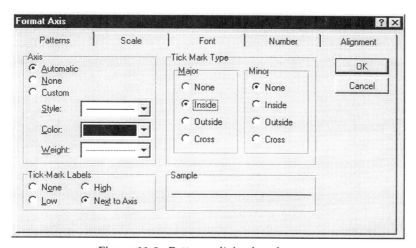

Figure 11-8. Patterns dialog box for axes.

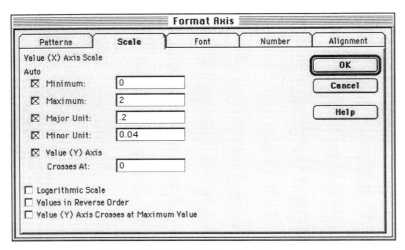

Figure 11-9. **Scale** dialog box for axes.

Excel Tip (Excel 4.0 only). *To change the number of decimal places displayed on a chart axis, change the number of decimal places in the data series in the spreadsheet. (Actually, you need only number-format the first data point in the series.)*

Excel Tip. *Sometimes it's difficult to select a chart element by clicking on it (for example, if two data series are almost superimposed). Instead of selecting with the mouse pointer, you can select chart elements by using the up and down arrows on the keyboard. This allows you to select each element in turn (Chart, Plot, Axis, etc.); the name of the selected chart element is displayed in the Reference Area of the formula bar. Using the left and right arrows allows you to select chart elements within a group (Series 1, Series 2, etc.). Once the desired chart element has been selected, you can format it using the **Format** menu.*

PLOTTING EXPERIMENTAL DATA POINTS
AND A CALCULATED CURVE

Plotting experimental data points and a smooth calculated curve is one of the most common applications of custom formatting. To do this you need to plot two Y-data series — the experimental data points and a series of points to describe the calculated curve. The Y_{obsd} data will be plotted as a series of symbols with no connecting line, the Y_{calc} data as a line with no symbols, as in Figure 11-10. To generate a smooth calculated curve, you'll need to have the Y_{calc} points fairly close together. But since too many points make a worksheet slow to recalculate, youll have to try to strike a balance between the two requirements.

Of course, to plot a calculated curve you need to have an equation that fits the data. It may be the least-squares straight line (obtained from LINEST) that best fits the data, or a curve produced by an equation appropriate for the data. To draw a smooth curve through data points without the use of a theoretical relationship, use a power series (see Chapter 16) or the cubic spline (see Chapter 14).

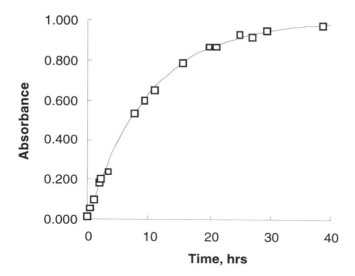

Figure 11-10. Chart with Y_{obsd} and Y_{calc}.

Figure 11-11 illustrates a portion of a data table showing experimental data points (time, absorbance) from a kinetics experiment and part of the table of calculated absorbance values, where the absorbance was calculated using the relationship =A_0*EXP(-k_*t)+const. The numerical values of the parameters were obtained from the experimental data.

	A	B	C
	t	A(obs)	A(calc)
6			
7	0.1	0.010	
8	0.5	0.055	
9	1.1	0.100	
10	1.9	0.180	
11	2.2	0.197	
12	3.3	0.240	
13	7.7	0.530	
14	9.3	0.600	
15	11.1	0.650	
16	15.7	0.790	
17	19.9	0.870	
18	21.1	0.870	
19	27	0.920	
20	24.9	0.929	
21	29.5	0.948	
22	38.7	0.975	
23	0		0.000
24	1		0.095
25	2		0.181
26	3		0.259
27	4		0.330
28	5		0.393
29	6		0.451

Figure 11-11. Spreadsheet for plotting Y_{obsd} and Y_{calc}.

To Plot Y_{obsd} and Y_{calc}

1. In the same column as the experimental X values, create a suitable range of X values (the best way is to use AutoFill) to calculate the theoretical curve. The X values can be appended either at the end or the beginning of the experimental X values.

2. Enter the expression for Y_{calc} in a separate column and **Fill Down** to produce the calculated values.

3. Select the three data series (X, Y_{obsd} and Y_{calc}) and create an X–Y chart. If you created an embedded chart, you'll have to double-click on it before you can proceed to customize it.

4. The chart will have the experimental and calculated values more or less superimposed. Click on one of the experimental points. You'll see the appropriate data series displayed in the formula bar. It may take a bit of searching and clicking to find the experimental data series.

5. Choose **Patterns** from the **Format** menu. In the Patterns dialog box (Figure 11-12), choose Line = None and Marker = Automatic. If you want to change the style of data marker, select from the Style box. To change a solid symbol to an open one, choose Foreground = Black, Background = White. The symbol that you've selected is displayed in the Sample box in the lower right corner.

6. Now click on one of the data points for the theoretical curve. Again, you'll see the appropriate data series displayed in the formula bar.

7. Choose **Patterns** from the **Format** menu. Choose Marker = None and Line = Custom, then choose the style in the Style box.

ADDING ERROR BARS TO AN X–Y CHART (EXCEL 5.0)

Error bars are often an important part of the graphical presentation of scientific data. In Excel 5.0 you can add error bars to either the Y-axis, the X-axis or both axes of X–Y charts. There are several ways to specify the magnitude of

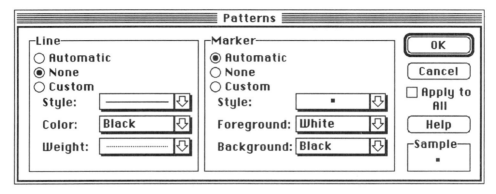

Figure 11-12. Excel 4.0 for the Macintosh **Patterns** dialog box for data series.

	A	B	C	D
1	Kinetic data for an acid–catalyzed reaction			
2				
3	[H+]/10^{-3} M	k(obsd)/ sec^{-1}	std.dev	k(calc)
4	0.00			0.0006
5	0.06	0.00073	0.00018	0.00070
6	0.12	0.00085	0.00015	0.00076
7	0.30	0.00099	0.00019	0.00096
8	0.50	0.00119	0.0002	0.00118
9	0.75	0.00152	0.00018	0.00145
10	1.00	0.00144	0.00035	0.00173
11	1.38	0.00191	0.0005	0.00215
12	1.80	0.00291	0.00038	0.00261

Figure 11-13. Spreadsheet layout, with standard deviations, for a chart with error bars.

the error bars; the following describes the method to use if you've calculated standard deviations (or other measures of scatter) for each data point in your worksheet. Almost always, you'll be adding error bars in the Y-axis, and this is what is described in the following.

Figure 11-13 shows a spreadsheet containing kinetic data for an acid-catalyzed reaction, together with standard deviations of the k_{obsd} values.

After creating and formatting an X–Y chart of the data (k_{obsd} and k_{calc} plotted vs. [H$^+$]), select the k_{obsd} data series, to which error bars are to be added, by clicking on any data point. Choose **Error Bars...** from the **Insert** menu to display the Error Bars dialog box (Figure 11-14) and choose the Y Error Bars tab. Choose one of the four types of error bar to be displayed (almost always the default, error bars above and below the data point), then choose Custom. Click in the Plus input box and then select the range of standard deviations in the worksheet. Repeat for the Minus input box, then click OK.

Figure 11-15 shows the formatted chart (k_{obsd} data points and k_{calc} line) with error bars corresponding to ± 1 σ.

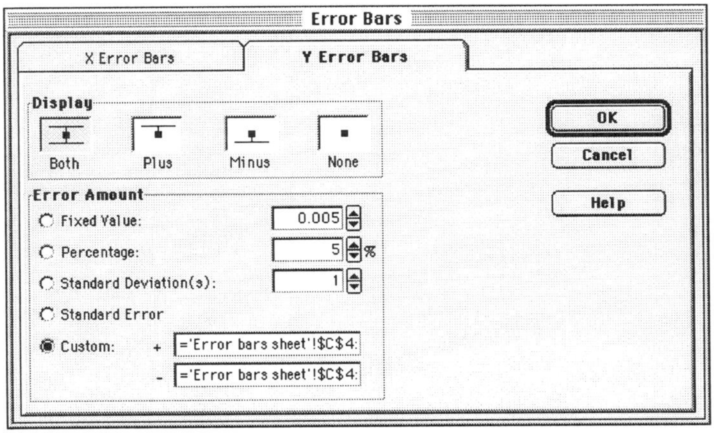

Figure 11-14. The Error Bars dialog box (Excel 5.0 only).

ADDING ERROR BARS TO AN X–Y CHART (EXCEL 4.0)

Although Excel 4.0 provides the capability of including error bars (actually, high and low values for stock quotes) in several kinds of line charts, there are no built-in X–Y chart formats with error bars. You can use the technique described in the first section of this chapter (see "Plotting Two Different Sets of X and Y Values in the Same Chart") to add error bars to an X–Y chart. In the data layout shown in Figure 11-16, the error bars for each point are calculated from the standard deviations; for example, cell D14 contains the formula =B5+C5, cell D15 the formula =B5-C5. Only a portion of the error bar data is shown.

The "horizontal bar" plotting symbol (the symbol for Series 12 in Figure 11-4) was selected for the data series in column D. There must be an empty row between each pair of identical X values; otherwise the error bars will be connected by a continuous line. A blank row is necessary between the original X data and the first pair of X values, for a similar reason. The sample chart with error bars is shown in Figure 11-17.

It is a laborious task to enter the formulas for the error bars. If you plan on creating a number of charts with error bars, it will be worth the effort to write a macro to perform the task. Figure 11-18 illustrates a portion of the *ErrorBar* macro, which creates the error bar data table.

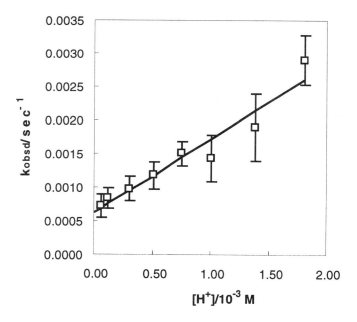

Figure 11-15. An X–Y chart with error bars added (Excel 5.0)

	A	B	C	D
1	Kinetic data for an acid–catalyzed reaction			
2				
3	[H$^+$]/10^{-3} M	k(obsd)/sec^{-1}	std.dev	k(calc)
4	0.00			0.0006
5	0.06	0.00073	0.00018	0.00070
6	0.12	0.00085	0.00015	0.00076
7	0.30	0.00099	0.00019	0.00096
8	0.50	0.00119	0.0002	0.00118
9	0.75	0.00152	0.00018	0.00145
10	1.00	0.00144	0.00035	0.00173
11	1.38	0.00191	0.0005	0.00215
12	1.80	0.00291	0.00038	0.00261
13				
14	0.06			0.00091
15	0.06			0.00055
16				
17	0.12			0.001
18	0.12			0.0007
19				
20	0.3			0.00118
21	0.3			0.0008
22				

Figure 11-16. Spreadsheet layout for error bars (only a portion of the error bar data is shown).

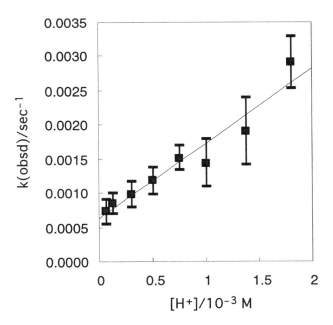

Figure 11-17. An error bar chart produced using the   ErrorBar macro.

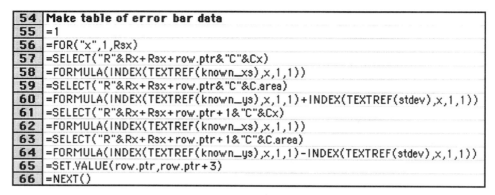

Figure 11-18. Portion of the *ErrorBar* macro.

THE MACRO. The following variables were defined earlier in the macro: Rx and Cx are the row and column numbers of the X data, C.area the column selected to receive the error bar maximum and minimum values; Rsx is the number of rows of X data; known_xs, known_ys and stdev are array variables containing the data.

Cell B55 contains the pointer to the row in which data is to be entered. In cell B56 we begin the loop in which data for each X value will be calculated. In cell B57 the appropriate cell of the worksheet, in the column containing the X values and in the row defined by Rx + Rsx + row.ptr, is selected as the destination cell for the X value of the error bar. In cell B58 the FORMULA function returns the appropriate value from the array known_xs, selected using the INDEX function. The TEXTREF function converts the reference known_xs, stored as text, to an absolute reference so that it can be used with the INDEX function. In cells B59 - B64, similar operations are performed for the upper and lower Y-values of the error bars. Cell B65 increments the row pointer by three so that there will be a space between the pairs of values for each error bar.

CHARTS SUITABLE FOR PUBLICATION

To produce a suitable Y_{calc} curve, you'll need to make sure that the X values are close enough together to produce a smooth curve. You can choose a sufficiently small increment for X to produce a smooth curve over the whole range of X, or you can manually insert smaller X increments only in those regions where the curvature requires them.

Usually X–Y charts for publication in journals have a border around the chart area, as in the example shown in Figure 11-19. Select the Plot area by clicking anywhere within it, then choose **Patterns** from the **Format** menu and choose Border = Automatic.

Publication-quality charts typically have data point symbols that are about 1/50th of the size of the chart. Since Excel's plotting symbols are 2 mm in size, irrespective of the size of the chart, it follows that the chart size should be about 100 mm, or 4 inches. The chart can be measured directly on the screen with a ruler and sized with the size box. It can be printed as "Size shown on screen" using the **Page Setup** command from the **File** menu, and it will be copied as exactly the size shown on screen to another application — Microsoft Word, for example.

USING SET PREFERRED TO APPLY CUSTOM CHART FORMATTING (EXCEL 4.0)

The Excel 4.0 default chart type is the format 1 column chart (format #1 in the **Gallery** of column charts). If you choose a different chart type, such as the X–Y chart, you can use the **Set Preferred** command from the **Gallery** menu to change the default chart type. Subsequent charts that you create will be in the new default format. (If you use the **ChartWizard**, use the >> button in the "Step 1 of 5" dialog box to go directly to the completed chart.)

There are potential problems, though, in using **Set Preferred**. As well as the chart type, **Set Preferred** saves any custom settings you have made by means of **Patterns, Font, Scale,** etc. in the **Format** menu. If you have made any changes to the X- or Y-axes, or to the major or minor tick mark settings, or have added axis labels or a title, then these will be applied to all subsequent charts. (The "Automatic" check boxes are unchecked if you manually enter values for Minimum, Maximum, Major Unit, etc. See the Axis Scale dialog box, Figure 11-9.) If all the charts you create are similar, the settings will be O.K., but if charts

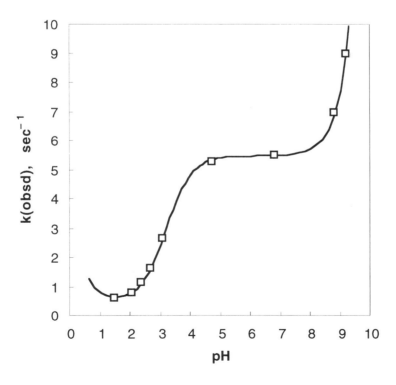

Figure 11-19. A chart suitable for publication.

with different scales are produced, then "Automatic" is to be preferred over **Preferred**.

The **Preferred** chart settings remain in effect until you quit Excel; when you open Excel the next time, the default settings will be in effect again. However, if the chart is part of a workbook, the preferred chart settings are saved in the workbook file.

CHANGING THE DEFAULT CHART FORMAT (EXCEL 5.0)

Excel 5.0 also uses the column chart type as the default chart format, and there's no way to change the default to another of the built-in chart formats (most likely you'd want the X–Y chart type to be the default). However you can use the procedure in the following box to define a custom chart format and make that the default format.

To Make a Custom Chart Format the Default Chart Format

1. Create a chart in the desired format. Make the format as general as possible (do not add title, or format the axis scales, etc.).

2. Activate the chart. Choose **Options** from the **Tools** menu and choose the **Chart** tab.

3. In the Default Chart Format box, press the Use the Current Chart button. You will be asked to enter a name for the chart format. The current chart format is now the default format.

You have now converted the ChartWizard into a one-step Wizard. After you press the ChartWizard button, just press the Finish button in the ChartWizard Step 1 of 5 dialog box to go directly to the completed chart.

The procedure to delete unwanted chart formats is shown in the box below.

To Delete a Chart Format

1. Activate any chart.

2. Choose **AutoFormat** from the **Format** menu.

3. Press the User-Defined button in the AutoFormat dialog box.

4. Select the format to be deleted, then press the Customize... button.

5. Press the Delete button, then press OK to exit.

	B
23	**Auto_Open_SetChartFormat**
24	=NEW(5)
25	=SET.PREFERRED("Default XY Chart")
26	=RETURN()

Figure 11-20. An Auto_Open macro to switch to a custom chart format.

The new chart format will remain in effect until you quit Excel. When you open Excel the next time, the built-in chart format will be in effect again. You can create a simple Auto_Open macro that will switch to your custom chart format each time Excel is opened. The macro shown in Figure 11-20 was created and saved in the Personal Macro Workbook.

3-D CHARTS

Occasionally you'll find it useful to be able to display graphical data in three dimensions, when a dependent variable Z depends on the values of two independent variables X and Y.

USING EXCEL'S BUILT-IN 3-D CHART FORMAT

The 3-D Surface chart type in Excel's gallery of charts is not a true 3-D chart, but rather a Line chart in two dimensions. The X- and Y-axes are Categories — only the Z-axis is proportional to the data plotted. Thus if you attempt to produce a 3-D chart using Excel, you'll have to take this limitation into account and work within it. The values on the X- and Y-axes must be equally spaced.

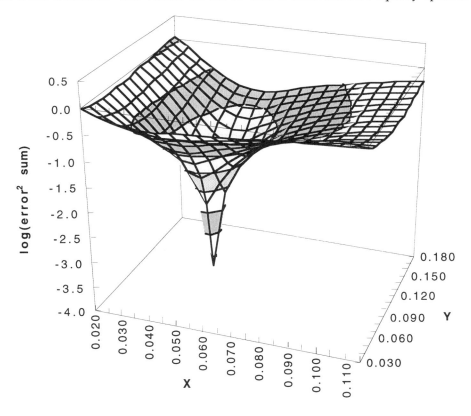

Figure 11-21. A 3-D chart.

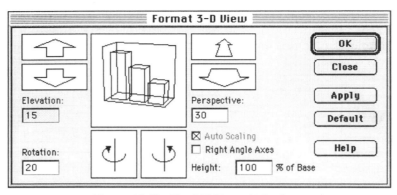

Figure 11-22. The 3-D View dialog box.

Since only one Z-axis value can be charted for any pair of X- and Y-axis values, you can't produce a plot of a closed surface, such as a sphere. But this type of chart can be useful, for example, to show the effect of changing two variables on the yield of a process.

There are two useful format options within the **Gallery** of 3-D Surface charts: format #1, the 3-D Surface chart, and format #2, the 3-D Wireframe chart. The 3-D Surface chart uses colors to indicate areas on the surface with different ranges of Z-axis values. The 3-D Wireframe chart is identical except that colors are not used. Figure 11-21 shows an example of a 3-D Surface chart.

You can change the view of a 3-D chart by choosing **3-D View...** from the **Format** menu (Figure 11-22). Use the Rotation, Elevation and Perspective buttons to view the chart from different angles. Often a more pleasing or informative chart can be produced by changing the view. To return to the original viewing angle, press the Default button.

COMBINATION CHARTS

You've seen how Excel can plot more than one set of Y-data on a single chart. However, all the data series are plotted on a single Y-axis scale. *Combination charts* such as Figure 11-23 permit the graphing of sets of data with different X- and/or Y-axis scales. Excel provides a whole menu of combination charts; for example, a bar chart can be combined with a line chart. For plotting scientific data, we'll be interested in producing combination X–Y charts. When a second chart with different axis scales is added to a chart, the second chart is referred to as an *overlay chart*. For X–Y combination charts, there can be two different Y-axis scales plotted using the same X-axis, or two different Y-axis scales and two different X-axis scales. The second Y-axis is plotted along the right side of the chart, the second X-axis along the top of the chart.

COMBINATION X–Y CHARTS IN EXCEL 5.0

Although Excel has a Combination category in its gallery of chart types, none of the built-in combination charts are X–Y charts. You'll have to create an X–Y combination chart from "scratch".

There are two ways to produce an X–Y combination chart in Excel 5.0. One way is to create a chart using the data series for both charts, then designate one of the Y data series as the series to be plotted on the *secondary axis*. The procedure is described in the following box.

To Create an X–Y Combination Chart in Excel 5.0

(two different Y-axis scales and the same X-axis)

1.   Select all data series to be plotted (the X-axis data series, two Y-axis data series).

2.   Create an X–Y chart. Activate the chart by double-clicking on it.

3.   Click on the data series whose axis you want to change.

4.   Choose **Selected Data Series...** from the **Format** menu and choose the Axis tab (see Figure 11-24).

5.   Press the Secondary Axis button. The form of the combination chart will be displayed. If the chart is suitable, press the OK button.

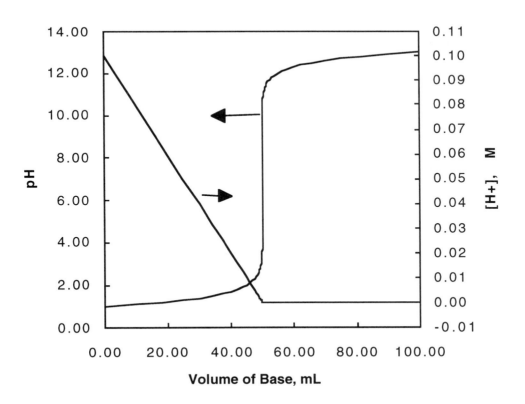

Figure 11-23. Combination chart with two Y-axis scales.

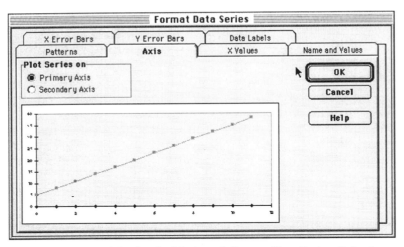

Figure 11-24. Excel 5.0 for the Macintosh Format Data Series dialog box .

The other way to produce an X–Y combination chart, with two different X-axes and two different Y-axes, is to create a chart with one set of X-and Y-data series, then paste the second set of data in the chart, as described in the following box.

> **To Add an X–Y Overlay to an X–Y Chart in Excel 5.0**
> (two different Y-axis scales and two different X-axis scales)
> 1. Create the first chart in the usual way.
> 2. Select the second set of X- and Y-data series, and **Copy**.
> 3. Activate the chart by double-clicking on it.
> 4. Choose **Paste Special...** from the **Edit** menu. In the Paste Special dialog box (see Figure 11-25), press buttons for New Series and X-Values in First Column, then OK.
> 5. The new series will be added to the chart. Select the new series by clicking on it.
> 6. Choose **Selected Data Series** from the **Format** menu. Choose the Axis tab and press the Plot Series on Secondary Axis button.
> 7. Choose **Axes...** from the **Insert** menu. The Axes dialog box is the same as in Excel 4.0 (see Figure 11-26). Check the X-axis and Y-axis buttons in the Secondary Axis box, then OK. The chart will now have different axes on all four sides.

COMBINATION X–Y CHARTS IN EXCEL 4.0

There are two ways to produce a combination X–Y chart in Excel 4.0. One is to select the data series for both charts, then choose **Combination** from the **Gallery** menu of chart types. The procedure is described in the box on the following page. Excel 4.0 provides a series of menu choices including several types of combination X–Y chart.

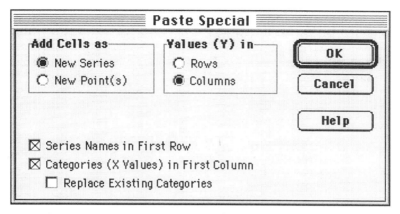

Figure 11-25. Adding a second chart data series in Excel 5.0.

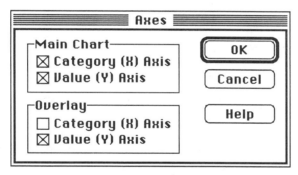

Figure 11-26. Excel 4.0 for the Macintosh Axes dialog box for combination chart.

To Create an X–Y Combination Chart in Excel 4.0

(two different Y-axis scales and the same X-axis)

1. Select all data series to be plotted (the X-axis data series, two Y-axis data series).

2. Create an X–Y chart.

3. From the **Gallery** menu, choose **Combination**.

4. The default type of overlay chart is a line chart. Choose **Overlay Chart** from the **Format** menu and change the chart type to an X–Y chart.

5. The overlay chart data will be plotted using the same axis scale (the left axis) as the main chart. Choose **Axes** from the **Chart** menu. In the Axes dialog box, check the Y-axis box for the Overlay chart. The axis for the overlay chart will appear as the right axis, and can be formatted in the usual way.

As with other charts, the first data series is the X-values. The rest are the Y-values. Excel partitions the Y-data series between the two charts. If there is an even number of Y-data series, the first half are plotted in the Main Chart, the second in the Overlay Chart. If there is an odd number, e.g., five, then the Main Chart gets the first three and the Overlay Chart the last two. If both X- and Y-axis boxes are checked in the Axes dialog box, then Excel partitions the selected data series in the same way, except that the first data series in each partition is the X-values.

The second way to produce a combination chart is to create the main chart in the usual way and then add the second data series as an overlay chart. The procedure is described in the following box.

To Add an X–Y Overlay to an X–Y Chart in Excel 4.0

(two different Y-axis scales and the same X-axis,
or two different Y-axis scales and two different X-axis scales)

1.   After creating the first chart in the usual way, choose **Add Overlay** from the **Chart** menu. (If you created the chart as an embedded chart, you'll first have to double-click on it to make it a separate document.)

2.   The default type of overlay chart is a line chart, so choose **Overlay Chart** from the **Format** menu and change the chart type to X–Y chart.

3.   Now choose **Edit Series** from the **Chart** menu and add a new series: first give the new series a name in the "Name" box (Figure 11-27). Then click the cursor in the "Y values" box, use the **Window** menu to activate the worksheet containing the chart data, select the Y-axis data series for the overlay chart, then press RETURN or click the OK button.

4.   Choose **Axes** from the **Chart** menu. In the **Axes** dialog box, check the Y-axis box for the Overlay chart. The right axis will appear and can be formatted in the usual way.

5.   To create a chart with two different X-axes as well as two different Y-axes, choose **Edit Series** and enter both X-axis and Y-axis data series in the appropriate boxes.

Excel Tip. *Use* ***Patterns****, not the Axes dialog box, to remove the X- or Y-axis labels.*

AN ADDITIONAL EXAMPLE OF AN OVERLAY CHART

The combination chart in Figure 11-28, containing a calculated curve, experimental data points and an overlay chart with a separate border and zero axis, illustrates a common way to show deviations of experimental points from a calculated curve. Both Y-series Tick Labels are on the left side.

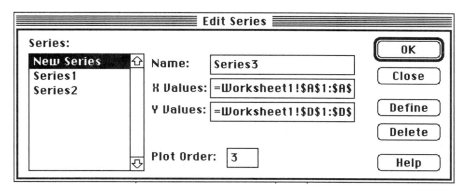

Figure 11-27. Excel 4.0 for the Macintosh Edit Series dialog box.

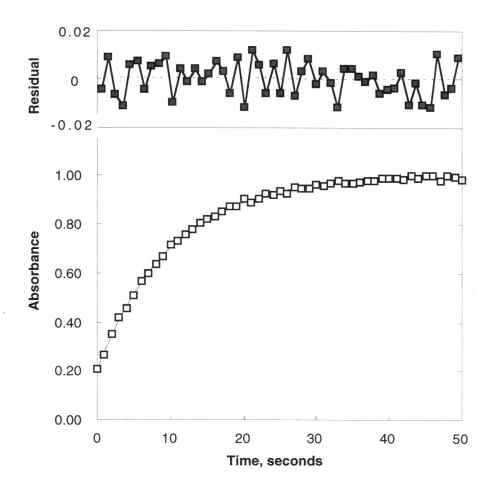

Figure 11-28. A customized combination chart.

To create this chart required some special techniques of customizing. Here's how Figure 11-28 was produced: (i) The experimental points and calculated curve were plotted and formatted as described earlier. (ii) The residuals data series was added as an overlay chart. (iii) The Y1 axis scale was formatted to provide space above the curve (initially, the maximum was 1.2; it was increased to 1.6). (iv) The Y2 axis scale for the overlay chart was formatted so that there were the same number of subdivisions as in the Y1 axis scale, i.e., eight. Initially the maximum was 0.015 and the minimum was −0.015; these were changed to 0.02 and −0.14 to place the residuals data at the top of the chart. (v) The X and Y1 axis labels were added; a little experimentation was necessary to get the correct number of spaces in front of "Absorbance" and between "Absorbance" and "Residuals" to position the Y1 axis label correctly. (vi) The border around the residuals data was produced by selecting Chart and choosing Border = Automatic (this provides the top border) and using an Arrow, minus the head, for the bottom border. (vii) A second arrow, minus the head and with Style changed to broken, provided the "zero" line. A little jockeying was required here also to get the line in position and perfectly horizontal. (viii) **Patterns** was used to remove the Y2-axis labels (choose Tick Labels = None in the dialog box). (ix) Text boxes were produced containing "0.02", "0" and "−0.02" and these were positioned over the Y1-axis Tick Labels 1.20, 1.40 and 1.60. (When created, text boxes have transparent background and don't cover up whatever is underneath them. Use **Patterns** to change Foreground and Background to White — this will cover up the Tick Labels underneath.)

A CHART WITH AN INSET

On the screen, you can easily create a chart with a smaller inset chart by superimposing the smaller chart on the main chart, but it's a little more difficult to transfer the result to paper. Once you've positioned the inset chart on the main chart, if you try to use Excel's **Copy** command, you can select and copy only one or the other of the two charts. There are two ways to produce a graphic containing both charts: you can either make a "screen shot" (see "Making a 'Screen Shot'" in Chapter 10) of the two superimposed charts, or you can copy the inset chart as a picture and paste it on the main chart.

Here's how to create a chart with an inset, similar to the example shown in Figure 11-29, which illustrates the separation of the two first-order rate constants of a consecutive process $A \rightarrow B \rightarrow C$.

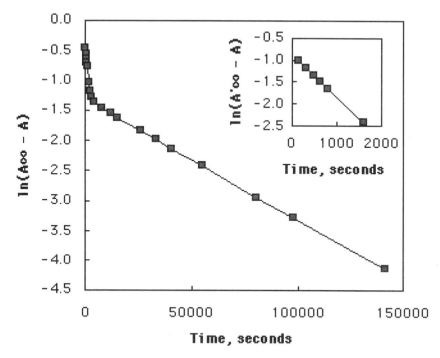

Figure 11-29 A chart with an inset.

To Create a Chart with an Inset

1. Create the main chart as an embedded chart or as a separate chart sheet. Customize it and size it as desired.

2. Create the inset chart as an embedded chart, and size it appropriately to fit in the main chart.

3. With the inset chart selected, hold down the SHIFT key while choosing the **Edit** menu, and choose **Copy Picture....** In the Copy Picture dialog box, choose As Shown On Screen, then press OK. Activate the main chart and **Paste** the picture. You can select the inset chart and move and size it as any graphic object. Position the inset chart on the main chart.

4. The black "handles" indicate the area of the main chart that will be obscured by the overlying inset chart. To remove the frame around the inset chart, select it, choose **Patterns** from the **Format** menu and set Border = None. If you set Background = None, you will be able to position the inset over the main chart without obscuring it.

Excel Tip. If you make an embedded chart transparent by setting Border = None and Background = None, it becomes very difficult to select the chart. If you click on the chart area, you will probably select the underlying worksheet cell or chart instead. To select such a chart embedded on a worksheet, move the mouse pointer slowly from the worksheet area (where the pointer will be the usual "plus-sign" pointer, as in Figure 11-30A) toward the chart. When you reach the invisible edge of the chart, the mouse pointer will become the arrowhead pointer, as in Figure 11-30B, and you can click to select the chart.

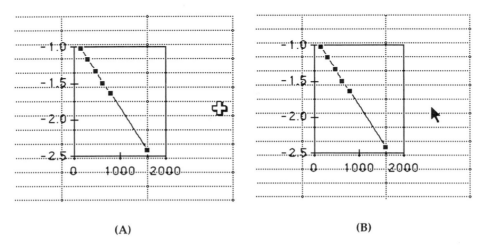

(A)   (B)

Figure 11-30. (A) Searching for, and (B) locating, the edge of an embedded chart with transparent background.

LINKING CHART TEXT ELEMENTS TO A WORKSHEET

Any text element in a chart can be linked to a worksheet cell, causing the text in the worksheet to be displayed in the chart. (Some of this can be done automatically, as, for example, category labels.) But chart titles or unattached text boxes can also be linked to the worksheet. In this way titles can be generated automatically, or explanatory notes in text boxes can include information that will change as the data in the worksheet is modified

Basically, to link a chart text element to a worksheet cell, you enter a formula as the chart text. The syntax of the formula is:

=worksheet_name!absolute_cell_reference.

Follow the procedure outlined in the following box.

To Link Worksheet Text to a Chart

1.   The worksheet must be open.

2.   The chart must be activated. If the chart is an embedded chart, you must activate it by double-clicking on it.

3.   With the chart as the active document, select the chart text element (Chart Title, X-axis Title, Y-axis Title or Unattached Text) by selecting it by using the **Chart** menu or by double-clicking on the chart text element if it already exists in the chart.

4.   In the formula bar, type

     =worksheet_name!absolute_cell_reference,

     e.g., ='Voltammetric curve'!$A$7. If the worksheet name contains spaces, it must be enclosed in single quotes.

5.   Alternatively, you can have Excel supply the worksheet name and cell reference. In the formula bar, type "=". Activate the worksheet by clicking on it or by selecting it from the **Window** menu. The external reference to the worksheet will appear in the formula bar. Select the desired cell in the worksheet. The absolute reference to the cell will appear in the formula bar.

The formula entered as chart text must be only a cell reference. If you want the chart text to be text concatenated with a number, e.g. to produce a title such as Half-wave Potential = -0.74V, where the potential is a value in cell B2 of the worksheet, the complete formula ="Half-wave potential = "&B2&"V" must appear in the worksheet cell, then the worksheet cell is linked to the chart.

EXTENDING A DATA SERIES OR ADDING A NEW SERIES (EXCEL 5.0)

Occasionally you may want to add additional data points to a data series in a chart, or add a new series. Of course you can simply delete the chart and create a new one. But it's possible to add to an existing chart.

To add extra data points to a series or add a new series in Excel 5.0, just **Copy** the extra data, switch to the chart sheet or activate an embedded chart by double-clicking on it, and **Paste**. The chart will be updated with the extra data.

EXTENDING A DATA SERIES OR ADDING A NEW SERIES (EXCEL 4.0)

To add extra data points in Excel 4.0, first switch to the chart sheet or activate an embedded chart by double-clicking on it. Then do either of the following:

- Choose **Edit Series** from the **Chart** menu. Select Series 1 (for example); the Name, external reference to the X-values and external reference to the Y-values are displayed in separate text input boxes, as in Figure 11-31. Use the TAB key to select the X-values text box. The worksheet will be displayed with a marquee around the X-values data series. Select the new range with the mouse, then press the Define button. Tab to the Y-values input box, and repeat.

  or...

- Click on the desired data series in the chart. The definition of the data series, in the format =SERIES(*name*, *x-values_ref*, *y-values_ref*, *plot_order*, for example =SERIES("Series2", Sheet4!$A$1:$A$11, Sheet4!$C$1:$C$11, 2) will appear in the formula bar. Using the usual editing techniques, change the references for both X- and Y-values to include the additional data points; each must be a single contiguous reference, of course.

To add a new data series in Excel 4.0, first switch to the chart sheet or activate an embedded chart by double-clicking on it. Then do either of the following:

- Choose **Edit Series** from the **Chart** menu. Select New Series. Use the TAB key to select the Y-values text box (the X-values will be the same for this new series); the worksheet will be activated. Select the new data series with the mouse, then press OK.

  or...

- Click the mouse pointer anywhere in the plot area. Type in the expression for the new series, in the form =SERIES(*name*, *x-values_ref*, *y-values_ref*, *plot_order*). Instead of typing the external references for the X- and Y-values, you can select the appropriate ranges on the worksheet. *Plot_order* should be an integer one greater than the last series in the chart. When you press RETURN or click the Check Box, the new series will appear in the chart.

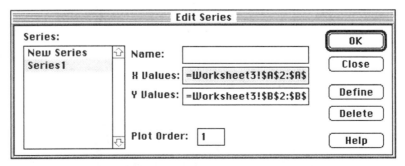

Figure 11-31. Edit Series dialog box.

TO SWITCH PLOTTING ORDER IN AN X–Y CHART (EXCEL 4.0 OR 5.0)

Excel always uses the leftmost column (of those that you selected) as the X values when it creates an X-Y chart. To change the plotting order, so that values from a column other than the leftmost one are used as the X-values, you must first create a chart in the normal way, then:

- Click on the desired data series in the chart. The series formula will appear in the formula bar, e.g.,

=SERIES(,Demo1!$A$1:$A$11,Demo1!$B$1:$B$11,1).

Manually edit the data series to reverse the X- and Y-values in the series, e.g.,

=SERIES(,Demo1!$B$1:$B$11,Demo1!$A$1:$A$11,1)

12

USING EXCEL AS A DATABASE

This chapter demonstrates how to create and use a database on an Excel worksheet. Although database software such as Filemaker is superior to Excel, it is sometimes useful to be able to combine the features of a database with Excel's superior calculation and data analysis features.

THE STRUCTURE OF A LIST OR DATABASE

A rectangular range of data in an Excel worksheet is termed a *list* or *database*. A database consists of a number of *records*, each of which contains a number of *fields*. For example, a compilation of physical properties of organic compounds, such as the one in the *CRC Handbook of Chemistry and Physics*, is a database; the row of data for a particular compound is a record and the values for the melting point, boiling point, solubility, etc. are the data fields within the record. In Excel, a list or database must be arranged in tabular form, with row or column labels; that's the only requirement.

Most fields in a database will contain values that have been entered as text, numbers or dates. A database may also contain *calculated fields*, containing values that are calculated by using Excel formulas.

SORTING A LIST

One of the most common operations performed on a list is to sort it in increasing or decreasing order with respect to the value in one of its fields. Use the **Sort...** command in the **Data** menu to do this.

Figure 12-1 illustrates a portion of a list of the elements, together with their symbols, atomic weights and electronic configurations. The list (which was imported into Excel by using a scanner, as described in Chapter 13) is arranged in order of increasing atomic number. It may seem redundant to include a column of atomic numbers in the list, but this column is necessary if you want to return the sorted list to its original order.

To sort the list according to any of its fields, select the complete list (be sure to select all rows and columns of the list, or it may become irreversibly scrambled). For a large list such as this one, it's convenient to use the method for selecting a block of cells described in Chapter 1: select the first row of cells, place the mouse pointer on the bottom edge of the selected row (the mouse pointer

	A	B	C	D	E
1	Table of Atomic Weights and Electron Configurations				
2					
3	Name	Symbol	At. Num.	At.Wt.	Elec. Config.
4	Hydrogen	H	1	1.00797	1s1
5	Helium	He	2	4.0026	1s2
6	Lithium	Li	3	6.939	[He] 2s1
7	Beryllium	Be	4	9.0122	[He] 2s2
8	Boron	B	5	10.811	[He] 2s2 2p1
9	Carbon	C	6	12.01115	[He] 2s2 2p2

Figure 12-1. A portion of a list of atomic weights and symbols of the elements.

turns into an arrow pointer), hold down the SHIFT key and double-click to select all cells to the bottom of the block. Then choose **Sort...** from the **Data** menu to display the Sort dialog box.

Excel proposes the data field where you began the selection (the cell with reverse highlighting) as the field to sort by, and displays the field name (sometimes called a *sortkey*) in the Sort By input box (Figure 12-2). If you want to sort according to a different field, type the field name or reference in the input box; for example, to sort by atomic symbol, you can enter either symbol (field names are not case-sensitive) or $B$3 in the Sort By input box. Then press OK. Figure 12-3 shows a portion of the list, sorted in ascending order according to column 1. If the list is not sorted as you want it, press the Undo button or choose **Undo Sort** from the **Edit** menu.

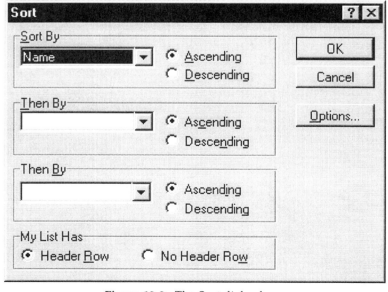

Figure 12-2. The **Sort** dialog box.

	A	B	C	D	E
3	**Name**	**Symbol**	**At. Num.**	**At.Wt.**	**Elec. Config.**
4	Actinium	Ac	89	[227]	[Rn] 7s2 6d1
5	Aluminum	Al	13	26.9815	[Ne] 3s2 3p1
6	Americium	Am	95	[243]	[Rn] 7s2 5f7
7	Antimony	Sb	51	121.75	[Kr] 5s2 4d10 5p3
8	Argon	Ar	18	39.948	[Ne] 3s2 3p6
9	Arsenic	As	33	74.9216	[Ar] 4s2 3d10 4p3

Figure 12-3. A portion of a list, sorted according to element name.

SORT ORDER

In an ascending sort, Excel orders dates and times in order of their serial number (see "Date and Time Functions" in Chapter 3). Text values are sorted in the order: 0 1 2 3 4 5 6 7 8 9 (space) ! " # $ % & ' ( ) * + , - . / : ; < = > ? @ A B C D E F G H I J K L M N O P Q R S T U V W X Y Z [\ ] ^ _ ` { | } ~. Blank cells are always sorted last, in both ascending and descending sorts.

SORTING ACCORDING TO MORE THAN ONE FIELD

As you can see from the Sort dialog box, it is possible to sort by up to three separate fields. For example, you can sort a list of chemistry students by descending order according to year of graduation (freshman, sophomore, junior, senior) and by ascending order alphabetically within each year.

SORT OPTIONS

Pressing the Options... button in the Sort dialog box displays the Sort Options dialog box, and allows you to change the default sorting options.

Excel assumes that your data fields are in columns and sorts your list by rearranging rows. If you want to sort a list horizontally, i.e., to rearrange the columns of the list rather than the rows, press the Sort Left to Right button in the Orientation box (Figure 12-4).

You can choose case-sensitive sorting, in which case lowercase letters follow uppercase letters in an ascending sort (AaBbCc..., not ABC...abc...). If you choose this option, the sortkeys will also be case-sensitive.

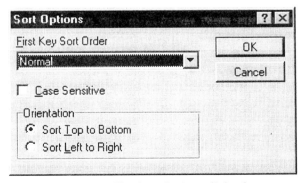

Figure 12-4. The **Sort** Options dialog box.

The Sort Options dialog box contains four lists for custom sorts (other than ascending or descending). For example, you can sort in the order Jan, Feb, Mar, Apr, etc. You can create a custom sort order for your own specialized application (see "Creating a custom sort order" in Excel's On-line Help for details).

OBTAINING A SUBSET OF A LIST USING AUTOFILTER (EXCEL 5.0)

Often you'll want to examine a subset of a list — only those records for which one or more data fields match certain criteria. The process of abstracting a subset of a list is known as using a *data filter* or *querying* the database. The **AutoFilter** command, introduced in Excel 5.0, makes it easy to obtain a subset of a list.

Figure 12-5 illustrates a portion of a list of polymer research samples and some of their physical properties. To use **AutoFilter**, first select the complete list or a partial area to be examined; if the list is separated from the rest of the worksheet data by blank cells, you need only select any cell within the list. Choose **Filter** from the **Data** menu and then choose **AutoFilter** from the submenu. Excel adds drop-down arrow buttons to row 1, which ideally should contain column labels.

To use a data filter on molecular weight (column B), click the arrow button in that column. Excel displays a drop-down list of all values in the column, plus

	A	B	C	D	E	F	G	H	I
	Sample Number	Molecular Weight	MW Dispersity (Mw/Mn)	Specific Gravity (STM D792)	Refractive Index	Dielectric Constant @ 100 Hz (STM D150)	Rockwell Hardness (STM D785)	Tensile Strength PSI/1000 (STM D638)	Melting Point (°C)
2	91976	14.6	2.4	1.135	1.497	2.88	110	15.4	232
3	91977	14.6	2.3	1.149	1.502	2.88	110	16.9	231
4	91978	11.5	2.6	1.105	1.500	2.88	107	16.9	209
5	91979	12.0	2.8	1.117	1.503	2.94	115	14.2	208
6	91980	12.8	2.9	1.181	1.496	2.89	108	13.8	207
7	91981	12.2	2.6	1.172	1.497	2.94	111	13.9	205
8	92007	13.6	2.2	1.174	1.501	2.88	114	14.4	215
9	92039	13.6	2.1	1.184	1.500	2.92	112	15.6	212
10	92071	14.5	3.4	1.135	1.496	2.91	113	16.3	185
11	92103	15.2	3.6	1.143	1.500	2.87	109	16.7	174
12	92135	11.5	3.7	1.148	1.504	2.89	110	17.4	218
13	92167	12.0	3.4	1.139	1.498	2.94	107	15.3	210
14	92199	12.8	3.0	1.178	1.495	2.89	106	16.7	215
15	92231	12.2	2.9	1.177	1.499	2.89	113	13.1	212

Figure 12-5. A portion of a list with AutoFilter buttons displayed.

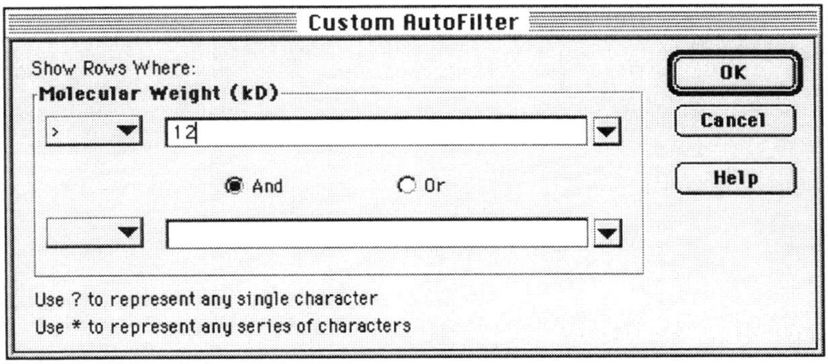

Figure 12-6. The Custom AutoFilter dialog box.

"custom". To display all records that match one of the values in the selected field, you can select it from the list. To perform other logical comparisons, click Custom to display the Custom AutoFilter dialog box (Figure 12-6).

Click the comparison operator drop-down arrow and choose >. Tab over to the text box and enter 12, then press the OK button. The records in the list for samples with molecular weight greater than 12,000 (12 kD) are displayed (Figure 12-7).

To restore the complete list, simply choose **AutoFilter** from the **Data** menu again. Notice that the **AutoFilter** command is checked, indicating that the list has been queried.

You can add a second data filter on data in the same field using the Custom

	A	B	C	D	E	F	G	H	I
1	Sample Number	Molecular Weight (kD)	MW Dispersity (Mw/Mn)	Specific Gravity (ASTM D792)	Refractive Index	Dielectric Constant @ 100 Hz (ASTM D150)	Rockwell Hardness (ASTM D785)	Tensile Strength PSI/1000 (ASTM D638)	Melting Point (°C)
2	91976	14.6	2.4	1.135	1.497	2.88	110	15.4	232
3	91977	14.6	2.3	1.149	1.502	2.88	110	16.9	231
6	91980	12.8	2.9	1.181	1.496	2.89	108	13.8	207
7	91981	12.2	2.6	1.172	1.497	2.94	111	13.9	205
8	92007	13.6	2.2	1.174	1.501	2.88	114	14.4	215
9	92039	13.6	2.1	1.184	1.500	2.92	112	15.6	212
10	92071	14.5	3.4	1.135	1.496	2.91	113	16.3	185
11	92103	15.2	3.6	1.143	1.500	2.87	109	16.7	174
14	92199	12.8	3.0	1.178	1.495	2.89	106	16.7	215
15	92231	12.2	2.9	1.177	1.499	2.89	113	13.1	212
16	92263	13.6	3.9	1.190	1.497	2.92	109	13.1	211

Figure 12-7. A portion of a list that has been filtered.

	A	B	C	D	E	F	G	H	I
1	Sample Number ▼	Molecular Weight ▼	MW Dispersity (Mw/Mn) ▼	Specific Gravity (ASTM D792) ▼	Refractive Index ▼	Dielectric Constant @ 100 Hz (ASTM D150) ▼	Rockwell Hardness (ASTM D785) ▼	Tensile Strength PSI/1000 (ASTM D638) ▼	Melting Point (°C) ▼
2	91976	14.6	2.4	1.135	1.497	2.88	110	15.4	232
3	91977	14.6	2.3	1.149	1.502	2.88	110	16.9	231
25	92358	13.0	3.0	1.188	1.500	2.88	106	13.0	218

Figure 12-8. A list that has been filtered on two fields.

AutoFilter dialog box. The second filter can be either an *And filter* (samples with molecular weight greater than 12 kD and less than 15 kD, for example) or an *Or filter*.

USING MULTIPLE DATA FILTERS

You can also add a second filter by querying data in a second field. For example, you can display all samples with melting point greater than 215°C. This is also an And filter. Figure 12-8 shows the subset of records having both molecular weight greater than 12 kD and melting point greater than 215°C.

To "undo" a particular filter, click the drop-down list button in that column and choose (all) from the list.

Although only three records are displayed, you can't calculate the average molecular weight of the three samples by entering the formula =AVERAGE() in a worksheet cell and selecting the displayed cells in column B that contain the molecular weight information. The range that will be entered by selecting will be (in this example) B2:B25, and the average will not be the desired one. The most convenient way to apply SUM, AVERAGE, STDEV, etc., to a subset is to **Copy** the desired cells, **Paste** them in a convenient location and then perform the calculation.

DEFINING AND USING A DATABASE (EXCEL 4.0)

Excel 4.0 provides a number of menu commands and functions that permit you to use a list as a database. You can use menu commands to **Find**, **Extract** or **Delete** records from the database that match criteria that you define.

DEFINING A DATABASE

To use the database commands and functions, you must first define a list as a database. The top row of the list must contain the *field names*, which you will use to identify the information stored in each field. Select the entire range of cells in the list, including the field names. Then choose **Set Database** from the **Data** menu.

Excel assigns the name Database to the selected range. You can also define a database by assigning the name Database to the selected range of cells using the Define Name command.   You can define only one list in a worksheet as Database; if you have more than one list in a worksheet, you will have to redefine the database range in order to switch databases.

The name Database is one of Excel's *built-in names*. Excel recognizes the name Database as the reference to use in database functions, the Data Form, and some others.   Other built-in names include Auto_Open, Auto_Close, Auto_Activate, Auto_Deactivate, Data_Form, Criteria, Extract, Sheet_Title, Print_Area, Print_Titles and Recorder. Don't use any of these names as variable names.

ADDING OR DELETING RECORDS OR FIELDS

You can add new records to a database either by inserting new rows within the database, or by entering new information below the last existing record. If you insert additional rows within the range, Excel changes the definition of the range name Database to conform to the new range. If you enter information below the last row of the defined range, you will have to redefine the range of the database using Excel 4.0's **Set Database** command.

If you prefer to keep the database records in the order of their entry, you can include a dummy row at the end of the database, and always **Insert** new rows just above the dummy row. You can add new data fields (columns) in exactly the same way. Alternatively, you can use Excel's Data Form to add new records.

UPDATING A DATABASE USING DATA FORM (EXCEL 4.0 OR EXCEL 5.0)

You can edit existing records or enter new data in a database by using Excel's Data Form. The database range must already have been defined, either by using **Set Database** (Excel 4.0) or by selecting a cell within the block of cells that comprise the database (Excel 5.0). Choose **Form**... from the **Data** menu. The Data Form dialog box will appear, with the name of the worksheet in the Title Bar, as in Figure 12-9.

Each field name in the database appears in the dialog box, along with the entries from the first record in the database. The current record, and the number of records in the database, are displayed in the upper right corner of the dialog box. Entries in *editable fields* display their values in a text entry box, while values that do not appear in a text box are either in calculated fields or in fields whose contents are protected.

You can move from record to record using either the Find Prev and Find Next buttons or the Up Arrow and Down Arrow, or the scroll bar. The first field of each record is highlighted. Move forward from field to field within a record by using the TAB key. You can also select a field for editing by clicking the mouse pointer in it.

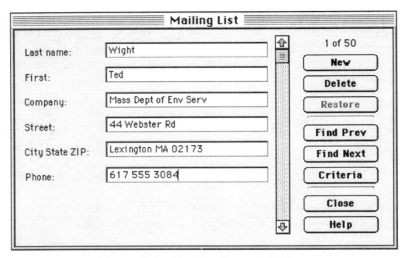

Figure 12-9. The Data Form dialog box.

Edit characters in a text box in the usual way. Once a change has been made in a field, the Restore button is enabled; you can use it to restore the original data if you've made an error. To edit a single field in all records, tab forward to that field or click the mouse pointer in it; then use the Up or Down arrows to move to the same field in other records.

To add a new record, press the New button or drag the scroll button to the bottom. New Record will be displayed in the upper right corner of the dialog box. The new record will be added to the bottom of the database. Formulas of calculated fields are also included in the new record. If the new record will overwrite existing information outside the database, you'll get the "Cannot extend database" message.

In Excel 4.0, but not in Excel 5.0, the range name Database will be updated automatically when you add or remove records. See "To Use Database in Excel 5.0" later in this chapter.

FINDING RECORDS THAT MEET CRITERIA (EXCEL 4.0)

You can use three commands from Excel 4.0's **Data** menu to locate information in a database: **Find** locates, one at a time, those records that match selection criteria that you define; **Delete** deletes all records that match the selection criteria; **Extract**... allows you to find the records that match the selection criteria and duplicate them in another location in the worksheet.

DEFINING AND USING SELECTION CRITERIA

You define selection criteria by setting up a table in the worksheet with one or more *field name*s in a row and the desired criteria below the field names. The Criteria field names must be the same as the database field names. For example, in Figure 12-10, the range K2:K3 is a criteria range to select all records having molecular weight greater than 12 kD. Now select the range of cells containing

Figure 12-10. A simple criteria table.

the field name and the criterion and choose **Set Criteria** from the **Data** menu. Excel assigns the name Criteria to the range. At any time, only one range in a worksheet can be defined as the Criteria range.

It's good practice to copy the complete row of database field names and use it to create a Criteria range, as in Figure 12-11. You can then enter single or multiple criteria in the appropriate cells.

USING MULTIPLE CRITERIA

You can select records on the basis of two or more criteria. Multiple criteria can be combined to produce a logical AND, a logical OR, or a combination of AND and OR.

If you enter two criteria in the same row, you have created an AND criterion, e.g., molecular weight >12 kD and melting point >215°C. If you enter two criteria in separate rows, you have created an OR criterion, e.g., molecular weight > 12 kD or melting point <200°C (see Figure 12-12).

Figure 12-11. Criteria tables are best made by copying the complete row of database field names.

	A	B	C	D	E	F	G	H	I
34				*Criteria range*					
35	Sample Number	Molecular Weight	MW Dispersity(M▸	Specific Gravity (▸	Refractive Index	Dielectric Consta▸	Rockwell Hardness?	Tensile Strength	Melting Point (°C)
36		>12							
37									<200

Figure 12-12. A criteria table with OR logic.

If the AND criteria apply to the same field (e.g., all samples with molecular weight >12 kD and <13 kD), you must duplicate the field name in the Criteria range (see Figure 12-13).

SPECIAL CRITERIA FOR TEXT ENTRIES

When you use a number value in a Criteria range, Excel returns only the records that match the field value exactly. But when you use a text value as a criterion, Excel returns all records that contain text values in the specified field that *begin* with the text value. For example, in a database of alumni information (LastName, FirstName, StreetAddress, City, State, ZIP, YearOfGraduation, etc.), using the letter N as criterion in the LastName criteria field will return all names that begin with N.

You can use wildcard characters in text criteria. The ? wildcard character represents any single character, the * character represents any number of characters. Thus, for example, the criterion *phen* will extract, from a database of chemicals, any entry that contains the character string "phen", such as phenol, 1,10-phenanthroline and benzophenone.

	A	B
31	Molecular Weight (kD)	Molecular Weight (kD)
32	>12	<13

Figure 12-13. A criteria table with AND logic applied to the same field.

To extract only the records that have an exact match between a text criterion and a field value, enter the criterion in the form

    ="criterion_text"

where criterion_text is the text string you want to find.

FINDING RECORDS

Once you've defined both the Database range and the Criteria range, you can use the **Find** command from the Excel 4.0 **Data** menu to select, one by one, the records in the database that match your criteria. When you choose **Find**, the first matching record in the database is highlighted. As you click the scroll arrows in the vertical scroll bar, Excel will scroll to the next matching record (the scroll box displays a bar pattern to indicate that you are in **Data Find** mode). To exit from **Find** and return to normal scrolling, choose **Exit Find** from the **Data** menu.

EXTRACTING RECORDS

To extract a copy of all records that meet specified criteria, use the **Extract...** command. To use this command, first define Database and Criteria ranges as described earlier. Then create a destination range for the extracted records, in the following way: **Copy** the database field names and **Paste** them in a suitable area of the worksheet as labels for the extracted information. Now select the row of field names and enough empty rows below the column labels for the extracted information. Define this range as the Extract range by choosing **Set Extract** from the **Data** menu. Now choose **Extract...** from the **Data** menu. The database records that match the criteria will be copied into the Extract range (Figure 12-14).

*Excel Tip.   Be sure to select a range of empty cells as the Extract range. If you select only the field names as the Extract range, the **Extract** command will clear all cells below the Extract field names to the bottom of the worksheet, whether values are extracted into them or not, erasing any previous information. The **Undo** command will not reverse the action and restore the missing information.*

	A	B	C	D	E	F	G	H	I
40				*Extract range*					
41	Sample Number	Molecular Weight	MW Dispersity(M)	Specific Gravity (	Refractive Index	Dielectric Constar	Rockwell Hardnes:	Tensile Strength	Melting Point (°C)
42	91976	14.6	2.4	1.135	1.497	2.88	110	15.4	232
43	91977	14.6	2.3	1.149	1.502	2.88	110	16.9	231
44	92358	13.0	3.0	1.188	1.500	2.88	106	13.0	218

Figure 12-14. An extract of a database.

	A	B	C	D	E
40	*Extract range*				
41	Sample Number	Molecular Weight	Specific Gravity	Refractive Index	Melting Point (°C)
42	91976	14.6	1.135	1.497	232
43	91977	14.6	1.149	1.502	231
44	92358	13.0	1.188	1.500	218

Figure 12-15. An extract containing only selected fields.

You can create an Extract range that contains only selected fields of each record. Only the fields included in the Extract range will be returned, as shown in Figure 12-15.

USING ADVANCED FILTER (EXCEL 5.0)

Excel 5.0's Advanced Filter is just a slightly different implementation of Excel 4.0's Database, Criteria and Extract. If you have skipped the earlier section entitled "Finding Records that Meet Criteria" you should go back and read it now.

To use Advanced Filter you must establish a Criteria range somewhere on your worksheet, as shown in Figure 12-10, -11, -12 or -13. You can give it a range name if you wish, by using Assign Name. Any name is acceptable, but Criteria is best since Excel will recognize it. You can also establish an Extract range and give it the name Extract. You do not need to assign the name Database to the list to be filtered. Just select any cell within the list and Excel will recognize the range as a list.

Now choose **Advanced Filter...** from the **Filter...** submenu of the **Data** menu. Excel will display the Advanced Filter dialog box (Figure 12-16). If you selected a cell within the list, the reference to the list will appear in the List Range box. If you previously assigned the name Criteria to the criteria table, the reference will appear in the Criteria Range box.

You can choose to Filter The List In-Place or Copy To Another Location. Filter The List In-Place hides the rows in the list that do not meet the criteria; Copy To Another Location copies the rows that meet the criteria to another location in the worksheet or to another worksheet. If you press the button for Copy To Another Location, and you have assigned the name Extract, Excel will automatically enter the reference to the Extract range. When you press OK, the records that meet the criteria will be copied to the Extract range. (Remember that all cells from the bottom of the Extract range to the bottom of the worksheet will be cleared.)

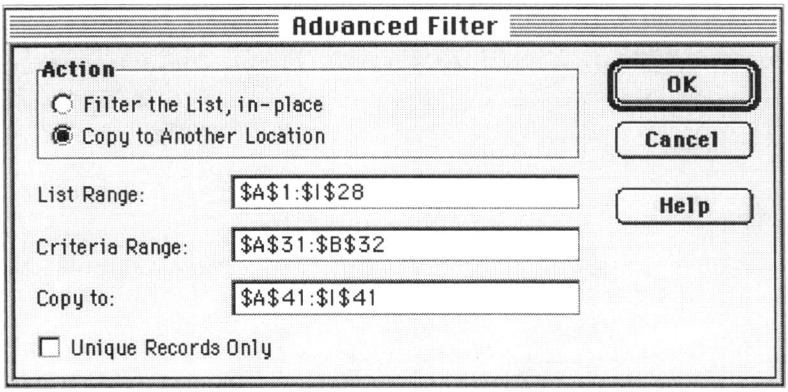

Figure 12-16. The Advanced Filter dialog box.

USING DATABASE FUNCTIONS (EXCEL 4.0 OR EXCEL 5.0)

There are 12 *database functions* that return information about the records in a database: DAVERAGE, DCOUNT, DCOUNTA, DGET, DMAX, DMIN, DPRODUCT, DSTDEV, DSTDEVP, DSUM, DVAR and DVARP. Some of these are particularly useful and are described here. See *Microsoft Excel Function Reference* or On-line Help for further information.

DAVERAGE returns the average of the values in the specified field of all records that match the criteria.

DMAX or DMIN returns, respectively, the maximum or minimum value in the specified field of all records that match the criteria.

DSTDEV returns the standard deviation of the values in the specified field of all records that match the criteria.

All 12 database functions have the syntax (*database*, *field*, *criteria*). *Database* is either the reference to the database or the name assigned to it. *Field* is either the name of the field, as text, or a number indicating the position of the field within the database. *Criteria* is either the reference to the criteria range or the name assigned to it.

For example, to obtain the average refractive index of all samples in the database that have molecular weight >12 and melting point >215, enter the database function DAVERAGE(Database, 5, Criteria). The formula returns the value 1.49948135, the average for the three samples that were extracted in the example shown in Figure 12-14.

You can use the formula =DAVERAGE(Database, "Refractive Index", Criteria) in place of the formula =DAVERAGE(Database, 5, Criteria). The field name must be identical to the field name in the database range. If the name used in the function is not identical to the name of one of the fields in the database, you'll get the #VALUE! error value.

For Excel 4.0 users, DCOUNT, DSUM and DSTDEV can be used to return the same values that are returned by the functions COUNTIF, SUMIF and STDEVIF, which are available only in Excel 5.0.

TO USE DATABASE IN EXCEL 5.0

For Excel 5.0 to recognize a list as a database, you simply place the mouse pointer anywhere within the list. It is not necessary to use the **Set Database** command to define the database, as in Excel 4.0. However, this means that a reference to the database used in other Excel worksheet functions, such as VLOOKUP, will not be automatically updated if you use the **Form...** command from the **Data** menu to add additional records to the list.

There are two ways to ensure that the reference to the database will be automatically updated. The first way is simply to select the complete database, including the title row, and then use **Define Name** from the **Insert** menu to give the range the name Database. Once you've done this the range name Database will be updated if you use the Data Form to add or delete records.

Or, if you use Database regularly in Excel 5.0, you may wish to install the Excel 4.0 **Set Database** command in the Excel 5.0 **Data** menu, so that you can define a list as a database. Then, when **Form...** is used to add records, the database reference will be automatically updated. To install the command, enter, on a macro sheet in the workbook containing the database, the macro shown in Figure 12-17.

Cell D12 (not shown) contains the following Status Bar message text: "(Excel 4.0 menu command) Defines selected cells as database".

When you run the macro the **Set Database** menu command will be installed in the **Data** menu, below the **Subtotals...** command. Use the command to define the database by selecting the complete database, including the title row, then choosing the command from the menu. You need to use the **Set Database** command only once to define the database.

You can now use worksheet formulas such as:

=VLOOKUP(A4,Sheet1!Database,9,0)

to operate on the database range.

	A	B
1		**Install_SetDatabase**
2	ActiveBar	=GET.BAR()
3		=IF(ISERROR(GET.BAR(ActiveBar,"Data","Set Database")),,RETURN())
4	MenuPosition	=GET.BAR(ActiveBar,"Data","Subtotals...")
5		=ADD.COMMAND(ActiveBar,"Data",$A$10:$D$11,MenuPosition+1)
6		=RETURN()
7	DoCommand	=SET.DATABASE()
8		=RETURN()
9	Command Table	
10	-	
11	Set Database	DoCommand

Figure 12-17. A macro to insert the **Set Database** command in Excel 5.0's Data menu.

13

GETTING EXPERIMENTAL DATA INTO EXCEL

Since much of what you use Excel for is the analysis of experimental data, the problem of how to get that data into Excel is crucial. Certainly you don't want to spend time transcribing data from a piece of paper into an Excel worksheet. There are several ways that you can transfer data to Excel. In order of decreasing preference, they are:

- directly from an instrument to an Excel worksheet,
- from a data file on a diskette, or from a data file received electronically, to an Excel worksheet,
- from data on paper, via a scanner, to an Excel worksheet.

DIRECT INPUT OF INSTRUMENT DATA INTO EXCEL

Software programs are available to accept data from instruments and transfer it in realtime to Excel. For example, SoftwareWedge (TAL Technologies, Inc., 2027 Wallace St., Philadelphia PA 19130) captures RS-232 serial I/O data, parses and filters it to user specifications, and transfers it to Excel for Windows via DDE (Dynamic Data Exchange). Data can be received from several serial ports simultaneously. The user can also define output strings which can be sent to an instrument to control it directly from Excel. A 32-bit version is available for use with Windows 95. Discussion of this and similar software is beyond the scope of this book; specific information can be obtained from the manufacturer.

TRANSFERRING FILES FROM OTHER APPLICATIONS TO EXCEL

Most modern spectrophotometers are controlled by a computer and can write a data file to disk. Most are run by PC's. Here's how to import spectrophotometric data, saved to disk as a text file, into Excel.

USING DATA PARSE (EXCEL 4.0)

In the following example, data from a Hewlett-Packard diode-array spectrophotometer interfaced to a PC was collected between 350 and 820 nm at 2 nm intervals and saved to disk as a comma-delimited text file (Figure 13-1). We

	A
1	350,4.580689E-02
2	352,4.936218E-02
3	354,.0541687
4	356,6.047058E-02
5	358,6.842041E-02
6	360,7.719421E-02
7	362,8.996582E-02
8	364,.1021271

Figure 13-1. A comma-delimited text file imported into Excel.

now want to read the data file into Excel and get the data fields into separate column. (Data from a PC disk can be read by most Macintosh computers as a PC Exchange Document using the Apple File Exchange utility program.)

If Excel is open, when you open the text file the data will be automatically imported into an Excel worksheet. Each data record consists of (wavelength, comma, absorbance) in cells A1, A2, etc.

Select the range of data in column A, then choose **Parse** from the **Data** menu. The Parse dialog box will be displayed, and Excel will attempt to partition the record into separate fields by placing delimiters ([and] characters) around each field, as in Figure 13-2.

You will occasionally need to reposition the delimiters or add additional ones. In this case, add additional delimiter characters to indicate separate fields for the wavelength, the comma and the absorbance, as illustrated in Figure 13-3.

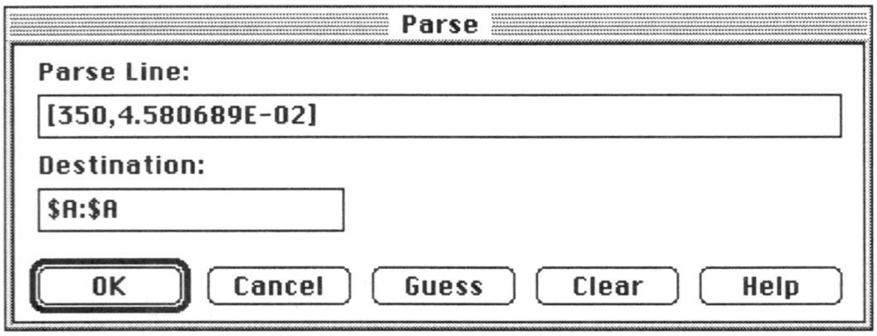

Parse

Parse Line:

[350,4.580689E-02]

Destination:

$A:$A

[OK] [Cancel] [Guess] [Clear] [Help]

Figure 13-2. Excel 4.0 for the Macintosh Parse dialog box with initial display of line to be parsed.

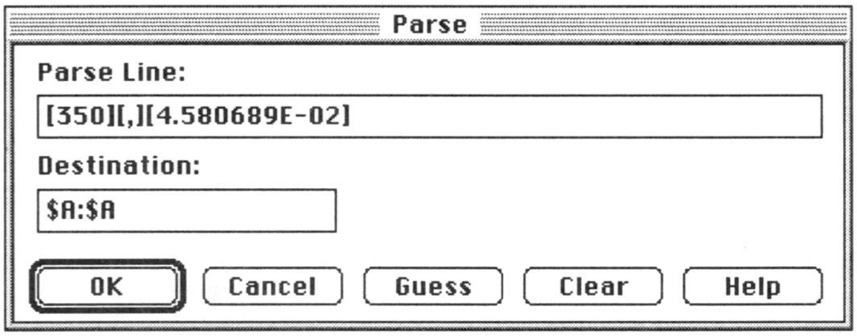

Parse

Parse Line:

[350][,][4.580689E-02]

Destination:

$A:$A

[OK] [Cancel] [Guess] [Clear] [Help]

Figure 13-3. The Parse dialog box with additional delimiters added.

	A	B	C
1	350	,	4.58E-02
2	352	,	4.94E-02
3	354	,	0.0541687
4	356	,	6.05E-02
5	358	,	6.84E-02
6	360	,	7.72E-02
7	362	,	9.00E-02
8	364	,	0.1021271

Figure 13-4. The text file after parsing.

When the fields are delimited correctly, press the OK button. If the data was not parsed correctly, use Undo and try again.

In this example, delete column B, which contains the delimiter characters (see Figure 13-4) , before you **Save** the data.

The Save dialog box will show that the Excel document is in Text format. Choose Normal format (Excel 4.0) or Microsoft Excel Workbook format (Excel 5.0).

USING TEXT TO COLUMNS (EXCEL 5.0)

The **TextWizard** is actuated by means of the **Text to Columns** command in the **Data** menu. Excel 5.0's TextWizard is a real improvement over Excel 4.0's **Parse** command. If you open the text file by double-clicking on the document, Excel will automatically place the data in a worksheet; if you open the document by using the **Open** command from the **File** menu, the TextWizard will intercept the data and allow you to parse it before it is displayed on a worksheet.

The procedure for using the TextWizard is essentially identical to that used for **Parse** in Excel 4.0, except that delimiters are automatically recognized and used to separate the fields. Either Delimited or Fixed Width text files can be parsed. The Fixed Width option is useful for columnar data separated by spaces, as illustrated in Figure 13-5.

Excel usually recognizes whether the text file is delimited or fixed width, and displays the choice in the Step 1 dialog box. You can override Excel's choice and manually select either Delimited or Fixed Width option, as shown in Figure 13-6.

	A
1	CH11101 1 L GENERAL CHEM LAB I M 2-5
2	CH11102 1 L GENERAL CHEM LAB I T 8 45-11 45
3	CH11103 1 L GENERAL CHEM LAB I T 1 30-4 30
4	CH11104 1 L GENERAL CHEM LAB I W 2-5
5	CH11105 1 L GENERAL CHEM LAB I TH 8 45-11 45

Figure 13-5. A fixed-width text file imported into Excel, before parsing.

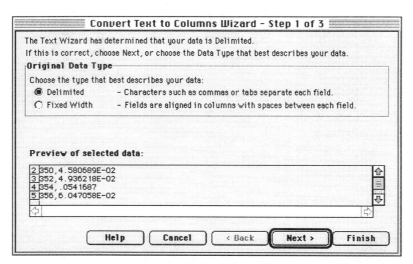

Figure 13-6. The **Text to Columns** Step 1 dialog box with preview of data to be parsed.

Next, the Step 2 dialog box allows you to select the type of delimiter and see a preview of the parsed data (Figure 13-7).

If you chose Fixed Width in Step 1, the Step 2 dialog box (Figure 13-8) will display a ruler on which you can position vertical bars indicating the field widths.

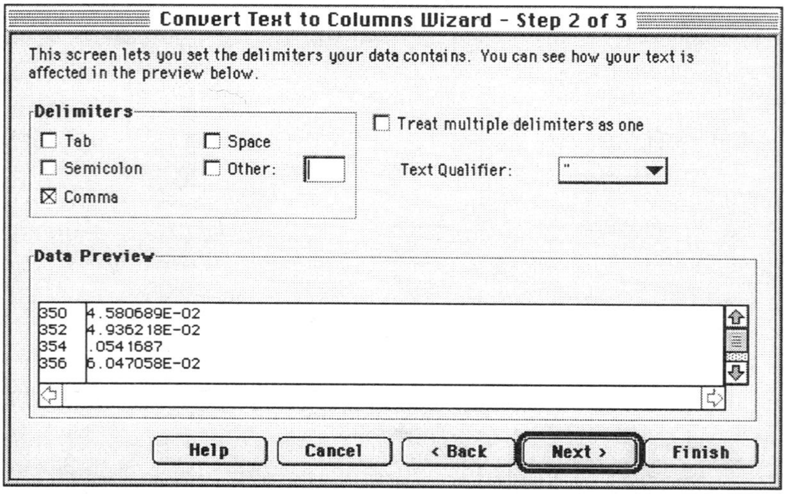

Figure 13-7. The **Text to Columns** Step 2 dialog box with preview of data fields.

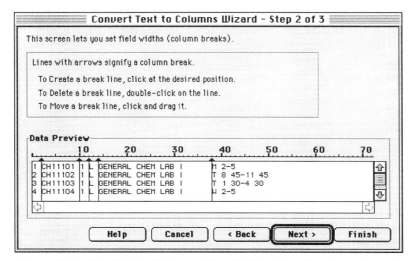

Figure 13-8. The **Text to Columns** Step 2 dialog box showing ruler and column break delimiters.

The Step 3 dialog box (Figure 13-9) allows you to number-format each column and also to specify where you want the parsed data to be placed. The default option parses the data into columns to the right of the original data column.

The parsed data is shown in Figure 13-10. The delimiter characters (in this case the commas) are automatically deleted. Again, the data is in a text file; **Save** it with Normal format (Excel 4.0) or Microsoft Excel Workbook format (Excel 5.0).

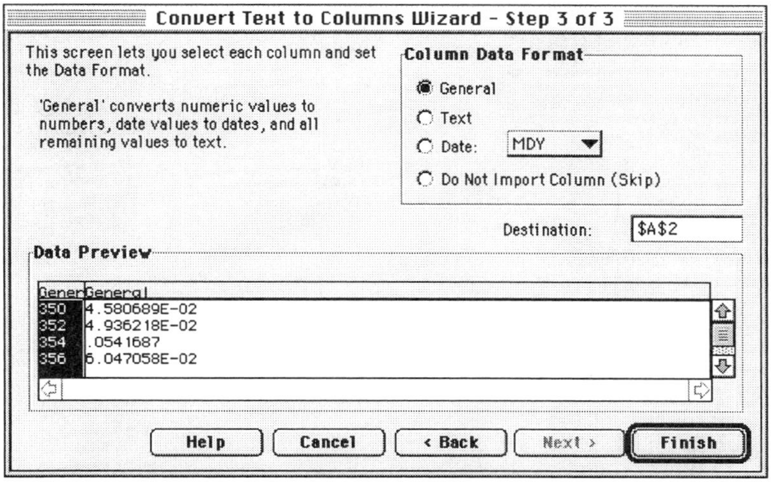

Figure 13-9. The **Text to Columns** Step 3 dialog box.

	A	B
1	350	4.58E-02
2	352	4.94E-02
3	354	0.0541687
4	356	6.05E-02
5	358	6.84E-02
6	360	7.72E-02
7	362	9.00E-02
8	364	0.1021271

Figure 13-10. The text file after parsing.

FROM HARD COPY (PAPER) TO EXCEL

Occasionally you'll have data in hard copy tabular form (from a book, report, correspondence, etc.). Unless the data set is very small, in which case it will be more efficient to enter the data manually, you'll want to use some semi-automatic way to import the data from paper copy to Excel.

USING A SCANNER TO TRANSFER NUMERIC DATA TO EXCEL

You can use a scanner and a software program that performs optical character recognition (e.g., OmniPage Professional from Caere Corp., Los Gatos CA 95030) to create a data file from hard copy. If the data is in a table, it may have to be manipulated in Excel to get it in useful form.

As an example of how to record a simple macro to convert scanned data with blank lines and other undesired features into useful columnar data, consider Figure 13-11, which shows scanned spectral data (absorbance values at 10-nm intervals from 500 to 680 nm for a metal ion at varying concentrations of a ligand) copied from a monograph and imported into Excel. Because the data in the original copy was double-spaced, the scanner put the data in alternate rows of

	A	B	C	D	E	F	G	H
1	2.51859E-4	0.0	1.21621E-4	1.42871E-2	0.0	10.0		
2								
3	0.465	0.489	0.506	0.517	0.522	0.519	0.502	0.473
4								
5	0.428	0.377	0.324	0.273	0.223	0.181	0.145	0.117
6								
7	0.093	0.074						
8								
9	2.51859E-4	0.0	1.21621E-4	2.85743E-2	0.0	10.0		
10								
11	0.569	0.581	0.576	0.555	0.529	0.496	0.459	0.416
12								

Figure 13-11. Spectral data imported from a monograph (Arthur E. Martell and Ramunas J. Motekaitis, *The Determination and Use of Stability Constants*, VCH Publishers, New York, 1988.)

the spreadsheet. The data consists of (i) in Row 1, Row 9, etc., a row of data giving concentrations and path length, not necessary in our data table, (ii) in Row 3, Row 11, etc., eight values of absorbance measurements (for 500 - 570 nm) for one concentration of ligand, (iii) in Row 5, etc., a second eight values of absorbance (580 - 660 nm), and in Row 7, etc., the final two values of absorbance (670 and 680 nm).

Of the 120 lines in data in the sheet, more than half need to be deleted. With a little bit of experience in recording macros, you can manually fix up the first lines of the data set while recording a macro, then let the macro do the rest. here's how this example was handled:

1. Choose **Record** from the **Macro** menu. The macro was saved in an .XLM sheet with C O N T R O L +a (Windows) or OPTION+COMMAND+a (Macintosh) as the shortcut key.

2. Choose **Relative Record**.

3. Choose **Record** from the **Macro** menu and perform the following operations: (i) **Delete** the first two rows, (ii) **Cut** the second row of data and **Paste** it at the end of the first row of data, (iii) **Cut** the third row of data (two cells) and **Paste** at the end of the first row of data, (iv) select five rows and **Delete**, (v) select the next row of data, so that operation will be repeated on the correct row when automated by using the macro.

4. Choose **Stop Recorder**.

The recorded macro is shown in Figure 13-12. Comments have been added in column B and the HLINE and HPAGE commands (recorded when the mouse was used to scroll back for the next operation), have been disabled by deleting the = sign.

	A	B
1	Record1 (a)	
2	=SELECT("R :R[1]")	Select the current row & the one below it.
3	=EDIT.DELETE(2)	Delete them.
4	=SELECT("R[2]C :R[2]C[7]")	Select the second row of data.
5	=CUT()	Cut.
6	=SELECT("R[-2]C[8]")	Select the cell 2 rows above, column 8.
7	=PASTE()	Paste the data.
8	=SELECT("R[4]C[-8]:R[4]C[-7]")	Select the two cells in the third row.
9	=CUT()	Cut...
10	HLINE(12)	(not needed)
11	=SELECT("R[-4]C[16]")	Select the destination, 4 rows above, column 16.
12	=PASTE()	& Paste.
13	HLINE(-1)	(not needed)
14	HPAGE(-2)	(not needed)
15	=SELECT("R[1]:R[5]")	Now select the 5 empty rows below the data.
16	=EDIT.DELETE(2)	& Delete them.
17	=SELECT("RC")	(Probably not needed)
18	=RETURN()	

Figure 13-12. Recorded macro, with comments added.

	A	B	C	D	E	F	G	H	I	J
1	0.465	0.489	0.506	0.517	0.522	0.519	0.502	0.473	0.428	0.377
2	0.569	0.581	0.576	0.555	0.529	0.496	0.459	0.416	0.365	0.313
3	0.668	0.668	0.640	0.592	0.533	0.473	0.413	0.357	0.301	0.251

Figure 13-13. Reformatted data.

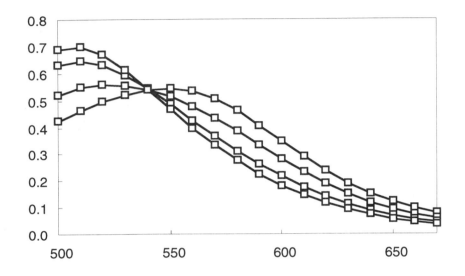

Figure 13-14. Chart created from imported data.

The macro can be actuated by using OPTION+COMMAND+a. If the cursor is placed on the first line of the original data, and OPTION+COMMAND+a pressed, the data is automatically formatted and the cursor moved down to the next row. Thus, repeatedly pressing OPTION+COMMAND+a will successively format each line of data.

The macro can be modified so that it automatically goes through the data table until all rows have been processed, simply by inserting the command =IF(SELECTION()<>"",GOTO(A2)) above row 18. The complete data table is thus processed by a single OPTION+COMMAND+a keystroke. A spreadsheet fragment showing some of the reformatted data is shown in Figure 13-13; Figure 13-14 is a chart of a portion of the imported data.

USING A SCANNER TO TRANSFER GRAPHICAL DATA TO EXCEL

You can use a scanner to convert graphs from strip chart recorders, published graphs or spectra, etc., to digitized X-Y data, using special software (e.g., Un-Scan-It from Silk Scientific, Inc., Orem UT 84059). First you scan the image using a scanner. Then you import the scanned image into the digitizing program. The program converts the scanned image into an X-Y ASCII file.

SELECTING EVERY NTH DATA POINT

If you have imported a data file with a large number of data records, e.g., a spectrum covering the range 300-900 nm with an absorbance measurement taken every 1 nm, you may wish to work with a reduced data set, say every tenth point. Any of the four worksheet formulas described below can be used to select every Nth data point from a data table. All four utilize the INDEX(array, row_num,column_num) function to select an element from the array of data, but use different methods to calculate the pointers row_num and column_num into the reference that contains the data.

- The first method uses names for the X and Y data ranges; a separate worksheet formula must be entered for each column of data. The worksheet formula

  =INDEX(XData,Nth*(ROW()-ROW(XData))+1)

  when entered in cell T5 of Figure 13-15 and **Filled Down** yields the results shown in Figure 13-15. A similar formula is entered in cell U5 and **Filled Down**. XData and YData were defined as named ranges; the variable Nth contains the value 5.

- The second method uses a single worksheet formula that employs a mixed reference. The formula can then be copied into multiple columns. The formula

  =INDEX(R$5:R$18,Nth*(ROW()-ROW(R$5:R$18))+1)

  was entered in cell T5 and **Filled Right** into cell U5 to yield the formula

  =INDEX(O$5:O$18,Nth*(ROW()-ROW(O$5:O$18))+1)

  Upon application of **Fill Down**, the values shown in Figure 13-15 are produced.

- The third method uses an array formula that, because it involves both rows and columns, must be entered into the complete range of cells. The following array formula was entered in cells T5 :U7:

  {=INDEX(Data,Nth*(ROW()-MIN(ROW()))+1,COLUMN()-MIN(COLUMN())+1)}

- The fourth method uses an array formula that can be entered into a single row of cells (here T5:U5) and then copied into the complete range of cells using either **Fill Down** or AutoFill:

  {=INDEX(Data,Nth*(ROW()-ROW(Data))+1,COLUMN()-MIN(COLUMN())+1)}

	R	S	T	U
3	<-- Data -->			
4	XData	YData		
5	350	0.0458	350	0.0458
6	351	0.0476	355	0.0574
7	352	0.0494	360	0.0772
8	353	0.0518	*etc.*	*etc.*
9	354	0.0542		
10	355	0.0574		
11	356	0.0605		
12	357	0.0645		
13	358	0.0684		
14	359	0.0728		
15	360	0.0772		
16	361	0.0836		
17	362	0.0900		
18	*etc.*	*etc.*		

Figure 13-15. Selecting every Nth data point by using a worksheet formula.

PART III

SPREADSHEET METHODS FOR CHEMISTS

14

SOME MATHEMATICAL TOOLS FOR SPREADSHEET CALCULATIONS

This chapter describes some mathematical methods that are useful for spreadsheet calculations. For the most part, they are methods that are applicable to tables or arrays of data. (It is assumed that the reader is familiar with basic algebra, logarithms, exponential functions and trigonometric functions.)

INTERPOLATION METHODS

Spreadsheets are ideal for examining and manipulating large arrays of x, y data points. Often it's necessary to estimate the value of y for some value of x intermediate between two adjacent data points — the process of *interpolation*. If the separation between x values is small, and the y values do not change rapidly with x, then linear interpolation may be adequate. To calculate the value of y at a value x_i that is intermediate between x_0 and x_1, use the linear interpolation formula (equation 14-1).

$$y_i = y_0 + \frac{(x_i - x_0)}{(x_1 - x_0)} (y_1 - y_0) \qquad (14\text{-}1)$$

For more general application, a *spline function* is often used for interpolation. A spline function uses the values of n adjacent data points to evaluate a polynomial of order n that passes through the data points. One of the most commonly used spline functions is the *cubic spline*, in which the values of four adjacent data points are used to evaluate the coefficients of the cubic equation $y = a + bx + cx^2 + dx^3$.

A compact and elegant implementation of the cubic spline in the form of an Excel 4.0 Macro Language custom function was provided by Orvis[*]. A slightly modified version is provided here (Figure 14-1). The syntax of the custom function is CUBIC.SPLINE(*known_xs, known_ys, x_value*).

The cubic spline function can be used to produce a smooth curve through data points. Figure 14-2 illustrates a portion of a spreadsheet with experimental spectrophotometric data taken at 5 nm intervals in columns A and B (the data is

---

[*]   William J. Orvis, *Excel 4 for Scientists and Engineers*, Sybex Inc., Alameda, CA, 1993.

	A	B
1		**Cubic Spline Macro**
2		Syntax: array of x values, array of y values, x
3		
4		=ARGUMENT("xvalues",64)
5		=ARGUMENT("yvalues",64)
6		=ARGUMENT("x",1)
7	row	=MATCH(x,xvalues)
8		=IF(row<2,SET.VALUE(row,2))
9		=IF(row>ROWS(xvalues)-2,SET.VALUE(row,ROWS(xvalues)-2))
10	y	=0
11		=FOR("I",row-1,row+2)
12	q	=1
13		=FOR("J",row-1,row+2)
14		=IF(I<>J,SET.VALUE(q,q*(x-INDEX(xvalues,J))/(INDEX(xvalues,I)-INDEX(xvalues,J))))
15		=NEXT()
16		=SET.VALUE(y,y+q*INDEX(yvalues,I))
17		=NEXT()
18		=RETURN(y)

Figure 14-1. Cubic spline custom function macro.

in rows 6 through 86), and a portion of the interpolated values, at 1 nm intervals, in columns C and D. The formula in cell D24 is

='Cubic Spline Macro'!Macro1($A$6:$A$86,$B$6:$B$86,C24).

The smoothed curve through the data points is illustrated in Figure 14-3.

The cubic spline forces the curve to pass through all of the known data points, known as *knots*. If there is any experimental scatter, the result will not be too pleasing. A better approach for data with scatter is to find the coefficients of a least-squares line through the data points, as described in Chapters 16 or 17.

	A	B	C	D
4	**Original Data**		**Interpolated Data**	
5	x	y	x	y
24	390	0.552	390	0.552
25	395	0.582	391	0.559
26	400	0.598	392	0.566
27	405	0.600	393	0.572
28	410	0.586	394	0.577
29	415	0.559	395	0.582
30	420	0.521	396	0.586
31	425	0.473	397	0.590
32	430	0.419	398	0.593
33	435	0.362	399	0.596
34	440	0.305	400	0.598

Figure 14-2. Interpolation by using a cubic spline function macro.

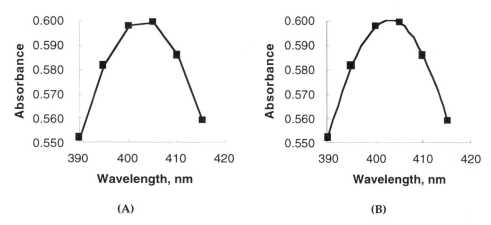

Figure 14-3. (A) Chart created using data points only. (B) Chart with smooth curve interpolated using Cubic Spline.

NUMERICAL DIFFERENTIATION

The process of finding the derivative or slope of a function is the basis of *differential calculus*. Since you will be dealing with spreadsheet data, you will be concerned, not with the algebraic differentiation of a function, but with obtaining the derivative of a data set or the derivative of a worksheet formula by numeric methods.

Often a function depends on more than one variable. The *partial derivative* of the function $F(x,y,z)$, e.g., $\partial F/\partial x$, is the slope of the function with respect to x, while y and z are held constant.

FIRST AND SECOND DERIVATIVES OF A DATA SET

The simplest method to obtain the first derivative of a function represented by a table of x, y data points is to calculate $\Delta y / \Delta x$. The first derivative or slope of the curve at a given data point x_n, y_n can be calculated using either of the formulas

$$\text{slope} = \frac{\Delta y}{\Delta x} = \frac{y_{n+1} - y_n}{x_{n+1} - x_n} \tag{14-2}$$

or

$$\text{slope} = \frac{y_n - y_{n-1}}{x_n - x_{n-1}} \tag{14-3}$$

The second derivative, d^2y/dx^2, of a data set is calculated in a similar manner, namely by calculating $\Delta(\Delta y/\Delta x)/\Delta x$.

Calculation of the first or second derivative of a data set tends to emphasize the "noise" in the data set; that is, small errors in the measurements become relatively much more important.

2	A V/mL	B pH	C ΔV	D ΔpH	E V(avge)	F ΔpH/ΔV
22	1.90	4.981	0.100	0.229	1.850	2.29
23	1.95	5.157	0.050	0.176	1.925	3.52
24	2.00	5.389	0.050	0.232	1.975	4.64
25	2.05	5.928	0.050	0.539	2.025	10.78
26	2.08	7.900	0.030	1.972	2.065	65.73
27	2.10	9.115	0.020	1.215	2.090	60.75
28	2.15	9.604	0.050	0.489	2.125	9.78
29	2.20	9.856	0.050	0.252	2.175	5.04
30	2.30	10.125	0.100	0.269	2.250	2.69

Figure 14-4. First derivative of titration data, near the end-point.

Points on a curve of x, y values for which the first derivative is either a maximum, minimum or equal to zero are often of particular importance and are termed *critical points*.

The calculation of the first derivative of a data set is illustrated, using pH titration data, in the spreadsheet shown in Figure 14-4.

Since the derivative has been calculated over the finite volume $\Delta V = V_{n+1} - V_n$, the most suitable volume to use when plotting the $\Delta pH/\Delta V$ values, when plotting the $\Delta pH/\Delta V$ values, as shown in Figure 14-4, is:

$$V_{\text{average}} = \frac{V_{n+1} + V_n}{2} \tag{14-4}$$

The maximum in $\Delta pH/\Delta V$ indicates the location of the inflection point of the titration (Figure 14-5). The second derivative, $\Delta(\Delta pH/\Delta V)/\Delta V$, which is calculated by means of the spreadsheet shown in Figure 14-6, can be used to

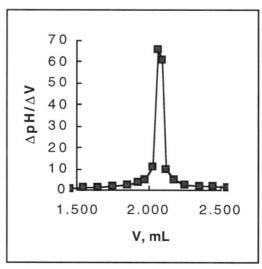

Figure 14-5. First derivative of titration data, near the end-point.

2	E	F	G	H	I	J
	V(avge)	ΔpH/ΔV	ΔV	Δ(ΔpH)	V(avge)	Δ(ΔpH)/ΔV
22	1.850	2.29	0.100	0.57	1.800	5.7
23	1.925	3.52	0.075	1.23	1.888	16.4
24	1.975	4.64	0.050	1.12	1.950	22.4
25	2.025	10.78	0.050	6.14	2.000	122.8
26	2.065	65.73	0.040	54.95	2.045	1373.8
27	2.090	60.75	0.025	-4.98	2.078	-199.3
28	2.125	9.78	0.035	-50.97	2.108	-1456.3
29	2.175	5.04	0.050	-4.74	2.150	-94.8
30	2.250	2.69	0.075	-2.35	2.213	-31.3

Figure 14-6. Second derivative of titration data, near the end-point.

locate the inflection point more precisely. The second derivative passes through zero at the inflection point. Linear interpolation can be used to calculate the point where the second derivative is zero (Figure 14-7).

There are more sophisticated equations for numerical differentiation. These equations calculate the derivative using three, four or five points instead of two points. Since they usually require equal intervals between points, they are of less generality. Their main advantage is that they minimize the effect of "noise".

DERIVATIVES OF A FUNCTION

The first derivative of a formula in a worksheet cell can be obtained with a high degree of accuracy by evaluating the formula at x and at $x + \Delta x$. Since Excel carries 15 significant figures, Δx can be made very small. Under these conditions $\Delta F / \Delta x$ approximates dF/dx very well.

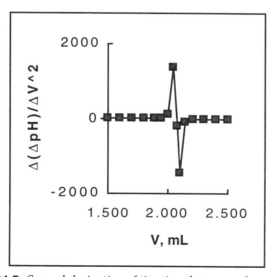

Figure 14-7. Second derivative of titration data, near the end-point.

	A	B	C	D	E
1		**Numerical Differentiation**			
2		$F(x) = tx^3 + ux^2 + vx + w$			
3		First derivative $F'(x) = 3tx^2 + 2ux + v$			
5		t	1		
6		u	-3		
7		v	-130		
8		w	150		
9		Δ (delta)	1.00E-09		
10				*Using differences*	*Using calculus*
11	x	F(x)	F(x+Δ)	F'(x)	F'(x)
12	-10	150	150	230.0	230.0
13	-9	348	348	167.0	167.0
14	-8	486	486	110.0	110.0
15	-7	570	570	59.0	59.0
16	-6	606	606	14.0	14.0
17	-5	600	600	-25.0	-25.0
18	-4	558	558	-58.0	-58.0
19	-3	486	486	-85.0	-85.0

Figure 14-8. Calculating the first derivative of a function.

The spreadsheet fragment shown in Figure 14-8 illustrates the calculation of the first derivative of a function ($F = x^3 - 3x^2 - 130x + 150$) by evaluating the function at x and at $x + \Delta x$. Here a value of Δx of 1×10^{-9} was used; alternatively Δx could be obtained by using a worksheet formula such as =1E-9*x. For comparison the first derivative was calculated from the expression from differential calculus: $F' = 3x^2 - 6x - 130$.

The Excel formulas in cells B12, C12, D12 and E12 are

= t*x^3+u*x^2+v*x + w

= t*(x+delta)^3 +u*(x+delta)^2 +v*(x+delta) + w

=(C12-B12)/delta

=3*t*x^2+2*u*x+v

A chart of the function and its first derivative is shown in Figure 14-9.

NUMERICAL INTEGRATION

A common use of numerical integration is to determine the area under a curve. Three methods for determining the area under a curve will be described here: the rectangle method, the trapezoid method and Simpson's method. Each involves approximating the area of each portion of the curve delineated by adjacent data points; the area under the curve is the sum of these individual segments.

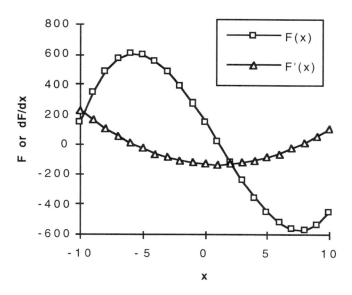

Figure 14-9. The function $F = x^3 - 3x^2 - 130x + 150$ and its first derivative.

The simplest approach is to approximate the area by the rectangle whose height is equal to the value of one of the two data points, illustrated in Figure 14-10.

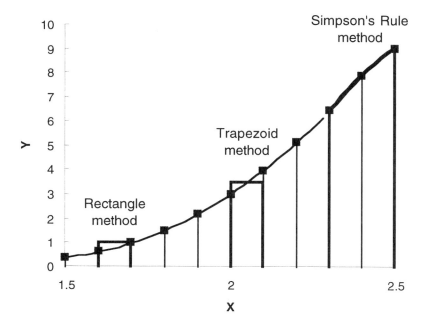

Figure 14-10. Graphical illustration of methods of calculating the area under a curve.

As the x increment (the interval between the data points) decreases, this rather crude approach becomes a better approximation to the area. The area under the curve bounded by the limits $x_{initial}$ and x_{final} is the sum of the individual rectangles, as given by equation 14-5.

$$\text{area} = \sum y_i (x_{i+1} - x_i) \tag{14-5}$$

A better approximation is to use the average of the two y values as the height of the rectangle. This is equivalent to approximating the area by a trapezoid rather than a rectangle. The area under the curve is given by equation 14-6.

$$\text{area} = \sum \frac{y_i + y_{i+1}}{2} (x_{i+1} - x_i) \tag{14-6}$$

Simpson's rule approximates the curvature of the function by means of a quadratic equation. To evaluate the coefficients of the quadratic requires the use of the y values for three adjacent data points. The x values must be equally spaced.

$$\text{area} = \sum \frac{y_i + 4y_{i+1} + y_{i+2}}{6} (x_{i+1} - x_i) \tag{14-7}$$

AN EXAMPLE: FINDING THE AREA UNDER A CURVE

The curve shown in Figure 14-11 is the sum of two Gaussian curves, with position and standard deviation $\mu = 90$, $s = 10$ and $\mu = 130$, $s = 20$, respectively. The equation used to calculate each Gaussian curve is

$$y = \frac{\exp[-(x - \mu)^2/2\sigma^2]}{\sigma\sqrt{2\pi}} \tag{14-8}$$

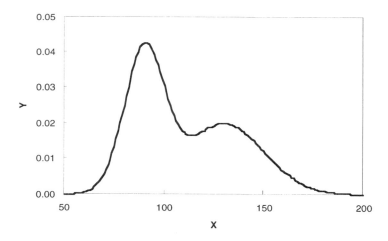

Figure 14-11. A curve that is the sum of two Gaussian curves.

	E	F	G	H	I
6			*Rectangular*	*Trapezoidal*	*Simpson's*
7	x	Y	*Approximation*	*Approx.*	*Rule*
8					
9	50	0.00002	0.00020	0.00010	0.00095
10	60	0.00049	0.00487	0.00253	0.01265
11	70	0.00562	0.05621	0.03054	0.08007
12	80	0.02507	0.25073	0.15347	0.24751
13	90	0.04259	0.42594	0.33834	0.37687
14	100	0.03067	0.30673	0.36633	0.30464

Figure 14-12. Portion of a spreadsheet for calculating the area under a curve.

The area of each Gaussian curve is equal to 1.000; thus the total area under the curve shown in Figure 14-11 is 2.000. The Excel formula used to calculate y is as follows (m is the position μ and s is the standard deviation σ):

=EXP(-0.5*((x-curv1   m)/curv1   s)^2)/(SQRT(2*PI())*curv1   s)+EXP(-0.5* ((x-curv2 m)/curv2 s)^2)/(SQRT(2*PI())*curv2 s)

The area under the curve, between the limits $x = 50$ and $x = 200$, was calculated by using each of the above equations (14-5, 14-6 and 14-7): the rectangular approximation, the trapezoidal approximation and Simpson's rule. In each case a constant x increment of 10 was used. A portion of the spreadsheet is shown in Figure 14-12.

The formulas in row 9, used to calculate the area increment, are as follows:

=10*F9   (rectangular approximation)

=10*(F8+F9)/2   (trapezoidal approximation)

=10*(F8+4*F9+F10)/6   (Simpson's rule)

The area increments were summed and the area under the curve, calculated by the three methods, is shown in Figure 14-13. All three methods of calculation appear to give acceptable results in this case.

DIFFERENTIAL EQUATIONS

Certain chemical problems, such as ones involving chemical kinetics, can be expressed by means of differential equations. For example, the coupled reaction scheme

$$A \underset{k_2}{\overset{k_1}{\rightleftharpoons}} B \underset{k_4}{\overset{k_3}{\rightleftharpoons}} C$$

results in the simultaneous equations

$$\frac{d[A]}{dt} = -k_1[A] + k_2[B]$$

$$\frac{d[B]}{dt} = k_1[A] - k_2[B] - k_3[B] + k_4[C]$$

$$\frac{d[C]}{dt} = k_3[B] - k_4[C]$$

	G	H	I
6	*Rectangular*	*Trapezoidal*	*Simpson's*
7	*Approximation*	*Approx.*	*Rule*
25			
26	1.9999	1.9997	1.9998

Figure 14-13. Area under a curve, calculated by three different methods.

To "solve" this system of simultaneous equations, we want to be able to calculate the value of [A], [B] and [C] for any value of t. For all but the simplest of these systems of equations, obtaining an exact or analytical expression is difficult or sometimes impossible. Such problems can always be solved by numerical methods, however. Numerical methods are completely general. They can be applied to systems of differential equations of any complexity, and they can be applied to any set of initial conditions. Numerical methods require extensive calculations but this is easily accomplished by spreadsheet methods.

In this chapter we will consider only ordinary differential equations, that is, equations involving only derivatives of a single independent variable. As well, we will discuss only initial-value problems — differential equations in which information about the system is known at $t = 0$. Two approaches are common: Euler's method and the Runge–Kutta (RK) methods.

EULER'S METHOD

Let us use as an example the simulation of the first-order kinetic process A → B with initial concentration $C_0 = 0.2000$ mol/L and rate constant $k = 5 \times 10^{-3}$ s^{-1}. We'll simulate the change in concentration vs. time over the interval from $t = 0$ to $t = 600$ seconds, in increments of 20 seconds.

The differential equation for the disappearance of A is $d[A]/dt = -k[A]$. Expressing this in terms of finite differences, the change in concentration $\Delta[A]$ that occurs during the time interval from $t = 0$ to $t = \Delta t$ is $\Delta[A] = -k[A]\Delta t$. Thus, if the concentration of A at $t = 0$ is 0.2000 M, then the concentration at $t = 0 + \Delta t$ is $[A] = 0.2000 - (5 \times 10^{-3})(0.2000)(20) = 0.1800$ M. The formula in cell B7 is

=B6-k*B6*DX.

The concentration at subsequent time intervals is calculated in the same way.

The advantage of this method, known as *Euler's method,* is that it can be easily expanded to handle systems of any complexity. Euler's method is not particularly useful, however, since the error introduced by the approximation $d[A]/dt = \Delta[A]/\Delta t$ is compounded with each additional calculation. Compare the Euler's method result in column B of Figure 14-14 with the analytical expression for the concentration, $[A]_t = [A]_0 e^{-kt}$, in column C. At the end of approximately one half-life (seven cycles of calculation in this example), the error has already increased to 3.7%. Accuracy can be increased by decreasing the size of Δt, but only at the expense of increased computation. A much more efficient way of increasing the accuracy is by means of a series expansion. The most commonly used formula is the Runge–Kutta method, which is described in the following section.

	A	B	C	D
1	Simulation of first order kinetics by Euler's Method			
2		rate constant =	5.0E-03	(k)
3		time increment =	20	(DX)
4				
5	t	A(t)	e^(-kt)	
6	0	0.2000	0.2000	
7	20	0.1800	0.1810	
8	40	0.1620	0.1637	
9	60	0.1458	0.1482	
10	80	0.1312	0.1341	
11	100	0.1181	0.1213	
12	120	0.1063	0.1098	
13	140	0.0957	0.0993	

Figure 14-14. Euler's method.

THE RUNGE–KUTTA METHODS

The Runge–Kutta methods for numerical solution of the differential equation $dy/dx = F(x, y)$ involve, in effect, the evaluation of the differential function at intermediate points between x_i and x_{i+1}. The value of y_{i+1} is obtained by appropriate summation of the intermediate terms in a single equation. The most widely used Runge-Kutta formula involves terms evaluated at x_i, $x_i + \Delta x/2$ and $x_i + \Delta x$. The *fourth-order Runge-Kutta* equations for $dy/dx = F(x, y)$ are

$$y_{i+1} = y_i + \frac{T_1 + 2T_2 + 2T_3 + T_4}{6} \tag{14-9}$$

where
$$T_1 = F(x_i, y_i)\, \Delta x \tag{14-10}$$

$$T_2 = F\left(x_i + \frac{\Delta x}{2}, y_i + \frac{T_1}{2}\right) \Delta x \tag{14-11}$$

$$T_3 = F\left(x_i + \frac{\Delta x}{2}, y_i + \frac{T_2}{2}\right) \Delta x \tag{14-12}$$

$$T_4 = F(x_i + \Delta x, y_i + T_3)\, \Delta x \tag{14-13}$$

If more than one variable appears in the expression, then each is corrected by using its own set of T_1 to T_4 terms.

The spreadsheet in Figure 14-15 illustrates the use of the RK method to simulate the first-order kinetic process A → B with initial concentration $[A]_0 = 0.2000$ and rate constant $k = 5 \times 10^{-3}$. The differential equation is $d[A]_t/dt = -k[A]_t$. This equation is one of the simple form $dy/dx = F(y)$ and thus only the y_i terms of T_1 to T_4 need to be evaluated. The RK terms are (note that T_1 is the Euler method term):

$$T_1 = -k[A]_t\, \Delta x \tag{14-14}$$

$$T_2 = -k([A]_t + T_1/2)\, \Delta x \tag{14-15}$$

$$T_3 = -k([A]_t + T_2/2)\, \Delta x \tag{14-16}$$

$$T_4 = -k([A]_t + T_3)\, \Delta x \tag{14-17}$$

	A	B	C	D	E	F	G
1		Runge-Kutta simulation of first order kinetics					
2			rate constant =		5.0E-03	(k)	
3			time increment =		20	(DX)	
4							
5	t	TA1	TA2	TA3	TA4	A(t)	e^(-kt)
6	0					0.2000	0.2000
7	20	-0.0200	-0.0190	-0.0191	-0.0181	0.1810	0.1810
8	40	-0.0181	-0.0172	-0.0172	-0.0164	0.1637	0.1637
9	60	-0.0164	-0.0156	-0.0156	-0.0148	0.1482	0.1482
10	80	-0.0148	-0.0141	-0.0141	-0.0134	0.1341	0.1341
11	100	-0.0134	-0.0127	-0.0128	-0.0121	0.1213	0.1213
12	120	-0.0121	-0.0115	-0.0116	-0.0110	0.1098	0.1098
13	140	-0.0110	-0.0104	-0.0105	-0.0099	0.0993	0.0993

Figure 14-15. The Runge–Kutta method

The RK equations in cells B7, C7, D7, E7 and F7, respectively, are:

=-k*F6*DX

=-k*(F6+TA1/2)*DX

=-k*(F6+TA2/2)*DX

=-k*(F6+TA3)*DX

=F6+(TA1+2*TA2+2*TA3+TA4)/6.

If you use the names TA1,···, TA4, TB1,···, TB4, etc., you'll find that (i) the nomenclature is expandable to systems requiring more than one set of Runge–Kutta terms, (ii) the names are accepted by Excel, whereas T1 is not a valid name, and (iii) you can generate the column labels TA1,···, TA4 using AutoFill.

Compare the RK result in column F of Figure 14-15 with the analytical expression for the concentration, $[A]_t = [A]_0 e^{-kt}$, in column G. After one half-life (row 13) the RK calculation differs from the analytical expression by only 0.00006%. (Compare this with the 3.6% error in the Euler method calculation at the same point.) Even after ten half-lives (not shown), the RK error is only 0.0006%.

If the spreadsheet is constructed as shown in Figure 14-15, it is not possible to use a formula in which a name is assigned to the concentration array in column F, since the formula in B7, for example, will use the concentration in F7 instead of the required F6. An alternative arrangement that permits using a name for the concentration $[A]_t$ is shown in Figure 14-16. Each row contains the concentration at the beginning and at the end of the time interval. The name A_t can now be assigned to the array of values in column B; the former formulas (now in cells C7 through G7) contain A_t in place of F6, and cell B7 contains the formula =G6.

	A	B	C	D	E	F	G	H
1			Runge–Kutta simulation of first order kinetics					
2			rate constant =		5.0E–03	(k)		
3			time increment =		20	(DX)		
4								
5	t	A(t)	TA1	TA2	TA3	TA4	A(t+DX)	e^(–kt)
6	0	0.2000	–0.0200	–0.0190	–0.0191	–0.0181	0.1810	0.1810
7	20	0.1810	–0.0181	–0.0172	–0.0172	–0.0164	0.1637	0.1637
8	40	0.1637	–0.0164	–0.0156	–0.0156	–0.0148	0.1482	0.1482
9	60	0.1482	–0.0148	–0.0141	–0.0141	–0.0134	0.1341	0.1341
10	80	0.1341	–0.0134	–0.0127	–0.0128	–0.0121	0.1213	0.1213

Figure 14-16. Alternative spreadsheet layout for the Runge-Kutta method.

In essence, the fourth order Runge–Kutta method performs four calculation steps for every time interval. In the solution by Euler's method, decreasing the time increment to 5 seconds, so as to perform four times as many calculation steps, still only reduces the error to 0.9% after 1 half-life.

Summary of Steps to Implement the RK Method

1. Write down the differential equations that describe the system.

2. From the set of differential equations, enter formulas in spreadsheet cells using Euler's Method, **Fill Down** and make sure that the results make sense.

3. Insert five columns to the right of each variable. Label them TA1,···, TA4 and A(t), TB1,···, TB4 and B(t), etc.

4. **Copy** the Euler's method formula from the formula bar and **Paste** it into the four cells for TA1,···, TA4. The Euler's method formula should be of the form = yo + F*yo*Dx. Edit it to leave only the = F*yo*Dx part. Modify the formulas to produce the formulas for TA1,···, TA4. TA1 remains F*yo*Dx, TA2 becomes F*(yo + TA1/2)*Dx, TA3 becomes F*(yo + TA2/2)*Dx, TA4 becomes F*(yo + TA3)*Dx.

5. Enter the formula for y(m): ym + (TA1 +2*TA2 + 2*TA3 + TA4)/6

6. **Fill Down** the TA1···TA4 & y(m) cells. y(m) should agree approximately with the Euler's method column.

7. **Delete** the columns containing the Euler's method values.

ARRAYS, MATRICES AND DETERMINANTS

Spreadsheet calculations lend themselves almost automatically to the use of arrays of values. As you've seen, arrays in Excel can be either one- or two-dimensional. For the solution of many types of problem, it is convenient to manipulate an entire rectangular array of values as a unit. Such an array is termed a *matrix*. A $n \times m$ matrix of values is illustrated below. In Excel, the terms "matrix" and "array" are virtually interchangeable.

$$\begin{vmatrix} a_{11} & a_{12} & \cdots & a_{1n} \\ a_{21} & a_{22} & \cdots & a_{2n} \\ \vdots & \vdots & \vdots & \vdots \\ a_{m1} & a_{m2} & \cdots & a_{mn} \end{vmatrix}$$

The values comprising the array are called *matrix elements*. Mathematical operations on matrices have their own special rules.

A *square matrix* has the same number of rows and columns. If all the elements of a square matrix are zero except those on the main diagonal (a_{11}, $a_{22}, \cdots, a_{nn}$), the matrix is termed a *diagonal matrix*. A diagonal matrix whose diagonal elements are all 1 is a *unit matrix*.

A *determinant* is simply a square matrix. There is a procedure for the numerical evaluation of a determinant, so that an $N \times N$ matrix can be reduced to a single numerical value. The value of the determinant has properties that make it useful in certain tests and equations. (See, for example, "Solving Sets of Simultaneous Linear Equations" in Chapter 15.)

AN INTRODUCTION TO MATRIX ALGEBRA

Matrix algebra provides a powerful method for the manipulation of sets of numbers. Many mathematical operations — addition, subtraction, multiplication, division, etc. — have their counterparts in matrix algebra. The discussion below will be limited to the manipulations of square matrices. For purposes of illustration, two 3×3 matrices will be defined, namely

$$\mathbf{A} = \begin{vmatrix} a & b & c \\ d & e & f \\ g & h & i \end{vmatrix} = \begin{vmatrix} 2 & 3 & 4 \\ 3 & 2 & -1 \\ 4 & 3 & 7 \end{vmatrix}$$

and

$$\mathbf{B} = \begin{vmatrix} r & s & t \\ u & v & w \\ x & y & z \end{vmatrix} = \begin{vmatrix} 2 & 0 & 2 \\ 0 & 3 & -3 \\ -3 & -2 & 1 \end{vmatrix}$$

The following examples illustrate addition, subtraction, multiplication and division using a constant.

Addition or subtraction of a constant:
$$\mathbf{A} + q = \begin{vmatrix} a+q & b+q & c+q \\ d+q & e+q & f+q \\ g+q & h+q & i+q \end{vmatrix}$$

Multiplication or division by a constant:
$$q\mathbf{A} = \begin{vmatrix} qa & qb & qc \\ qd & qe & qf \\ qg & qh & qi \end{vmatrix}$$

Addition or subtraction of two matrices (both must contain the same number of rows and columns):

$$\mathbf{A + B} = \begin{vmatrix} a & b & c \\ d & e & f \\ g & h & i \end{vmatrix} + \begin{vmatrix} r & s & t \\ u & v & w \\ x & y & z \end{vmatrix} = \begin{vmatrix} a+r & b+s & c+t \\ d+u & e+v & f+w \\ g+x & h+y & i+z \end{vmatrix}$$

Performing matrix algebra with Excel is very simple. Let's begin by assuming that the matrices **A** and **B** have been defined by selecting the 3R × 3C arrays of cells containing the values and naming them by using **Define Name**. To add a constant, e.g., 3, to matrix **A**, simply select a range of cells the same size as the matrix, enter the formula =A + 3, then press COMMAND+ENTER (Macintosh) or CONTROL+SHIFT+ENTER (Windows). Subtraction of a constant, multiplication or division by a constant, or addition of two matrices also is performed by using standard Excel algebraic operators.

The multiplication of two matrices is somewhat more complicated:

$$\mathbf{A\,B} = \begin{vmatrix} a & b & c \\ d & e & f \\ g & h & i \end{vmatrix}\begin{vmatrix} r & s & t \\ u & v & w \\ x & y & z \end{vmatrix} = \begin{vmatrix} ar+bu+cx & as+bv+cy & at+bw+cz \\ dr+eu+fx & ds+ev+fy & dt+ew+fz \\ gr+hu+ix & gs+hv+iy & gt+hw+iz \end{vmatrix}$$

Matrix multiplication is not commutative, that is **A B** ≠ **B A**.

Matrix multiplication can be accomplished easily by the use of one of Excel's worksheet functions for matrix algebra, MMULT(*matrix1, matrix2*). For the matrices **A** and **B** defined above,

$$\mathbf{A\,B} = \begin{vmatrix} -8 & 1 & -1 \\ 9 & 8 & -1 \\ -13 & -5 & 6 \end{vmatrix} \quad \mathbf{B\,A} = \begin{vmatrix} 12 & 12 & 22 \\ -3 & -3 & -24 \\ -8 & -10 & -3 \end{vmatrix}$$

The *transpose* of a matrix, indicated by a prime ('), is produced when rows and columns of a matrix are interchanged, i.e.,

$$\mathbf{A'} = \begin{vmatrix} a & d & g \\ b & e & h \\ c & f & i \end{vmatrix}$$

The transpose is obtained by using the worksheet function TRANSPOSE(*array*) or the Transpose option in the **Paste Special...** menu command (see "Using Paste Special to Transpose Rows and Columns" in Chapter 1).

The process of *matrix inversion* is analogous to obtaining the reciprocal of a number a. The matrix relationship that corresponds to the algebraic relationship $a \times (1/a) = 1$ is

$$\mathbf{A\,A^{-1} = I}$$

where $\mathbf{A^{-1}}$ is the inverse matrix and **I** is the unit matrix. The process for inverting a matrix "manually" (i.e., using pencil, paper and calculator) is complicated, but the operation can be carried out readily using Excel's worksheet function MINVERSE(*array*). The inverse of the matrix **A** above is:

$$\mathbf{A}^{-1} = \begin{vmatrix} -0.25 & -0.3333 & -0.5 \\ 0.75 & 0.6667 & 0.5 \\ 0.75 & 0.3333 & 0.5 \end{vmatrix}$$

The "pencil-and-paper" evaluation of a determinant of N rows $\times$ N columns is also complicated, but it can be done simply by using the worksheet function MDETERM(*array*). The function returns a single numerical value, not an array, and thus you do not have to use COMMAND+ENTER. The value of the determinant of $\mathbf{A}$, represented by $|\mathbf{A}|$, is 12.

POLAR TO CARTESIAN COORDINATES

You may occasionally need to chart a function that involves angles. Instead of using the familiar Cartesian coordinate system (x, y and z coordinates), such functions often use the polar coordinate system, in which the coordinates are two angles, θ and ϕ, and a distance r. The two coordinate systems are related by the equations $x = r \sin \theta \cos \phi$, $y = r \sin \theta \sin \phi$, $z = r \cos \theta$. Angle θ is the angle between the vector r and the Cartesian z-axis; ϕ is the angle between the projection of r on the x, y plane and the x-axis. Since Excel's trigonometric functions only consider x- and y-axes, the simplified relationships are, for angles in the x, y plane: $x = r \cos \phi$, $y = r \sin \phi$.

As an example of transformation of polar to Cartesian coordinates, we'll graph the wave function for the $d_{x^2 - y^2}$ orbital in the x, y plane. The angular component of the wave function in the x, y plane is

$$\Phi = \sqrt{\frac{15}{16\pi}} \cos 2\phi \tag{14-18}$$

and Φ can be equated to the radial vector r for the conversion of polar to Cartesian coordinates.

In the spreadsheet fragment shown in Figure 14-17, column A contains angles from 0 to 360 in 2 degree increments. Column B converts the angles to radians (required by the COS worksheet function) using the relationship =A4*PI()/180 in row 4. The formulas in cells C4, D4 and   E4 are:

=SQRT(15/(16*PI()))*COS(2*B4)

=C4*COS(B4)

=C4*SIN(B4).

The chart of the x and y values is shown in Figure 14-18.

	A	B	C	D	E
1	Representation of Angular Wave Function of dx^2-y^2 Orbital				
2					
3	Angle, deg	Angle, rad	d	x coord	y coord
4	0	0.0000	0.5463	0.5463	0.0000
5	2	0.0349	0.5449	0.5446	0.0190
6	4	0.0698	0.5410	0.5396	0.0377
7	6	0.1047	0.5343	0.5314	0.0559
8	8	0.1396	0.5251	0.5200	0.0731
9	10	0.1745	0.5133	0.5055	0.0891
10	12	0.2094	0.4990	0.4881	0.1038

Figure 14-17. Converting from polar to Cartesian coordinates

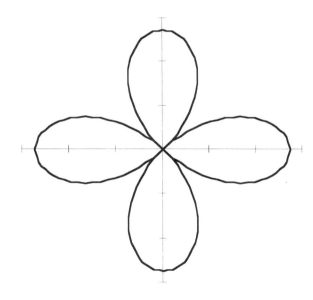

Figure 14-18. Angular wave function for $d_x2 - y^2$ orbital.

USEFUL REFERENCES

K. Jeffrey Johnson, *Numerical Methods in Chemistry*, Marcel Dekker, New York, 1980.

15

GRAPHICAL AND NUMERICAL
METHODS OF ANALYSIS

In this chapter you'll learn how to use graphical and numerical methods to solve chemical problems. The methods described in this chapter range from the simple (finding the roots of a polynomial from a graph of the function) to the complex (solving sets of simultaneous equations using matrix methods). Excel does most of the work, though, by virtue of its graphing capabilities and built-in numerical analysis capabilities.

FINDING ROOTS OF EQUATIONS

Sometimes a chemical problem can be reduced algebraically, by pencil and paper, to a polynomial expression for which the solution to the problem is one of the roots of the polynomial. Almost everyone remembers the quadratic formula for the roots of a quadratic equation, but finding the roots of a more complicated polynomial is more difficult. This first section describes three methods for finding the real roots of a polynomial.

THE GRAPHICAL METHOD

The roots of a polynomial $y = F(x)$ are the values of x that make $F(x) = 0$. One simple way to find those values is to create a spreadsheet table of x and corresponding y values, then create a chart from the values. The x-values where the curve crosses the x-axis (where $y = 0$) are the roots of the equation. Figure 15-1 is a graph of the function $y = x^3 - 3x^2 - 13x + 15$, for which the roots are 5, 1 and -3. The roots can't be read from this chart with any degree of accuracy, but it's a simple matter to create a chart of the region immediately around an intersection point and get a much more precise value. In general, though, the main use of this method is to gain an idea of the approximate value of the roots.

How many real roots does a given polynomial have? Generally, in chemical problems, x represents a quantity that can only be positive, such as a concentration or an equilibrium constant. *Descartes' Rule of Signs* states that, for any polynomial $F(x)$ written in decreasing powers of x, the number of positive roots of the equation $F(x) = 0$ is equal to the number of changes in sign between the coefficients of adjacent terms, or is less than this number by some positive even integer. For example, for the cubic equation $x^3 + (2.29 \times 10^2)x^2 - (2.29 \times 10^1)x - (9.14 \times 10^{-1}) = 0$, there is one change in sign, between the x^2 term and the x term. Thus there is only one positive root of the equation. Almost always, in

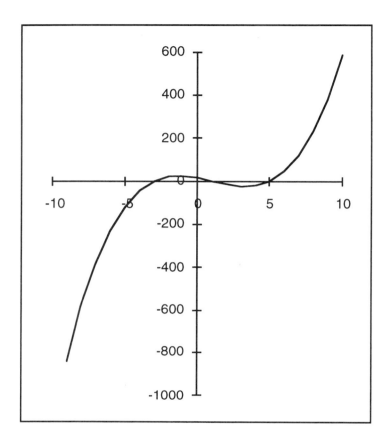

Figure 15-1. Graph of the equation $y = x^3 - 3x^2 - 13x + 15$. The roots are 5, 1 and –3.

chemical problems, such a polynomial will have only one positive real root, and this one root will be the solution to the problem.

THE METHOD OF SUCCESSIVE APPROXIMATIONS

The graphical method can give us only an approximate idea of the roots of a polynomial. To obtain a more accurate numerical result, the roots of $y = F(x)$ can easily be obtained by trial-and-error, finding the values of x that make the function y equal to zero.

As an example, let's calculate the solubility of barium carbonate in pure water. Barium carbonate, $BaCO_3$, is a sparingly-soluble salt with $K_{sp} = 5.1 \times 10^{-9}$:

$$BaCO_3(s) = Ba^{2+} + CO_3^{2-} \qquad\qquad K_{sp} = [Ba^{2+}][CO_3^{2-}] \quad (15\text{-}1)$$

If the hydrolysis of the carbonate ion to form bicarbonate is taken into account (equation 15-2, $K_b = 2.1 \times 10^{-4}$), solving the system of mass- and charge-balance

$$CO_3^{2-} + H_2O = OH^- + HCO_3^- \qquad\qquad K_b = \frac{[OH^-][HCO_3^-]}{[CO_3^{2-}]} \quad (15\text{-}2)$$

equations leads to the following pseudo-quadratic expression, in which x is the solubility of barium carbonate.

$$x^2 - \sqrt{K_b\,K_{sp}}\,x - K_{sp} = 0 \qquad\qquad (15\text{-}3)$$

Inserting the values of the constants, you can write

$$x^2 - 1.0 \times 10^{-6}\,x^{1/2} - 5.1 \times 10^{-9} = 0 \qquad\qquad (15\text{-}4)$$

You can locate the positive root in a systematic fashion by changing the value of x in increments and observing the value of the function y. When y exhibits a sign change, you'll know that it has passed through zero. Then decrease the size of the increment and repeat the process. In this way you can determine the value of x to any desired degree of accuracy.

You can start with any value of x. Eventually you'll get the right answer. But if you use an educated guess, you'll get there sooner. Since the K_{sp} of $BaCO_3$ is 5.1×10^{-9}, and the solubility can be approximated by $\sqrt{K_{sp}}$, use 7×10^{-5} as the initial guess. The solubility, as carbonate hydrolyzes to bicarbonate, will be higher than this, so increase it in increments of 1×10^{-5}.

Open a new worksheet and label cell A1 "X". Label cell B1 "Y". Enter 7E-5 in cell A2, 8E-5 in cell A3, then use Autofill to **Fill Down** about 10 cells. In cell B2 enter the formula =A2^2-1.0E-6*SQRT(A2)-5.1E-9, and use AutoFill to **Fill Down**. The result is shown in Figure 15-2.

The function y changes sign between rows 7 and 8 (between $x = 1.2\text{E-}04$ and $1.3\text{E-}04$). Now repeat the process beginning with $x = 1.2\text{E-}04$ and using a smaller increment. To do this, **Copy** cells A7:B7 and **Paste** into cell C7:D7. In cell C8 enter 1.21E-04. Select C7:C8 and use AutoFill to **Fill Down** about 10 cells, then use AutoFill to **Fill Down** column D. The result is shown in Figure 15-3. The sign change occurs between rows 15 and 16.

	A	B
1	X	Y
2	7.00E-05	-8.57E-09
3	8.00E-05	-7.64E-09
4	9.00E-05	-6.49E-09
5	1.00E-04	-5.10E-09
6	1.10E-04	-3.49E-09
7	1.20E-04	-1.65E-09
8	1.30E-04	3.98E-10
9	1.40E-04	2.67E-09
10	1.50E-04	5.15E-09
11	1.60E-04	7.85E-09
12	1.70E-04	1.08E-08
13	1.80E-04	1.39E-08

Figure 15-2. Stage 1 in solution by successive approximations

	C	D
7	1.20E-04	-1.65E-09
8	1.21E-04	-1.46E-09
9	1.22E-04	-1.26E-09
10	1.23E-04	-1.06E-09
11	1.24E-04	-8.60E-10
12	1.25E-04	-6.55E-10
13	1.26E-04	-4.49E-10
14	1.27E-04	-2.40E-10
15	1.28E-04	-2.97E-11
16	1.29E-04	1.83E-10
17	1.30E-04	3.98E-10

Figure 15-3. Stage 2 in solution by successive approximations

	E	F
15	1.280E-04	-2.97E-11
16	1.281E-04	-8.52E-12
17	1.282E-04	1.27E-11

Figure 15-4. Stage 3 in solution by successive approximations

You've now established that the solubility is between 1.28×10^{-5} and 1.29×10^{-5} M. Repeating the process a third time with an initial value of 1.280×10^{-4} and an increment of 1×10^{-7} yields a value for the solubility between 1.281 and 1.282×10^{-4} (Figure 15-4). From the magnitude of y in cell F16 compared to F15 and F17, you can say that the solubility is close to 1.281×10^{-4} M.

THE NEWTON-RAPHSON METHOD

Instead of simply changing the value of x in regular increments until the function y approaches zero, the *Newton-Raphson method* uses the slope of the function at an initial estimate of the root, x_1, to obtain an improved estimate of the root, x_2. Figure 15-5 illustrates an expanded portion of the graph of the function $y = x^3 - 3x^2 - 13x + 15$, between $x = 4$ and $x = 8$. The value $x = 7$ has been chosen as the initial estimate for finding a root of the function. Here's how an improved estimate for the root is calculated. The slope of the curve at x_1 is $y_1/(x_1-x_2)$. Therefore the improved estimate of the root is given by $x_2 = x_1 - (y_1/\text{slope})$. The process is repeated until no appreciable change in x occurs.

In pencil-and-paper calculations, the slope of the function would be obtained by calculating its first derivative, i.e., $dF/dx = 3x^2 - 6x - 13$ in this particular case. In spreadsheet calculations, however, the slope is conveniently obtained by numerical differentiation, i.e., slope $= \Delta F(x)/\Delta x$, as illustrated in the spreadsheet fragment shown in Figure 15-6.

The formula in cell E35 is

=(B36-B35)/(A36-A35)

and the formula in cell F35 is

=A35-(B35/E35).

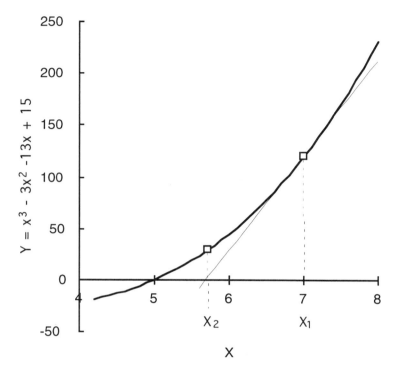

Figure 15-5. The Newton-Raphson method used to obtain the roots of the equation $y = x^3 - 3x^2 - 13x + 15$.

Unless Δx is very small, the slope calculated this way will not be the true slope, but the Newton-Raphson method will converge to the root anyway. The Newton-Raphson method converges rapidly; for the preceding example,

$x_1 = 7$,

$x_2 = 5.7$,

$x_3 = 5.13$,

$x_4 = 5.006$,

$x_5 = 5.00001$.

	A	B	E	F
5	**X**	**Y**	**slope**	**X(new)**
33	6.8	102.312		
34	6.9	110.979		
35	7.0	120	93.81	5.72081868
36	7.1	129.381		

Figure 15-6. Portion of a spreadsheet showing calculation of an improved estimate of a root of a function by the Newton-Raphson method.

If a function has more than one real root, the particular root to which the Newton-Raphson method converges will depend on the initial estimate chosen. In the example used here, if $x_1 = -5$ is chosen, the method converges to the root $x = -3$. Thus, to obtain the desired root, some guidance must be provided by the user.

A MACRO FOR THE NEWTON-RAPHSON METHOD

The macro shown in Figure 15-9 implements the Newton-Raphson method. This macro was written more as an exercise than as a tool, since Excel provides the **Goal Seek** option (described later in this chapter), which does the same thing. However, it demonstrates the techniques of reading values from a worksheet into a macro sheet and sending values from a macro sheet back to a worksheet.

To use the macro, you select a 1R × nC range in a worksheet. The leftmost cell in the range must contain the x value, and the rightmost cell the function. When the macro is run, the values of x and $F(x)$ change rapidly for a second or two until the value of $F(x)$ becomes zero. The value of x displayed is now the root of the function. Figures 7 and 8 illustrate the use of the macro to obtain the solubility of barium carbonate. In the worksheet you created earlier for the determination of the solubility of barium carbonate illustrated in Figure 15-4, copy the formula from cell **F17** into cell **F20**. As a starting estimate of x, enter the value 1 in cell **E20**.

Then select cells **E20:F20**, choose the **Macro** menu command, and double-click on the macro **Newton-Raphson!for_single_cell**. After a few cycles of iteration, the result $x = 0.000121814$ is displayed.

If the macro continues to run for many seconds, you have almost certainly entered the formula of a polynomial that has no real roots. Press COMMAND+period to stop the macro.

THE MACRO. Cells **B6-B8** save the row and column number of the leftmost cell of the selection, and the number of columns in the range. Cells **B9** and **B10** obtain the x_1 and y_1 values. Cell **B11** tests to see if y_1 is less than 1E-15; if so, the root has been found and the macro returns to the calling sheet. Otherwise, cell **B12** calculates the increment for numerical differentiation. If the increment is zero, it is set to 1E-10. Cell **B14** uses the FORMULA function to replace the

	E	F
19	X	Y
20	1	0.99999899

Figure 15-7. Cells selected for use with the Newton-Raphson macro. An arbitrary value of x has been chosen as the starting value.

	E	F
19	X	Y
20	0.00012814	3.4274E-16

Figure 15-8. Values of x and y resulting after the Newton-Raphson macro has been run.

	A	B
1		**Newton_Raphson**
2		Gets a single root by N-R method using numerical differentiation
3		Seeks positive root less than initial estimate.
4		
5		**for_single_cell**
6	col	=COLUMN(SELECTION())
7	cols	=COLUMNS(SELECTION())
8	row	=ROW(SELECTION())
9	X.1	=OFFSET(SELECTION(),0,0,1,1)
10	Y.1	=OFFSET(SELECTION(),0,cols-1,1,1)
11		=IF(ABS(Y.1)<0.000000000000001,RETURN())
12	incr	=0.01*OFFSET(SELECTION(),0,0,1,1)
13		=IF(OFFSET(SELECTION(),0,0,1,1)=0,SET.VALUE(incr,0.0000000001))
14		=FORMULA(X.1+incr,"R"&row&"C"&col)
15	Y.2	=OFFSET(SELECTION(),0,cols-1,1,1)
16	X.2	=X.1-Y.1/((Y.2-Y.1)/(incr))
17		=FORMULA(X.2,"R"&row&"C"&col)
18		=GOTO(B9)
19		

Figure 15-9. Newton-Raphson macro.

original value of x by the incremented value, and cell B15 retrieves the new value of y from which the slope of the curve will be calculated. Cell B16 calculates the improved value of x (the calculation of the slope and the calculation of x_2 are combined in one formula) and cell B17 returns the value to the spreadsheet.

SOLVING A PROBLEM USING GOAL SEEK...

Excel provides a built-in way to perform Newton-Raphson approximations: the **Goal Seek** command in the **Tools** menu (Excel 5.0) or the **Formula** menu (Excel 4.0). **Goal Seek** varies the value of a selected cell (the *changing cell*) to make the value of another cell (the *target cell*) reach a desired value.

To illustrate, let's repeat the calculation of the solubility of barium carbonate. **Goal Seek** allows you to obtain the same result much more easily. Open a new worksheet and in cell A1 enter the value 1. In cell B1, enter the formula =A1^2 - 1.0E-6*SQRT(A1)-5.1E-9. The task now will be to use **Goal Seek...** to find the value in cell A1 that makes the function (in B1) equal to zero. The accuracy of the result will depend on the magnitude of the Maximum Change parameter, which you can adjust by choosing the **Calculation** tab in the **Options** command of the **Tools** menu (Excel 5.0) or the **Calculation** command in the **Options** menu (Excel 4.0). The default value is 0.001 You'll see in a moment that adjusting the Maximum Change parameter is critical when using **Goal Seek.** But for now, set the value of Maximum Change to 1E-12.

Select **Goal Seek** from the **Formula** or **Tools** menu. As shown in Figure 15-11, enter $B$1 in the Set Cell box (the cell reference will appear there if you selected that cell before choosing **Goal Seek...**). Put the cursor in the To Value box and enter the desired value, zero. Put the cursor in the By Changing Cell box and enter $A$1 by selecting the cell or by typing. Then click on OK.

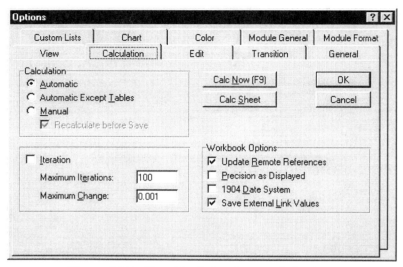

Figure 15-10. The Calculation Options dialog box.

After a few iteration cycles the Status dialog box (Figure 15-12) is displayed.

Adjusting the Maximum Change parameter is critical when using **Goal Seek.** That's because Excel stops iterating when the change in the result is less than the Maximum Change parameter. Therefore the Maximum Change parameter needs to be adjusted to match the value of the function. The data in Table 15-1 illustrate the importance of adjusting the Maximum Change parameter. For most chemical calculations it's a good idea to set Maximum Change to 1E-12 or 1E-15 as a matter of course.

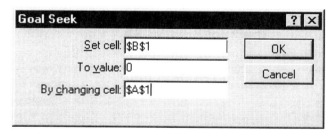

Figure 15-11. The Goal Seek dialog box.

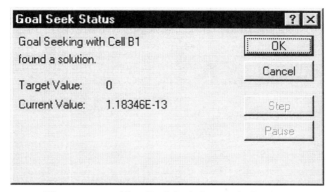

Figure 15-12. The Goal Seek Status dialog box.

Table 15-1. The Effect of Setting the Maximum Change Parameter

Trial #	Maximum Change	Final Value of F(x)*	Value of x
1	1.00E-03	5E-04	2.22E-02
2	1.00E-06	6E-07	7.82E-04
3	1.00E-09	3E-10	1.30E-04
4	1.00E-12	2E-13	1.281E-04
5	1.00E-15	4E-17	1.281E-04

* The final value of the function depends to some extent on the starting value of x.

For problems requiring the variation of two or more parameters, that is, varying the values of several cells to make the value of another cell reach a desired value, you must use the Solver, which is described in detail in Chapter 17.

SOLVING A PROBLEM BY CIRCULAR REFERENCE AND ITERATION

When a formula refers to itself, either directly or indirectly, it creates what is referred to as a *circular reference*. If a circular reference occurs, Excel issues a "Cannot resolve circular references" message (Figure 15-13).

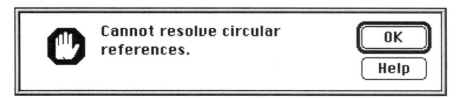

Figure 15-13. Circular References alert box.

Usually circular references occur unintentionally, because a user incorrectly entered a cell reference in an equation. But occasionally a problem can be solved by intentionally creating a circular reference.

An example of a problem that can be solved by circular reference and iteration is the calculation of the pH of a solution of a weak acid. Combining the equations for mass balance and K_a, we obtain the following equation for the $[H^+]$ of a weak acid of concentration C mol/L.

$$[H^+] = \sqrt{K_a (C - [H^+])} \qquad (15\text{-}5)$$

Provided the extent of dissociation is small (i.e., $[H^+] \ll C$), the $[H^+]$ term can be neglected and the equation reduces to:

$$[H^+] = \sqrt{K_a C} \qquad (15\text{-}6)$$

The majority of weak acid problems can be solved using the approximate equation. However, if the extent of dissociation is relatively large (typically, greater than 10%), the $[H^+]$ term cannot be neglected and equation 15-5, when multiplied and rearranged, yields a quadratic equation. These problems are usually solved by using the quadratic formula, but an alternative approach is to use the method of successive approximations. An initial estimate of $[H^+]$, obtained from equation 15-6, is subtracted from C, as in equation 15-5, to obtain an improved value of $[H^+]$. The process is repeated until there is no further significant change in $[H^+]$. Each cycle of calculation is referred to as an iteration.

As an example of a weak acid problem that can readily be solved by circular reference and iteration, consider the calculation of the pH of a 0.1000 M solution of sodium bisulfate, $NaHSO_4$. The bisulfate ion is a relatively strong weak acid with $K_a = 1.2 \times 10^{-2}$. (Roughly speaking, weak acids with K_a values greater than 10^{-4} will require the quadratic or successive approximations approach.) The spreadsheet fragment in Figure 15-14 illustrates a case where a cell contains a circular reference. Cell C5 contains the formula =SQRT($C$3*(A5-C5)).

An equivalent circular reference, in which a cell refers to itself indirectly, is shown in Figure 15-15. Cell C5 contains the formula =SQRT($C$3*B5) and cell B5 contains the formula =A5-C5.

	A	B	C
3	Bisulfate ion	Ka =	1.20E-02
4	[HA](initial)		[H⁺]
5	0.1		0.02916

Figure 15-14. Circular reference. Cell C5 refers to itself.

	A	B	C
3	Bisulfate ion	Ka =	1.20E-02
4	[HA](initial)	[HA](final)	[H⁺]
5	0.1	0.07084	0.02916

Figure 15-15. Circular reference. Cell C5 refers to a cell that refers to C5.

If the expression =SQRT($C$3*(A5-C5)) is typed into cell C5 in Figure 15-14, the "Cannot resolve circular references" message is displayed; when you click on OK in the dialog box, a zero value is displayed in the cell.

The **Calculation**... option allows you to select and control iteration of circular references. After displaying the Calculation Options dialog box (see Figure 15-10), click on the Iteration check box. Recalculation of cells containing circular references will occur either until the number of cycles of calculation shown in the Maximum Iterations box have been performed or until all cells change by an amount less than that shown in the Maximum Change box. (The default settings are Maximum Iterations = 100 and Maximum Change = .001.) Change the value in the Maximum Change box to 0.000001

When you press the OK button, the worksheet will be recalculated until (in this case) the change in the cells is less than 0.00001.

For many chemistry calculations, this limit will not be sufficiently small. You will often need to reduce the Maximum Change parameter; you will rarely have to increase the Maximum Iterations parameter.

The current iteration number is displayed in the formula bar. To terminate calculations, press COMMAND+(period) or ESC (Macintosh) or ESC (Windows).

SOLVING SETS OF SIMULTANEOUS LINEAR EQUATIONS

Sometimes a chemical system can be represented by a set of n linear equations in n unknowns, i.e.,

$$a_{11}x_1 + a_{12}x_2 + a_{13}x_3 + \cdots = c_1$$
$$a_{21}x_1 + a_{22}x_2 + a_{23}x_3 + \cdots = c_2$$
$$\vdots$$
$$a_{n1}x_1 + a_{n2}x_2 + a_{n3}x_3 + \cdots = c_n$$

where x_1 , x_2 , x_3 , $\cdots$ are the experimental unknowns, c is the experimentally measured quantity, and the a_{ij} are coefficients. The equations must be linearly independent; in other words, no equation is simply a multiple of another. These equations can be represented in matrix notation by

$$\mathbf{AX = C} \qquad\qquad (15\text{-}8)$$

A familiar example is the spectrophotometric determination of the concentrations of a mixture of n components by absorbance measurements at n different wavelengths. The coefficients a_{ij} are the ε, the molar absorptivities of the components at different wavelengths (for simplicity, the cell path length, usually 1.00 cm, has been omitted from these equations). For example, for a mixture of three species P, Q and R, where absorbance measurements are made at λ_1, λ_2 and λ_3, the equations are:

$$\varepsilon_{\lambda_1}^{P}[P] + \varepsilon_{\lambda_1}^{Q}[Q] + \varepsilon_{\lambda_1}^{R}[R] = A\lambda_1$$

$$\varepsilon_{\lambda_2}^{P}[P] + \varepsilon_{\lambda_2}^{Q}[Q] + \varepsilon_{\lambda_2}^{R}[R] = A\lambda_2$$

$$\varepsilon_{\lambda_3}^{P}[P] + \varepsilon_{\lambda_3}^{Q}[Q] + \varepsilon_{\lambda_3}^{R}[R] = A\lambda_3$$

Thus nine coefficients are required for the determination of three unknown concentrations.

CRAMER'S RULE

According to Cramer's rule, a system of simultaneous linear equations has a unique solution if the determinant D of the coefficients is non-zero.

$$D = \begin{vmatrix} a_{11} & a_{12} & a_{13} \\ a_{21} & a_{22} & a_{23} \\ a_{31} & a_{32} & a_{33} \end{vmatrix}$$

Thus, for example, for the set of equations

$$2x + y - z = 0$$
$$x - y + z = 6$$
$$x + 2y + z = 3$$

the determinant is

$$\begin{vmatrix} 2 & 1 & -1 \\ 1 & -1 & 1 \\ 1 & 2 & 1 \end{vmatrix}$$

The coefficients and constants lend themselves readily to spreadsheet solution, as illustrated in Figure 15-16.

Using the formula =MDETERM(A2:C4), the value of the determinant is found to be –9, indicating that the system is soluble.

	A	B	C	D
1	Coefficients			Constants
2	2	1	-1	0
3	1	-1	1	6
4	1	2	1	3

Figure 15-16. Spreadsheet data for three equations in three unknowns.

	A	B	C
8	0	1	-1
9	6	-1	1
10	3	2	1

Figure 15-17. The determinant for obtaining x.

The x values that comprise the solution of the set of equations can be calculated in the following manner: x_k is given by a quotient in which the denominator is D and the numerator is obtained from D by replacing the kth column of coefficients by the constants c_1, c_2, $\cdots$. The unknowns are obtained readily by copying the coefficients and constants to appropriate columns in another location in the sheet. For example, to obtain x, the determinant is shown in Figure 15-17 and $x = 2$ is obtained from the formula

=MDETERM(A8:C10)/MDETERM(A2:C4)

$y = -1$ and $z = 3$ are obtained from appropriate forms of the same formula.

SOLUTION USING MATRIX INVERSION

If equation 15-8 is multiplied by the inverse of $\mathbf{A}$, we obtain the relationship

$$\mathbf{X} = \mathbf{A}^{-1}\,\mathbf{C} \tag{15-9}$$

In other words, the solution matrix is obtained by multiplying the matrix of constants by the inverse matrix of the coefficients. To return the solution values shown in Figure 15-18, the array formula

{=MMULT(MINVERSE(A2:C4),D2:D4)}

was entered in cells E2:E4.

	A	B	C	D	E
	Coefficients			Constants	Solution
1					
2	2	1	-1	0	2
3	1	-1	1	6	-1
4	1	2	1	3	3

Figure 15-18. Solving a set of simultaneous equations by means of matrix methods.

16

LINEAR REGRESSION

Excel provides several ways to find the coefficients that provide the best fit of a function to a set of data points — a process sometimes referred to as *curve fitting*. The "best fit" of the curve is considered to be found when the sum of the squares of the deviations of the data points from the calculated curve is a minimum. In the field of statistics, finding the least-squares best-fit parameters that describe a data set is known as *regression analysis*. In this chapter you'll learn how to perform simple and multiple linear regression, using Excel's LINEST function.

LEAST-SQUARES CURVE FITTING

Regression analysis is a statistical technique used to determine whether experimental variables are interdependent, and to express the relationship between them, and the degree of correlation, in a quantitative manner. For most chemical systems, the mathematical form of the equation relating the dependent and independent variables is known; the goal is to obtain the values of the coefficients in the equation — the *regression coefficients*. In other cases, the data may be fitted by a power series, simply for purposes of graphing or interpolation. In any event, you must provide the form of the equation; regression analysis merely provides the coefficients.

A secondary but no less important goal is to obtain the standard deviations[*] of the regression parameters, and an estimate of the goodness of fit of the data to the model equation.

The *method of least-squares* yields the parameters which minimize the sum of squares of the residuals (the deviation of each measurement of the dependent variable from its calculated value).

$$SS_{resid} = \sum_{n=1}^{N}(y_{obsd} - y_{calc})^2 \qquad (16\text{-}1)$$

Linear regression is not restricted merely to straight-line relationships, but

---

[*] In Chapters 16 and 17, the symbol σ is used for the population standard deviation (i.e., when the sample size is large) and the symbol s for the sample standard deviation (when the sample size is small).

refers to any relationship that is *linear in the regression coefficients*, that is, any relationship of the form $y = a_0 + a_1 x_1 + a_2 x_2 + a_3 x_3 + \cdots$. The x_i can be different independent variables (e.g., pressure, temperature, time) or functions of a single independent variable (e.g., $[H^+], [H^+]^2, [H^+]^3$).

LEAST-SQUARES FIT TO A STRAIGHT LINE

One of the most common methods of treating a set of x, y data points is to draw a straight line through them. Although it is relatively easy to draw a straight line through a series of points if they all fall on or near the line, it becomes a matter of judgment if the data are scattered. The method of least-squares provides the best method for objectively determining the best straight line through a series of points. In its simplest form, the least-squares approach assumes that all deviations from the line are the result of error in the measurement of the dependent variable y. Excel provides several worksheet functions that return regression parameters of the least-squares best fit of the straight line $y = mx + b$ to a data set.

THE SLOPE AND INTERCEPT FUNCTIONS

The worksheet functions SLOPE(*known_ys, known_xs*) and INTERCEPT(*known_ys, known_xs*) return the slope, m, and intercept, b, respectively of the least-squares straight line through a set of data points. For example, Figure 16-1 illustrates some spectrophotometric calibration curve data (concentration of potassium permanganate standards in column B, absorbance of the standards in column C). The formula =SLOPE(C4:C8,B4:B8) in cell C10 was used to obtain the slope of the straight-line calibration curve. The SLOPE and INTERCEPT functions should be used with some caution, since they do not provide a measure of how well the data conforms to a straight line relationship. At the least, a chart of the data should be produced for visual inspection of the fit, as illustrated in Figure 16-2.

Column D in Figure 16-1 contains the absorbance values calculated from the slope and intercept; for example, cell D4 contains =$C$10*B4 + $C$11.

	A	B	C	D	E
1	Calibration curve of potassium permanganate solutions				
2					
3		C, M	Abs	A(calc)	
4		0.000E+00	0.000	0.002	
5		1.029E-04	0.257	0.258	
6		2.058E-04	0.518	0.513	
7		3.087E-04	0.771	0.769	
8		4.116E-04	1.021	1.025	
9					
10		slope =	2.484E+03		
11		intercept =	0.0022		
12					

Figure 16-1. Using SLOPE and INTERCEPT functions.

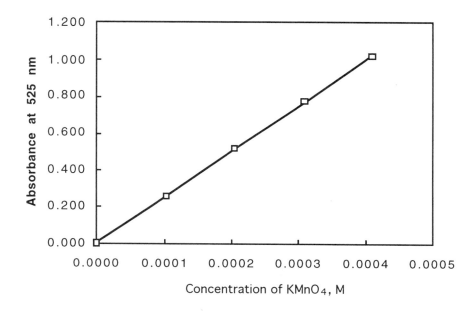

Figure 16-2. Least-squares best fit line through data points of a calibration curve.

LINEAR REGRESSION USING LINEST

The worksheet function LINEST performs linear regression analysis on a set of x,y data points. (LINEST stands for **LIN**ear **EST**imation, not **LINE ST**raight.) The general form of the linear equation that can be handled by LINEST is

$$y = m_1x_1 + m_2x_2 + m_3x_3 + \cdots + b \tag{16-2}$$

LINEST returns the array of regression parameters $m_n, \cdots, m_2, m_1, b$. The syntax is LINEST(**known_ys**, *known_xs, const_logical, stats_logical*). If *const_logical* is TRUE, 1 or omitted, the regression parameters include an intercept b; if *const_logical* is FALSE or 0, the fit does not include the intecept b. If *stats_logical* is TRUE or 1, LINEST returns an array of regression statistics in addition to the regression coefficients $m_n, \cdots m_1$ and b. The layout of the array of returned values is shown in Figure 16-3. A 1-, 2-, 3-, 4-, or 5-row array may be selected.

m(n)	m(n-1)	...	m(2)	m(1)	b
std.dev(n)	std.dev(n-1)	...	std.dev(2)	std.dev(1)	std.dev(b)
R^2	SE(y)				
F	df				
SS(reg)	SS(resid)				

Figure 16-3. Layout of regression results and statistics returned by LINEST.

Mathematical relationships between the regression parameters are given below (N = number of data points, k = number of regression coefficients to be determined):

$$df \text{ (degrees of freedom)} = N - k \tag{16-3}$$

$$SS_{regression} = \Sigma(y_{mean} - y_{calc})^2 \tag{16-4}$$

$$SS_{residuals} = \Sigma(y_{obsd} - y_{calc})^2 \tag{16-5}$$

$$R^2 = 1 - \frac{SS_{resid}}{SS_{regression}} \tag{16-6}$$

$$F = \frac{SS_{regression}}{SS_{resid}/df} \tag{16-7}$$

$$SE(y) = \sqrt{\frac{SS_{resid}}{N-k}} \tag{16-8}$$

The coefficient of determination, R^2 (or the correlation coefficient, R), is a measure of the goodness of fit of the data to (in this case) a straight line. If x and y are perfectly correlated (i.e., the difference between y_{obsd} and y_{calc} is zero) then $R^2 = 1$. In contrast, an R^2 value of zero means that there is no correlation between x and y. A value of R^2 of less than 0.9 corresponds to a rather poor fit of data to a straight line.

The $SE(y)$ parameter, the standard error of the y estimate, is sometimes referred to as the RMSD (root-mean-square deviation).

The F-statistic is used to determine whether the proposed relationship is significant (that is, whether y does in fact vary with respect to x). For most relationships observed in chemistry, there will be no question that there is a relationship. If it is necessary to determine whether the variation of y with x is statistically significant, or merely occurs by chance, you should consult a book on statistics.

The regression coefficients, the standard deviations of the coefficients, and R^2, the coefficient of determination, are the statistical parameters of most interest to chemists.

LEAST-SQUARES FIT OF $y = mx + b$

The most common application of LINEST is to find the best straight line through a set of data points, i.e., finding the regression parameters m and b of the best straight line $y = mx + b$ through the data points.

To use the LINEST function, select a 1-5 R × 2 C array on the spreadsheet, depending on the regression parameters that you want returned. The array of regression parameters and statistics returned by LINEST when a 3R × 2C array is selected is shown in Figure 16-4. Then enter the LINEST function, followed by the appropriate values for *known_ys* and *known_xs*. Enter TRUE or 1 for *constant_logical* and for *stats_logical*. Since LINEST is an array function, after typing the closing parenthesis of the function, press COMMAND+ENTER

m	b
std. dev of m	std. dev of b
R^2	SE(y)

Figure 16-4. Table of regression statistics returned by LINEST for slope and intercept of a straight line.

	B	C	D
13		*From LINEST:*	
14		m	b
15	*parameters*	2.484E+03	0.0022
16	*std. dev.s*	1.17E+01	0.0030
17	*R^2, SE(y)*	0.99993	0.00381

Figure 16-5. Slope and intercept of a straight line, with regression statistics.

(Macintosh) or CONTROL+SHIFT+ENTER (Windows). The regression parameters and statistics are returned, as shown in Figure 16-5. The returned values were formatted to display an appropriate number of significant figures.

REGRESSION LINE WITHOUT AN INTERCEPT

If the LINEST argument *const_logical* is set to FALSE, the m coefficient of the line of the form $y = mx$ that best fits the data is returned. Applying the LINEST function to the permanganate data yields the parameters shown in Figure 16-6.

LINEAR REGRESSION USING THE ANALYSIS TOOLPAK

Linear regression can also be performed using the Add-in package called the Analysis ToolPak. Choose **Data Analysis...** from the **Tools** menu in Excel 5.0 or **Analysis Tools...** from the **Options** menu in Excel 4.0. Choose Regression from the Analysis Tools list box. The Regression dialog box (Figure 16-7) will prompt you to enter the range of dependent variable (y) values, the range of independent variable (x) values, as well as whether the constant is zero, whether the first cell in each range is a label, and the confidence level desired in the output summary. Then select a range for the summary table. You need only select a single cell for this range; it will be the upper-left corner of the range. You can also request a table of residuals and a normal probability plot. If you select a cell or range such that the summary table will over-write cells containing values, you will get a warning message.

	B	C	D
13		*From LINEST:*	
14		m	b
15	*parameters*	2.491E+03	0
16	*std. dev.s*	6.38E+00	#N/A
17	*R^2, SE(y)*	0.99992	0.00359

Figure 16-6. Slope of a straight line through the origin, with regression statistics.

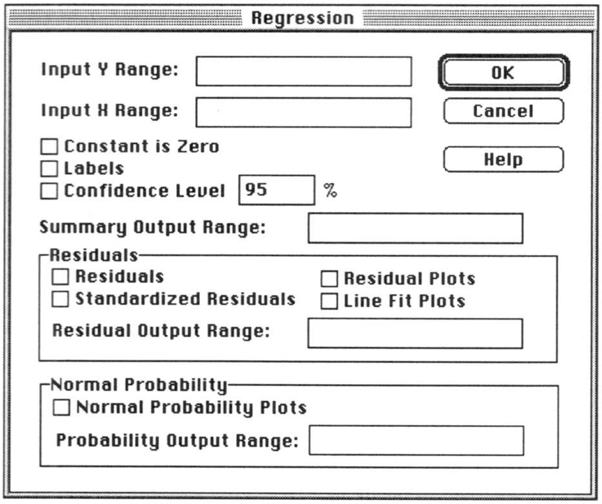

Figure 16-7 Regression dialog box.

In contrast to the results returned by LINEST, the output is clearly labeled, and some additional statistical data is provided. Regression data for the example shown in Figure 16-1 is shown in the three tables of Figure 16-8. Three tables are produced: regression statistics, analysis of variance, and regression coefficients. (The coefficients table has been broken into two parts to fit the page.)

Regression Statistics	
Multiple R	0.999966633
R Square	0.999933268
Adjusted R Square	0.999911024
Standard Error	0.003812261
Observations	5

(A)

Analysis of Variance					
	df	Sum of Squares	Mean Square	F	Significance F
Regression	1	0.6533136	0.6533136	44952.7706	2.3137E-07
Residual	3	4.36E-05	1.4533E-05		
Total	4	0.6533572			

(B)

	Coefficients	Standard Error	t Statistic
Intercept	0.0022	0.00295296	0.74501401
x1	2483.96501	11.715673	212.020684

(C)

P-value	Lower 95%	Upper 95%
0.49765621	-0.0071977	0.01159766
2.9688E-09	2446.68048	2521.24955

Figure 16-8. Data obtained by using Regression from the Analysis ToolPak: (A) Regression statistics, (B) Analysis of Variance,(C) Regression coefficients and statistics.

MULTIPLE LINEAR REGRESSION

As is clear from equation 16-2, LINEST can be used to find the least-squares coefficients for equations involving more than one x variable. In the example that follows, LINEST was used to find the correlation between the wavelength of the visible absorption band of square-planar copper(II) complexes of peptide, amino acid and polyamine ligands and the identity of the four ligand donor groups coordinated to the copper(II) ion. According to the "rule of average environment", v_{max}, the frequency or energy of the absorption band (as wavenumber), can be expressed as the sum of the ligand-field contributions of the four individual donor atoms in the complex. The wavenumber, in reciprocal centimeters, is used instead of wavelength since v_{max} is proportional to the ligand-field splitting caused by the donor atoms; $v_{max} = 10^8/\lambda_{max}$. The donor atoms considered in the study consisted of N(amino), N(peptide), O(carboxylate), O(peptide), H_2O and OH^-. Preliminary analysis indicated that the ligand-field strengths of O(peptide), H_2O and OH^- are identical, and these were combined in the treatment that follows. Thus

$$v_{max} = \sum_{i=1}^{4} n_i v_i \qquad (16\text{-}9)$$

and the regression equation is

$$v_{calc} = v_1 n_1 + v_2 n_2 + v_3 n_3 + v_4 n_4 \qquad (16\text{-}10)$$

where the n_i are the numbers of each type of donor atom in the complex (n_4 = total number of O(peptide), H_2O, OH^- donor atoms). The n_i are the independent variables and the v_i are the regression coefficients to be determined.

The data table (Figure 16-9) has been compressed by hiding rows containing some of the less-interesting data. To obtain the regression parameters, a 3R $\times$ 4C array was selected on the spreadsheet and the formula =LINEST(G5:G37,B5:E37,0,1) was entered. Since an array is being returned, the array formula was entered by using COMMAND+ENTER (Macintosh) or CONTROL+SHIFT+ENTER (Windows). . The array of returned values is shown in Figure 16-10.

	A	B	C	D	E	F	G	H
1	Analysis of Copper(II) Chromophores							
2		#N	#N	#O	#O			
3		(amino)	(peptide)	(carboxy)	(H_2O)	λ_{max},nm	ν_{max},cm^{-1}	λ_{max},nm
4	Complex	n_1	n_2	n_3	n_4	Observed	Observed	Predicted
5	CuGG$^+$	1	0	0	3	735	13.61	738
6	CuGGG$^+$	1	0	0	3	730	13.70	738
8	CuGa^{2+}	1	0	0	3	725	13.79	738
9	CuGGa^{2+}	1	0	0	3	725	13.79	738
11	Cu(Ga)$_2$	2	0	0	2	665	15.04	662
12	Cu gly$^+$	1	0	1	2	715	13.99	713
13	Cu(gly)$_2$	2	0	2	0	617	16.21	623
14	CuH$_{-1}$GG	1	1	1	1	635	15.75	630
15	CuH$_{-1}$GGG	1	1	0	2	660	15.15	650
17	CuH$_{-1}$GGa$^+$	1	1	0	2	660	15.15	650
19	Cu(Ga)(H$_{-1}$Ga)	2	1	0	1	605	16.53	590
20	Cu en^{2+}	2	0	0	2	660	15.15	662
22	CuH$_{-2}$GGG$^-$	1	2	1	0	555	18.02	564
24	CuH$_{-2}$GGa	1	2	0	1	580	17.24	580
26	CuH$_{-2}$GGGOH$^-$	1	2	0	1	578	17.30	580
27	CuH$_{-2}$GGaOH	1	2	0	1	575	17.39	580
28	CuH$_{-1}$GGOH	1	1	1	1	650	15.38	630
29	Cu dien^{2+}	3	0	0	1	611	16.37	600
30	Cu(H$_{-1}$Ga)$_2$	2	2	0	0	540	18.52	532
31	CuH$_{-3}$GGGG	1	3	0	0	520	19.23	524
34	Cu(en)$_2$$^{2+}$	4	0	0	0	549	18.21	549
35	Cu(1,2-DAP)$_2$$^{2+}$	4	0	0	0	545	18.35	549
37	Cu(tetmeen)$_2$$^{2+}$	4	0	0	0	547	18.28	549

Figure 16-9. Data table for multiple linear regression analysis of spectra of copper(II) complexes by LINEST. Ligand abbreviations: GG, glycylglycinate; Ga, glycinamide; gly, glycinate; en, ethylenediamine; dien, diethylenetriamine; 1,2-DAP, 1,2-diaminopropane; tetmeen, tetramethylethylenediamine. Data from E. J. Billo, *Inorg. Nucl. Chem. Lett.* **1974**, 10, 613.

The regression parameters can be used to predict the absorption maxima of other complexes, as illustrated in Figure 16-11.

37	Results from LINEST				
38		ν_4	ν_3	ν_2	ν_1
39		O(H_2O)	O(carboxy)	N(peptide)	N(amino)
40	coeffs	3.00	3.47	4.85	4.55
41	std.devs	0.03	0.09	0.04	0.03
42	R^2, SE(y)	0.985	0.24	#N/A	#N/A

Figure 16-10. Regression results and statistics returned by LINEST.

45	Predicted Absorption Maxima of Some Copper(II) Complexes (nm)					
46		n_1	n_2	n_3	n_4	calc	obsd
47	$Cu(H_2O)_4{}^{2+}$	0	0	0	4	834	808
48	$Cu(ala)_2$	2	0	2	0	623	620
49	$Cu(en)(Ga)^{2+}$	3	0	0	1	600	600
50	$Cu(en)(H_{-1}Ga)^+$	3	1	0	0	540	550
51	$Cu(en)(ala)^+$	3	0	1	0	584	582
52	$Cu(H_{-2}biuret)_2$	0	4	0	0	516	505

Figure 16-11. Using LINEST results to predict other values.

LINEAR REGRESSION ON EQUATIONS OF HIGHER ORDER

LINEST can also be used to find the regression coefficients for equations of higher order. It is sometimes convenient, in the absence of a suitable equation, to fit data to a power series. The equation can then be used for data interpolation. Often a power series $y = a + bx + cx^2 + dx^3$ is sufficient to fit data of moderate curvature. The lowest order polynomial that produces a satisfactory fit should be used; if there are N data points, the highest order polynomial that can be used is of order $(N - 1)$.

The example presented in Figure 16-12 fits the solubility of oxygen in water as a function of temperature over the range 0-100°C. Columns for the T^2 and T^3 independent variables were inserted and the solubility data in column D were fitted to a cubic equation using LINEST.

	A	B	C	D
1	Solubility of Oxygen in Water			
2	$(g\, O_2/100\, g\, H_2O)$			
3				
4	T, deg C	T^2	T^3	S, g
5	0	0	0	0.006945
6	5	25	125	0.006072
7	10	100	1000	0.005368
8	15	225	3375	0.004802
9	20	400	8000	0.004339
10	25	625	15625	0.003931
11	30	900	27000	0.003588
12	35	1225	42875	0.003315
13	40	1600	64000	0.003082
14	45	2025	91125	0.002858
15	50	2500	125000	0.002657
16	60	3600	216000	0.002274
17	70	4900	343000	0.001856
18	80	6400	512000	0.001381
19	90	8100	729000	0.00079
20	100	10000	1000000	0

Figure 16-12. Fitting O_2 solubility in water by a power series. (Data reprinted with permission from *CRC Handbook of Chemistry and Physics*. Copyright CRC Press, Boca Raton FL.)

22	*Results from LINEST*			
23	-1.3E-08	2.21E-06	-0.00016	0.00685682
24	5.65E-10	8.39E-08	3.4E-06	3.6788E-05
25	0.999456	4.93E-05	#N/A	#N/A

Figure 16-13. Least-squares parameters of a power series for O_2 solubility in water.

The regression parameters returned by LINEST are shown in Figure 16-13. All four parameters seem to be significant, since the standard deviations are, at the most, less than 5% of the parameter value.

Figure 16-14 shows the fit of the polynomial to the data points. Except for the data points for 80°C and above, the deviation of the calculated line from the data points is less than 2% .

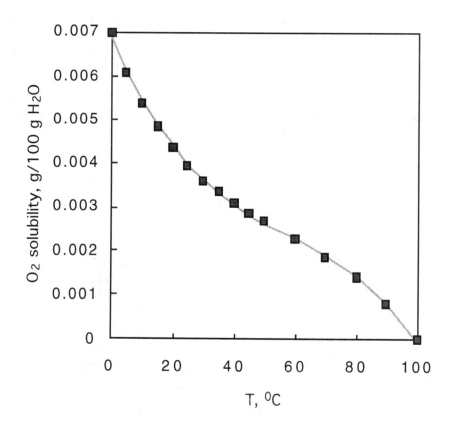

Figure 16-14. Fitting O_2 solubility in water by a power series.

WEIGHTED LEAST-SQUARES

If the y data to be fitted range over several orders of magnitude, it is sometimes advisable to use a *weighting factor* in the regression. If the error in the measurement is proportional to the magnitude of the measurement, the residuals of the largest measurements will have an overwhelming effect on the sum of squares. The weighting factor W is applied to each data point, so that the sum of squares is calculated according to equation 16-11.

$$\text{weighted } SS_{resid} = \sum_{n=1}^{N} W_n (y_n - y_{calc})^2 \qquad (16\text{-}11)$$

Some weighting functions that have been used include $W_n = 1/y_n$ and $W_n = 1/y_n^2$.

Weighting of the regression data can also be used if each y_n is an average of J_n observations. In this case the weighting factor is $W_n = J_n$.

USING THE REGRESSION STATISTICS

Regression statistics can be used to make decisions concerning the value or significance of regression coefficients — for example, whether the least-squares slopes of two lines are identical. This statistical procedure is known as *hypothesis testing*. The most common procedure is to set up a null hypothesis: for example, that the difference between the two slopes is zero. The answer — yes or no — must be stated within the context of a stated level of uncertainty, or confidence level. A confidence level of 95% is most commonly used; if the confidence level chosen, e.g., 99%, is too stringent, then a significant difference may be missed, while if the confidence level is too low, e.g., 50%, an insignificant difference may be accepted as real.

The t-test is used to test a null hypothesis. The computed t-statistic (calculated using one of the relationships given below) is compared with the value found in a table of t values corresponding to the appropriate number of degrees of freedom and at the desired confidence level. If the computed t-statistic exceeds the value from the table, the null hypothesis is rejected.

TESTING WHETHER AN INTERCEPT
IS SIGNIFICANTLY DIFFERENT FROM ZERO

For the permanganate calibration curve example examined earlier in this chapter, the intercept was found to be 0.0022 absorbance units with a standard deviation of 0.0030. Here it is obvious that the postulated zero intercept lies within the confidence interval for the intercept. But consider another case, where LINEST returns $b = 0.0011$ and $\sigma_b = 0.0005$; this calibration curve also contained $N = 5$ data points. Is the intercept statistically different from zero?

t-Table for Various Levels of Probability			
df	90%	95%	99%
1	6.31	12.70	63.70
2	2.92	4.30	9.92
3	2.35	3.18	5.84
4	2.13	2.78	4.60
5	2.02	2.57	4.03
10	1.81	2.23	3.17
15	1.75	2.13	2.95
20	1.72	2.09	2.84
30	1.70	2.04	2.75
60	1.67	2.00	2.66
∞	1.64	1.96	2.58

Figure 16-15. An abbreviated t-table.

To test whether the intercept b is equal to a given value b_0, compute the t-statistic

$$t = \frac{b - b_0}{\sigma_b} \tag{16-12}$$

and compare it to the t distribution with df degrees of freedom (for a straight line, $N - 2$ degrees of freedom).

For the second calibration curve

$$t = \frac{0.0011 - 0}{0.0005} = 2.2 \tag{16-13}$$

If you consult a table of t values in a statistics text (a portion of such a table is shown in Figure 16-15), you will find, for 95% probability and 3 degrees of freedom, that $t = 3.18$. Since the calculated t statistic is less than t-critical from the table, the intercept value of 0.0011 does not differ significantly from zero.

TESTING WHETHER TWO SLOPES ARE SIGNIFICANTLY DIFFERENT

To test whether the slopes of two different straight lines are equal, use the following equation:

$$t = \frac{m_2 - m_1}{\sigma_m(\text{pooled})} \tag{16-14}$$

The pooled estimate of the standard deviation of the slope is given as follows, provided the number of data points is large:

$$\sigma_m(\text{pooled}) = \sqrt{\sigma^2_{m1} + \sigma^2_{m2}} \tag{16-15}$$

If the numbers of data points for each line, n_1 and n_2, are not large, then you must use the following equation to calculate the pooled estimate of the standard deviation:

$s_m(\text{pooled}) =$

$$\sqrt{\frac{(n_1-2)\,SE^2_{y1} + (n_2-2)\,SE^2_{y2}}{(n_1-2)\,+(n_2-2)}}\;\sqrt{\frac{1}{(n_1-1)S^2_{x_1}}+\frac{1}{(n_2-1)S^2_{x_2}}} \qquad (16\text{-}16)$$

where S^2_{x1}, the variance of the x-values in data set 1, is given by:

$$S^2_{x1} = \frac{\sum x_1^2 - ((\sum x_1)^2\,/n_1)}{(n_1-1)} \qquad (16\text{-}17)$$

The variance can easily be obtained using the Excel function **VAR**. Equation 16-16 for the pooled estimate of the standard deviation is implemented, for example, using the Excel formula

=((N.1-2)*SEY1^2+(N.2-2)*SEY2^2)/(N.1+N.2-4)*(1/((N.1-1)*var1)+1/((N.2-1)*var2))

where N.1, SEY1 and var1 are the number of data points, the standard deviation of y, and the variance of the x-values of data set number 1, respectively.

TESTING WHETHER A REGRESSION COEFFICIENT IS SIGNIFICANT

In a multiple linear regression model using an empirical equation, you may want to know whether individual coefficients are significant, that is, whether they are useful in predicting the dependent variable. As in the previous examples, if the standard deviation of the coefficient is small relative to the coefficient, the coefficient is clearly significant, while if the standard deviation of the coefficient is larger than the coefficient itself, it is clear that the coefficient is much less significant.

If you divide the coefficient by its standard deviation, the result is the *t*-statistic for that coefficient. The *t*-statistic is used to test hypotheses about the value of the coefficient. In a multiple regression model, the value of a_n/SE_n shows the relative importance of each term in the model.

TESTING WHETHER REGRESSION COEFFICIENTS ARE CORRELATED

Occasionally, in a multiple linear regression model using an empirical equation, two or more independent variables are correlated among themselves, that is, one independent variable is a linear function of another independent variable. Intercorrelated variables are said to exhibit *multicollinearity*. A non-chemical example of multicollinearity might exist in the relationship between sales of a product and the demographics of the customer base. Perhaps it is found that both educational level and family income are good predictors of sales of the product, yet it is clear that these two independent variables are strongly correlated.

If two independent variables are perfectly correlated (i.e., one is an exact linear function of the other) then LINEST will return #NUM! in all cells. If they are only approximately correlated, then LINEST will return regression parameters that can be very misleading: R^2 can be very close to 1, but the coefficients may have no meaning. The SE's of one or more of the coefficients will probably be fairly large.

	A	B	C	D	E
1		*(amino)*	*(peptide)*	*(carboxy)*	*(H2O)*
2	*(amino)*	1			
3	*(peptide)*	-0.4132	1		
4	*(carboxy)*	-0.1193	-0.0863	1	
5	*(H2O)*	-0.5072	-0.4790	-0.2327	1

Figure 16-16. Degree of correlation of independent variables from the Analysis ToolPak.

THE MULTIPLE-CORRELATION COEFFICIENT

The Analysis ToolPak provides a convenient method for determining the correlation between independent variables. Choose Correlation from the Analysis Tools list box. The dialog box will prompt you to enter the range of independent variables; the dependent variables are not included. The output from Correlation, applied to the independent variable data from Figure 16-9, is shown in Figure 16-16. The output is a symmetrical matrix of correlation coefficients. The diagonal elements (correlation of a variable with itself) are all 1. The off-diagonal elements indicate the degree of correlation of two different independent variables. Unless a value is very close to 1.0, there is no significant correlation between independent variables.

CONFIDENCE INTERVALS FOR SLOPE AND INTERCEPT

The confidence limits for the slope of a straight line are given by $m \pm ts_m$, where t is obtained from the t-table for the desired confidence level and $(N - 2)$ degrees of freedom. Similarly the confidence limits for the intercept are $b \pm ts_b$. Using the LINEST results for the permanganate calibration curve, shown in Figure 16-5, and using $t = 3.18$ for 95% confidence level, the 95% confidence limits for the slope are $2.484 \times 10^3 \pm (3.18)(1.17 \times 10^1)$ or $(2.484 \pm 0.037) \times 10^3$.

CONFIDENCE LIMITS AND PREDICTION LIMITS
FOR A STRAIGHT LINE

The upper and lower *confidence limits* for y_{calc} at a particular value of x are given by equation 16-18. The confidence limits for y_{calc} are calculated by using a value of the t-statistic for a particular number of degrees of freedom and a particular probability level. Often the 90% probability level is used.

$$y_{calc} \pm t_{N-2} \sqrt{\frac{SS_{resid}}{N-2}} \sqrt{\frac{1}{N} + \frac{(x-\bar{x})^2}{(N-1)S^2_x}} \qquad (16-18)$$

The $\pm$ limits can be calculated using the Excel formula

=SQRT(SUM(d.sq)/(N-2))*SQRT(1/N+((X-X_bar)^2)/((N-1)*VAR(X)))

where d.sq is the residual squared, N is the number of observations, X_bar is the mean of the x-values, and VAR(X) returns the variance of the x-values.

The upper and lower *prediction limits* are given by the equation

$$y_{calc} \pm t_{N-2} \sqrt{\frac{SS_{resid}}{N-2}} \sqrt{1 + \frac{1}{N} + \frac{(x-\bar{x})^2}{(N-1)S^2_x}} \qquad (16\text{-}19)$$

which differs from equation 16-18 only in the extra 1 within the square root term. The prediction limits describe the confidence limits for predicting a single y-value for a particular x-value. The prediction limits contain two sources of error: the error in estimating y_{calc} and the error associated with a single measurement.

Very commonly, as in the calibration curve example, we use the measured y-value to estimate an x-value. Once the slope and intercept of a straight-line calibration curve have been established, it is easy to calculate an x-value (e.g., a concentration) from a measured y-value. The *estimation limits* of the estimated x-value are given by the equation

$$x_{meas} \pm t_{N-2} \frac{1}{b} \sqrt{\frac{SS_{resid}}{N-2}} \sqrt{1 + \frac{1}{N} + \frac{(y_{meas} - \bar{y})^2}{b^2 \Sigma (x_i - \bar{x})^2}} \qquad (16\text{-}20)$$

If y_{meas} is an average of M readings, then the equation becomes

$$x_{meas} \pm t_{N-2} \frac{1}{b} \sqrt{\frac{SS_{resid}}{N-2}} \sqrt{\frac{1}{M} + \frac{1}{N} + \frac{(y_{meas} - \bar{y})^2}{b^2 \Sigma (x_i - \bar{x})^2}} \qquad (16\text{-}21)$$

USEFUL REFERENCES

David G. Kleinbaum and Lawrence L. Kupper, *Applied Regression Analysis and Other Multivariable Methods*, Duxbury Press, Belmont, CA, 1978.

J. C. Miller and J. N. Miller, *Statistics for Analytical Chemistry*, 3rd ed., Ellis Horwood, New York , 1993.

17

NON-LINEAR REGRESSION USING THE SOLVER

In this chapter you'll learn how to use the Solver, Excel's powerful optimization package, to perform non-linear least-squares curve fitting.

NON-LINEAR FUNCTIONS

The function

$$y = F(x_1, x_2, \cdots, a_0, a_1, a_2, \cdots) \qquad (17\text{-}1)$$

where y and x_1, x_2, $\cdots$ are the dependent and independent variables, respectively, and a_0, a_1, a_2, $\cdots$ are coefficients, can be either *linear* or *non-linear* in the coefficients. A linear function is one of the type

$$y = a_0 + a_1 Z_1 + a_2 Z_2 + a_3 Z_3 + \cdots \qquad (17\text{-}2)$$

where the Z's are functions of the independent variables x_i. The word "linear" in linear regression does not mean that the function is a straight line, but that the partial derivatives with respect to each coefficient are not functions of other coefficients.

An example of a function that is linear in the coefficients is

$$y = a_0 + a_1 x + a_2 x^2 \qquad (17\text{-}3)$$

and it should be clear from reading Chapter 16 that LINEST can be used to obtain the regression coefficients for a data set that can be described by such an equation. However, if the function is one such as

$$y = \exp(a_0 + a_1 x) \qquad (17\text{-}4)$$

or

$$y = \exp(-a_1 x) - \exp(-a_2 x) \qquad (17\text{-}5)$$

then it is not linear in the coefficients, since the equation cannot be rearranged to obtain an expression containing separate $a_i Z_i$ terms. Some non-linear equations can be transformed into a linear form. Equation 17-4, for example, can be transformed by taking the logarithm to the base e of each side, to yield the equation

$$\ln y = a_0 + a_1 x \qquad (17\text{-}6)$$

which is linear. Others, such as 17-5, cannot be converted into a linear form and are said to be *intrinsically non-linear*.

USING THE SOLVER TO PERFORM NON-LINEAR LEAST-SQUARES CURVE FITTING

There are many published computer programs and commercial software packages that perform non-linear regression analysis. You can obtain the same results very easily using the Solver. When applied to the same data set, the Solver gives the same results as commercial software packages.

OPTIMIZATION USING THE SOLVER

The Solver is an optimization package that finds a maximum, minimum or specified value of a *target cell* by varying the values in one or several *changing cells*. It accomplishes this by means of an iterative process, beginning with trial values of the coefficients. The value of each coefficient is changed by a suitable increment, the new value of the function is calculated and the change in the value of the function is used to calculate improved values for each of the coefficients. The process is repeated until the desired result is obtained. The Solver uses gradient methods or the simplex method to find the optimum set of coefficients.

With the Solver you can apply constraints to the solution. For example, you can specify that a coefficient must be greater than or equal to zero, or that a coefficient must be an integer. Solutions to chemical problems will rarely use the integer option, and although the ability to apply constraints to a solution may be tempting, it can sometimes lead to an incorrect solution.

The Solver is an Add-in, a separate software package. To save memory, it may not automatically be opened whenever you start Excel. Open the Solver Add-In by choosing **Add-ins** from the **Tools** menu (Excel 5.0) or from the **Options** menu (Excel 4.0), check the box for the Solver, then press OK. If the Solver Add-in has been opened, you will see the **Solver...** command in the **Tools** menu (Excel 5.0) or the **Formula** menu (Excel 4.0). To use the Solver to perform multiple non-linear least-squares curve fitting, follow the procedure outlined in the accompanying box.

To Use the Solver to Perform
Non-Linear Least-Squares Curve Fitting

1. Start with a worksheet containing the data (independent variable x and dependent variable y_{obsd}) to be fitted.

2. Add a column containing y_{calc} values, calculated by means of an appropriate formula, and involving the x values and one or more coefficients to be varied.

3. Add a column to calculate the square of the residual ($y_{obsd} - y_{calc}$) for each data point.

4. Calculate the sum of squares of the residuals.

5. Use **Solver...** to minimize the sum of squares of residuals (the target cell) by changing the coefficients of the function (the changing cells).

Since the Solver operates by a search routine, it will find a solution most rapidly and efficiently if the initial estimates that you provide are close to the final values. Conversely, it may not be able to find a solution if the initial estimates are far from the final values. To ensure that the Solver has found a *global minimum* rather than a *local minimum*, it's a good idea to obtain a solution using different sets of initial estimates.

The least-squares regression coefficients that it returns may be slightly different, depending on the starting values that you provide.

USING THE SOLVER: AN EXAMPLE

The following example illustrates the ease with which the Solver can be used to perform non-linear least-squares curve fitting. Kinetics data (absorbance vs. time) from a biphasic reaction involving two consecutive first-order reactions $(A \rightarrow B \rightarrow C)$ are analyzed to obtain two rate constants and the molar absorptivity of the intermediate species B.

The equations for the concentrations of the species A, B and C in a reaction sequence of two consecutive first-order reactions can be found in almost any kinetics text. The expression for $[B]_t$ is

$$[B]_t = [A]_0 \frac{k_1}{k_2 - k_1} \{\exp(-k_1 t) - \exp(-k_2 t)\} \tag{17-7}$$

where $[A]_0$ is the initial concentration of the reactant species and k_1 and k_2 are the rate constants. This is an example of an intrinsically non-linear equation.

The kinetics data[*], for the reaction of a nickel(II) complex NiL_2^{2+} of a substituted bidentate diamine ligand (L = 2,3-dimethyl-2,3-diaminobutane) with cyanide ion, was obtained by using a stopped-flow spectrophotometer. The reaction is biphasic; one diamine ligand is replaced by two cyanide ligands, then the second diamine ligand is replaced:

$$NiL_2^{2+} + 2\,CN^- \rightarrow NiL(CN)_2 + L \tag{17-8}$$

$$NiL(CN)_2 + 2\,CN^- \rightarrow Ni(CN)_4^{2-} + L \tag{17-9}$$

The species $NiL(CN)_2$, which is formed and then decays during the reaction, absorbs at 243 nm, where NiL_2^{2+}, $Ni(CN)_4^{2-}$, CN^- or L does not absorb appreciably. The stopped-flow data (measured manually from an oscilloscope trace used to collect the data) are shown in columns A and B of Figure 17-1.

Using the expression for $[B]_t$ (equation 17-7) where $[A]_0$ is the initial concentration of the reactant species NiL_2^{2+}, and k_1 and k_2 are the rate constants for reactions in equations 17-7 and 17-8, together with Beer's law

$$A_{obsd} = \varepsilon b[B]_t \tag{17-10}$$

where A_{obsd} is the measured absorbance, ε is the molar absorptivity of the intermediate species $NiL(CN)_2$, and b is the path length of the stopped-flow cell, yields an expression for the absorbance as a function of time. The parameters ε, k_1 and k_2 are the regression parameters that we want to obtain by non-linear least-squares curve fitting.

---

[*]   From J. C. Pleskowicz and E. J. Billo, *Inorg. Chim. Acta* **1985**, *99*, 149.

In Figure 17-1, columns C and D contain the concentration of the intermediate species B and the absorbance, calculated using equations 17-9 and 17-10, respectively. The formulas in cells C12 and D12 are (the optical path length was 0.4 cm):

=C_A*k_1*(EXP(-k_2*t)-EXP(-k_1*t))/(k_1-k_2) and

=0.4*E_B*C12

In column E the squares of the residuals are calculated. The sum of squares is obtained in cell E27, using the Σ tool (select cell E27, click twice on the Σ tool).

The changing cells (B8, E5, E6) and the target cell (E27) are shown in bold. Initial values, estimated from the data, were 3000, 1 and 0.2, respectively.

To use the Solver, choose **Solver...** from the **Formula** menu. Since the Solver is an Add-in, a few seconds will be required while the Solver program is loaded. The Solver Parameters dialog box (Figure 17-2) will then be displayed.

	A	B	C	D	E
1	Rate of Reaction of Ni(tetmeen)$_2^{2+}$ with Cyanide				
2	Stopped-flow run #173-3 (page 72)				
3	Date:	4/8/96			
4	Operator:	JA Tyrell		*Rate constants:*	
5	Wavelength:	243 nm		$k_1 =$	**1.000**
6	Conc:	4.00E-05	(C_A)	$k_2 =$	**0.200**
7	E(A):	0			
8	E(B):	**3.00E+03**	(E_B)		
9	E(C):	0			
10	t, sec	A(obsd)	[B]	A(calc)	∂^2
12	0.2	0.0047	7.10E-06	0.0085	1.5E-05
13	0.6	0.0129	1.69E-05	0.0203	5.5E-05
14	1.0	0.0163	2.25E-05	0.0271	1.2E-04
15	1.4	0.0188	2.55E-05	0.0306	1.4E-04
16	1.8	0.0201	2.66E-05	0.0319	1.4E-04
17	2.2	0.0208	2.67E-05	0.0320	1.3E-04
18	2.6	0.0208	2.60E-05	0.0312	1.1E-04
19	3.0	0.0205	2.50E-05	0.0299	8.9E-05
20	4.0	0.0178	2.16E-05	0.0259	6.5E-05
21	5.0	0.0149	1.81E-05	0.0217	4.6E-05
22	6.0	0.0118	1.49E-05	0.0179	3.7E-05
23	7.0	0.0090	1.23E-05	0.0147	3.3E-05
24	8.0	0.0070	1.01E-05	0.0121	2.6E-05
25	9.0	0.0052	8.26E-06	0.0099	2.2E-05
26	10.0	0.0038	6.76E-06	0.0081	1.9E-05
27				$\Sigma(\partial^2) =$	1.03E-03

Figure 17-1. The spreadsheet before optimization of coefficients by the Solver.

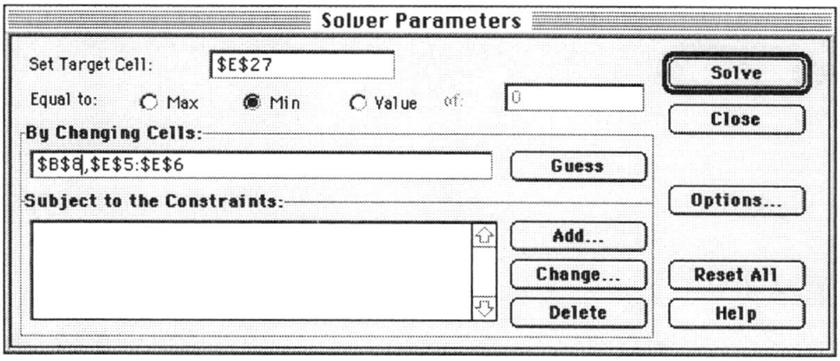

Figure 17-2. The Solver Parameters dialog box.

To solve the problem:

1. In the Set Cell box, type **E27**, or select cell **E27** with the mouse. (If you selected **E27** before running the Solver, $E$27 will appear in the Set Cell box.)

2. You want to minimize the sum of squares, so press the Min button.

3. Select cells **B8** and **E5:E6** so that they appear in the By Changing Cells box.

4. Do not enter any constraints.

5. Press the Solve button.

As the problem is set up and solved, messages will appear in the status bar at the bottom of the screen. The value of the target cell is displayed in the status bar after every iteration. In a few seconds the Solver finds a solution and displays the completion dialog box. You have the options of accepting the Solver's solution or restoring the original values. Press the Accept Solver Solution button. The spreadsheet will be displayed with the final values of the target cells (Figure 17-3).

> ***Excel Tip.*** *You should not introduce constraints (for example, to force a constant to be greater than or equal to zero) if you're using the Solver to obtain the least-squares best fit. The solution will not be the "global minimum" of the error-square sum, and the regression coefficients may be seriously in error.*

There are some additional controls in the Solver Parameters dialog box:

*   The Add..., Change... and Delete buttons are used to apply constraints to the model. Since the use of constraints is to be avoided, these buttons are not of much interest.

*   Pressing the Guess button will enter references to all cells that are precedents of the target cell. In the example above, pressing the Guess button enters the cell references A$12:B$26, $B$6, $B$8, $E$5, $E$6 in the By Changing Cells box.

*   The current Solver model is automatically saved with the worksheet. The Reset All button permits you to "erase" the current model and begin again.

	A	B	C	D	E
1	Rate of Reaction of Ni(tetmeen)$_2^{2+}$ with Cyanide				
2	Stopped-flow run #173-3 (page 72)				
3	Date:	4/8/96			
4	Operator:	JA Tyrell		*Rate constants:*	
5	Wavelength:	243 nm		$k_1 =$	**0.639**
6	Conc:	4.00E-05	(C_A)	$k_2 =$	**0.285**
7	E(A):	0			
8	E(B):	**2.53E+03**	(E_B)		
9	E(C):	0			
10	t, sec	A(obsd)	[B]	A(calc)	∂^2
12	0.2	0.0047	4.66E-06	0.0047	8.5E-11
13	0.6	0.0129	1.16E-05	0.0118	1.3E-06
14	1.0	0.0163	1.62E-05	0.0164	2.6E-09
15	1.4	0.0188	1.89E-05	0.0191	1.0E-07
16	1.8	0.0201	2.04E-05	0.0206	2.2E-07
17	2.2	0.0208	2.09E-05	0.0211	7.6E-08
18	2.6	0.0208	2.07E-05	0.0209	1.2E-08
19	3.0	0.0205	2.01E-05	0.0203	4.6E-08
20	4.0	0.0178	1.75E-05	0.0177	2.1E-08
21	5.0	0.0149	1.44E-05	0.0145	1.3E-07
22	6.0	0.0118	1.15E-05	0.0116	3.9E-08
23	7.0	0.0090	8.98E-06	0.0091	5.7E-09
24	8.0	0.0070	6.94E-06	0.0070	8.6E-11
25	9.0	0.0052	5.31E-06	0.0054	2.8E-08
26	10.0	0.0038	4.04E-06	0.0041	8.2E-08
27				$\Sigma(\partial^2) =$	**2.06E-06**

Figure 17-3. Regression coefficients (in bold) returned by the Solver.

Excel Tip. If, after you've made changes to a worksheet, a Solver model that had previously converged to a reasonable solution refuses to converge, and all attempts to find the problem have failed, use the Reset All button to erase the current model, then re-enter the Target Cell and the Changing Cells. This may solve the problem.

The fit of the curve to the data points is shown in Figure 17-4.

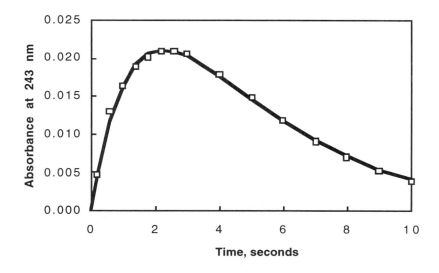

Figure 17-4. Calculated curve, using the coefficients obtained by the Solver

COMPARISON WITH A COMMERCIAL NON-LINEAR LEAST-SQUARES PACKAGE

In Figure 17-5, the Solver results are compared with the results obtained by using NLLSQ, a commercial non-linear regression analysis program (CET Research, Inc., Norman, OK 73070). It is clear that the Solver provides results that are essentially identical to those from the commercial software package. The slight differences (ca. 0.001% or less) arise from the fact that the coefficients are found by a search method; the "final" values will differ depending on the convergence criteria used in each program.

Other software packages which can be used to perform non-linear regression analysis are MathCad (MathSoft, Inc., 201 Broadway, Cambridge, MA 02139) and MatLab (Mathworks Inc., 24 Prime Park Way, Natick, MA 01760).

	A	B	C
33		Comparison of Results from the Solver and from NLLSQ	
34		*Solver*	*NLLSQ*
35	k_1	0.63877	0.63878
36	k_2	0.285394	0.285392
37	ε	2526.43	2526.39
38	$\Sigma(\partial^2) =$	2.0620E-06	2.0620E-06

Figure 17-5. Comparison of regression coefficients.

A major disadvantage of the Solver is that it does not provide the standard deviations of the coefficients. This problem is addressed later in this chapter (see "Statistics of Non-linear Regression").

SOLVER OPTIONS

The Options button in the Solver Parameters dialog box activates the Solver Options dialog box (Figure 17-6), and allows you to control the way Solver attempts to reach a solution. The default values of the options are shown in the dialog box.

The Max Time and Iterations parameters determine when the Solver will return a solution or halt. If either the Max Time (100 seconds) or Iterations (100) is exceeded before a solution has been reached, the Solver will pause and ask if you want to continue. For most simple problems, these limits will not be exceeded. In any event, you don't need to adjust the Max Time or Iterations parameter, since if either is exceeded, the Solver will pause and issue a "Continue anyway?" message.

At first glance the Precision parameter appears to correspond to the Maximum Change parameter in the Calculations Options dialog box (see Chapter 15, Figure 15-10), but both the Precision and Tolerance options apply only to problems with constraints. The Precision parameter determines the amount by which a constraint can be violated. The Tolerance parameter is similar to the Precision parameter, but applies only to problems with integer solutions. Since adding constraints to a model that involves minimization of the error-square sum is not recommended, neither the Precision nor the Tolerance parameter is of use in non-linear regression analysis.

There is no way to set the convergence (stopping tolerance) for the Solver. This parameter is "built-in". If the Solver refuses to converge to a reasonable solution (and you are positive that the model has been set up correctly), there are two things you can do to get around this problem: (i) check the Use Automatic Scaling checkbox (see below) or (ii) try to scale the model by hand, such that the

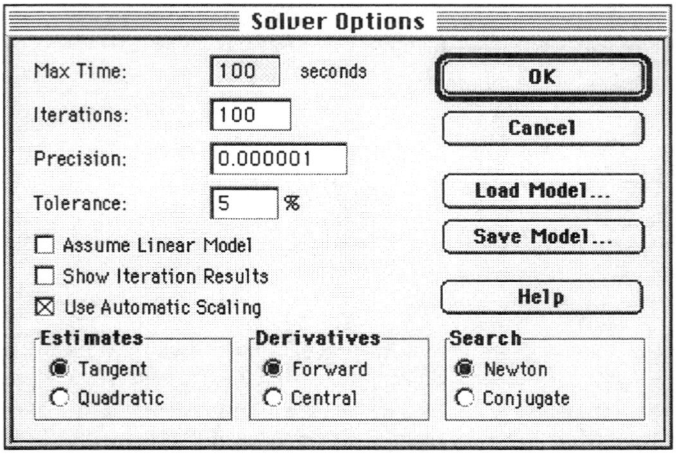

Figure 17-6. The Solver Options dialog box.

objective is a value in the neighborhood of 1. Future versions of Excel's Solver may provide the ability to control the convergence parameter.

If the function is linear, checking the Assume Linear Model box will speed up the solution process.

If the Show Iteration Results box is checked, the Solver will pause and display the result after each iteration.

The Use Automatic Scaling option should be checked if there are large differences in magnitude between the Target Cell and the Changing Cells, for example, if you are varying internuclear bond distance (on the order of 10^{-9} m) to obtain a minimum value of bond energy (on the order of 10^5 Joules).

The Estimates, Derivatives and Search parameters can be changed in order to optimize the solution process. The Search parameter specifies which gradient search method to use: the Newton method requires more memory but fewer iterations, the Conjugate method requires less memory but more iterations. The Derivatives parameter specifies how the gradients for the search are calculated: the Central derivatives method requires more calculations but may be helpful if the Solver reports that it is unable to find a solution. The Estimates parameter determines the method by which new estimates of the coefficients are obtained from previous values: the Quadratic method may improve results if the system is highly nonlinear.

The current Solver model is automatically saved with the worksheet. The Save Model... and Load Model... buttons permit you to save multiple Solver models.

STATISTICS OF NON-LINEAR REGRESSION

As you've seen, the Solver finds the set of least-squares regression coefficients very quickly and efficiently. However, it does not provide the standard deviations of the coefficients. Without these, the Solver's solution is essentially useless. The following illustrates how to obtain the standard deviations of the regression coefficients after obtaining the coefficients by using the Solver.

The standard deviation of the regression coefficient a_i is given by[*]

$$\sigma_i = \sqrt{P_{ii}^{-1}} \; SE(y) \qquad (17\text{-}11)$$

where P_{ii}^{-1} is the ith diagonal element of the inverse of the P_{ij} matrix,

$$P_{ij} = \sum_{n=1}^{N} \frac{\delta F_n}{\delta a_i} \frac{\delta F_n}{\delta a_j} \qquad (17\text{-}12)$$

$\delta F_n / \delta a_i$ is the partial derivative of the function with respect to a_i evaluated at x_n and

---

[*]   K. J. Johnson, *Numerical Methods in Chemistry*, Marcel Dekker, Inc., New York, 1980, p. 278.

$$SE(y) = \sqrt{\frac{SS_{resid}}{N-k}} \qquad (17\text{-}13)$$

The quantities SS_{resid}, N and k are as defined in Chapter 16.

The $(\delta F/\delta a_i)$ terms can be calculated for each data point by numerical differentiation. The term a_i is varied by a small amount from its optimized value while the other a_j terms are held constant. The differential $\delta F/\delta a_i \approx \Delta F/\Delta a_i = (F_{new} - F_{opt})/(a_{new} - a_{opt})$ is calculated for each data point. Since Excel carries 15 significant figures, the change in a_i can be made very small, so that $\delta F/\delta a_i = \Delta F/\Delta a_i$. This process is repeated for each of the k regression coefficients. Then the cross-products $(\delta F/\delta a_i)(\delta F/\delta a_j)$ are computed for each of the N data points and the $\Sigma(\delta F/\delta a_i)(\delta F/\delta a_j)$ terms obtained. The $\mathbf{P}_{ij}$ matrix of $\Sigma(\delta F/\delta a_i)(\delta F/\delta a_j)$ terms is constructed and inverted. The terms along the main diagonal of the inverse matrix are then used to calculate the standard deviations of the coefficients, using equation 17-11. This method may be applied to either linear or non-linear systems.

To illustrate the procedure, the method is applied to a small data set shown in Figure 17-7. The formula Y = 5*X + 3 + 0.1*(0.5-RAND()) was used to generate Y(data) in column B, with a small amount of experimental "noise" provided by the RAND() function. A linear function was chosen so that the standard deviations returned by LINEST could be compared with those provided by this method. Because the test equation is a simple linear one, the $\delta F/\delta a_i$ values ($\delta y/\delta_m$ and $\delta y/\delta_b$ in columns H and I of Figure 17-7) are very close to simple integer values. This is not the case when the procedure is applied to more complicated functions.

	A	B	C	D	E	F	G	H	I
1	*Method for Obtaining Standard Deviations of Solver Parameters*								
2	1. Regression parms obtained using Solver by minimizing Σd^2 (cell D17).								
3	2. Copy & Paste Y(calc) values to save them as Y(fixed) in column E.								
4	3. Solver parms Copied and Pasted to another location to use as ones to vary (C21:D21).								
5	4. Equations for $\partial Y/\partial m$ and $\partial Y/\partial b$ entered. Cell F12 contains (E12-C12)/($C$19-$C$21).								
6	They should display #DIV/0! initially.								
7	5. ($\partial Y/\partial m$) obtained by multipying parm value by 1.000001.								
8	6. Save column of $\partial Y/\partial m$ values (Copy & Paste Values), then restore original parm value.								
9	7. Repeat process to obtain ($\partial Y/\partial b$)								
10								Saved Values	
11	x	Y(data)	Y(calc)	d²	Y(fixed)	($\partial Y/\partial m$)	($\partial Y/\partial b$)	($\partial Y/\partial m$)	($\partial Y/\partial b$)
12	0	5.026	5.008	3.5E-04	5.008	#DIV/0!	#DIV/0!	0.0000	1.0000
13	1	7.965	8.006	1.6E-03	8.006	#DIV/0!	#DIV/0!	1.0000	1.0000
14	2	11.012	11.004	6.5E-05	11.004	#DIV/0!	#DIV/0!	2.0000	1.0000
15	3	14.033	14.002	9.4E-04	14.002	#DIV/0!	#DIV/0!	3.0000	1.0000
16	4	16.984	17.001	2.9E-04	17.001	#DIV/0!	#DIV/0!	4.0000	1.0000
17	N = 5		Σd^2 =	3.3E-03					
18			*Values from Solver:*						
19			2.99831	5.00755					
20			*Values to vary:*						
21			2.99831	5.00755					
22									

Figure 17-7. Method for obtaining regression statistics of coefficients obtained by the Solver (part 1).

	K	L	M	N
1	**Standard Deviations of Solver Parameters (continued)**			
2	8. Calculate the $(\partial Y/\partial m)$, $(\partial Y/\partial b)$ products and Σ terms.			
3		$(\partial Y/\partial m)^2$	$(\partial Y/\partial m)(\partial Y/\partial b)$	$(\partial Y/\partial b)^2$
4		0	0	1
5		1	1	1
6		4	2	1
7		9	3	1
8		16	4	1
9	$\Sigma =$	*30*	*10*	*5*
10				
11	9. Construct matrix from the Σ terms & obtain inverse matrix.			
12		**Matrix (P)**		
13		30	10	
14		10	5	
15		**Inverse matrix (P-1)**		
16		0.1	-0.2	
17		-0.2	0.6	
18				
19	10. Calculate standard deviation of the parms using SE(y)			
20		& i,i terms from inverse matrix.		
21		**SE(y)**	0.033048936	
22		Formula used: $\sqrt{\{\Sigma d^2/(N-2)\}}$		
23		$\sigma(m)$	$\sigma(b)$	
24		0.01045099	0.025599596	
25		Formula used: SQRT(P-1ii)*SE(y)		

Figure 17-8. Method for obtaining regression statistics of coefficients obtained by the Solver (part 2).

Compare the standard deviations using the above procedure (cells L24 and M24 in Figure 17-8) with those obtained from LINEST, which are shown in row 32 of Figure 17-9. Once again, it should be made clear that a linear problem was chosen so that the standard deviations of the regression coefficients could be compared with those from LINEST.

A MACRO TO PROVIDE REGRESSION STATISTICS FOR THE SOLVER

The preceding example was provided to illustrate the steps in the application of the method. The procedure is cumbersome to apply in practice. The macro SOLVSTAT.XLM on the diskette that accompanies this book was written to perform the calculations. It returns the standard deviations of the coefficients, the correlation coefficient and the SE(y) or RMSD; it can be applied to linear or non-linear regression.

	K	L	M
30	*From LINEST:*		
31	*coeff.s*	2.99831	5.00755
32	*std.dev.s*	0.01045099	0.025599596
33	R^2 , *SE(y)*	0.999964	0.033048936

Figure 17-9. Comparison of regression statistics returned by LINEST.

To calculate the standard deviations of the regression coefficients, the macro uses the same approach outlined in the preceding example. The known y's, the calculated y's, the Solver coefficients and the area to receive the returned statistical parameters are obtained using INPUT dialog boxes. The $SS_{residuals}$, $SS_{regression}$ and $SE(y)$ are calculated from the known y's and calculated y's. The partial differentials $\partial y/\partial a_i$ for each of the k regression coefficients are calculated for all N data points by a procedure similar to that used in the Newton-Raphson macro of chapter 15. A table of products of partial differentials is created and used to create a $k \times k$ matrix, and the matrix is inverted. The standard deviations are calculated from the diagonal elements of the inverted matrix, using equation 17-11. Finally, the standard deviations, correlation coefficient and $SE(y)$ are returned to the source worksheet.

USING THE MACRO. The SOLVSTAT.XLM macro is an Auto_Open macro; when you **Open** the document; it will appear on screen and then **Hide** itself. It installs a new menu command, **Solver Statistics...**, directly under the **Solver...** command in either the **Formula** menu (Excel 4.0) or **Tools** menu (Excel 5.0). If the Solver Add-in has not been loaded, the **Solver Statistics...** command will be at the top of the menu. The command will remain in the menu until you exit from Excel.

When you choose the **Solver Statistics...** command, a sequence of four dialog boxes will be displayed, and you will be asked to select four cell ranges: (i) the y_{obsd} data, (ii) the y_{calc} data, (iii) the Solver coefficients, (iv) a 3R × nC range of cells to receive the statistical parameters. The Step 1 dialog box is shown in Figure 17-10. The y values can each be in a single row or column. The Solver coefficients can be in non-adjacent cells.

The array of results returned by the macro is similar to that returned by LINEST, as shown in Figure 17-11.

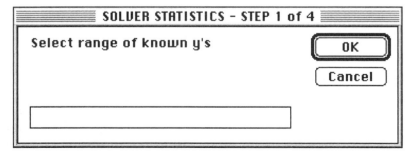

Figure 17-10. SOLVSTAT.XLM Step 1 of 4 dialog box.

parm(n)	parm(n-1)	...	parm(2)	parm(1)
std.dev(n)	std.dev(n-1)	...	std.dev(2)	std.dev(1)
R^2	SE(y)			

Figure 17-11. Layout of regression parameters and statistics returned by SOLVER.STATS macro.

	A	B	C
40	Comparison of results from the Solver and the SOLVSTAT macro with a commercial software package		
41		Solver	NLLSQ
42	k_1	0.63877	0.63878
43	std.dev. of k_1	0.04967	0.04965
44	k_2	0.285394	0.285392
45	std.dev. of k_2	0.01989	0.01987
46	ε	2526.43	2526.39
47	std.dev. of ε	144.18	144.07
48	corr. coeff. R^2	0.9964	0.9964
49	$\Sigma(\delta^2) =$	2.0620E-06	2.0620E-06

Figure 17-12. Regression parameters returned by SOLVSTAT.XLM .

The parameters *parm(1)* to *parm(n)* are not calculated by the macro; they are echoed simply to indicate which standard deviation is associated with which coefficient (since the Solver coefficients can be in non-adjacent cells).

If the SOLVSTAT.XLM macro is used with the kinetics data of Figure 17-3, the regression parameters shown in Figure 17-12 are returned.

AN ADDITIONAL BENEFIT FROM USING SOLVER.STATS

There is an additional major advantage in using the Solver Add-in and the SOLVSTAT.XLM macro, even for functions that are linear or can be rearranged to a linear form. For example, if the function is

$$y = \frac{abx}{1 + bx} \tag{17-14}$$

it can be re-cast as a linear function by taking the reciprocal of each side and rearranging to give

$$\frac{1}{y} = \frac{1}{abx} + \frac{1}{a} \tag{17-15}$$

Plotting $1/y$ vs $1/x$ will yield a straight line with slope $1/ab$ and intercept $1/a$. LINEST can be used to provide the regression coefficients $1/ab$ and $1/a$, and their associated standard deviations. The coefficients a and b can be obtained from the regression coefficients ($a = 1/$intercept, $b = $ intercept$/$slope). However, the standard deviations of a and b must be calculated from the standard deviations of $1/a$ and $1/ab$ by using relationships dealing with the propagation of error.

In contrast, with the Solver, the expression does not need to be rearranged, y_{calc} is calculated directly from equation 17-14, the Solver returns the coefficients a and b, and SOLVSTAT.XLM returns the standard deviations of a and b.

USEFUL REFERENCES

James F. Rusling and Thomas F. Kumosinski, *Nonlinear Computer Modeling of Chemical and Biochemical Data*, Academic Press, San Diego, CA, 1996.

PART IV

SOME APPLICATIONS

18

ANALYSIS OF
SOLUTION EQUILIBRIA

In this chapter you'll learn how to calculate species distributions of polyprotic weak acid species, how to apply Gran's method for the estimation of end-points in titrations, and a general method for the calculation of titration curves.

SPECIES DISTRIBUTION DIAGRAMS

To illustrate the variation in composition of an aqueous solution of a polyprotic weak acid species, it is useful to plot a species distribution curve such as the one shown in Figure 18-1 for citric acid. The parameter plotted versus pH for each species, H_3Cit, H_2Cit^-, $HCit^{2-}$ and Cit^{3-}, is α, the fraction of the total citric acid concentration in the form of a particular species.

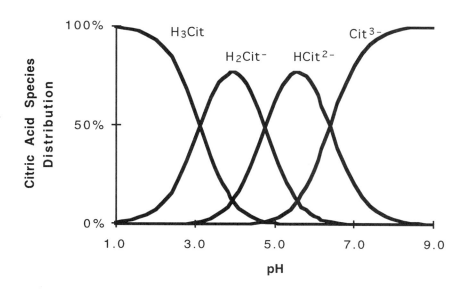

Figure 18-1. Species distribution diagram for citric acid.

For a weak acid H_KA, the fraction of the acid in the form containing j protons is:

$$\alpha_j = \frac{[H_jA]}{[H_KA]_T} \tag{18-1}$$

For example, the expression for α_0 for the Cit^{3-} species is:

$$\alpha_0 = \frac{[Cit^{3-}]}{[H_3Cit]_T} \tag{18-2}$$

In the equations that follow, protonation constants rather than dissociation constants are used. The stepwise protonation constant K^H of a weak base species B is the equilibrium constant for the following reaction (charges are omitted for clarity):

$$B + H^+ \rightleftharpoons HB$$

$$K^H = \frac{[HB]}{[B][H^+]}$$

and in general

$$K_j^H = \frac{[H_jB]}{[H_{j-1}B][H^+]}$$

K^H values are typically much greater than 1 and are often reported as $\log K^H$ values. K_a values are typically very small, and are often reported as pK_a values ($-\log K_a$). Since K^H is the reciprocal of K_a, it follows that numerically, $\log K^H$ is identical to pK_a for a given conjugate acid/conjugate base equilibrium.

It is also convenient to use the cumulative protonation constant β_j for the overall reaction

$$B + jH^+ \rightleftharpoons H_jB$$

$$\beta_j = \frac{[H_jB]}{[B][H^+]^j}$$

The relationship between K's and β_j is

$$\beta_j = K^H K_1^H ... K_j^H$$

The expression for the α_j value for a particular protonated species is easily calculated from the expression

$$\alpha_j = \frac{\beta_j [H^+]^j}{\displaystyle\sum_{j=0}^{K} \beta_j [H^+]^j} \tag{18-3}$$

where $\beta_0 = 1$. Thus the expression for α_2 for citric acid is

$$\alpha_2 = \frac{K_1^H K_2^H [H^+]^2}{1 + K_1^H [H^+] + K_1^H K_2^H [H^+]^2 + K_1^H K_2^H K_3^H [H^+]^3} \tag{18-4}$$

```
'ALPHA Function
'Written by        E. J. Billo
'Begun             1/25/96
'Last Modified     2/7/96


Function ALPHA(j, pH, pKs_or_logKs, pKa_logical)
'pKa_logical:      TRUE if table of pKs_or_logKs is pKa values
'                  FALSE if table of pKs_or_logKs is logK values

N = pKs_or_logKs.Count
If (j < 0) Or (j > N) Or (pH = "") Or (N = 0) Then ALPHA = "n/a":
GoTo 1000
logbeta = 0
denom = 1
numerator = 1
If (pKa_logical = False) Then            'Calculation using protonation
constants
   For x = 1 To N
   logbeta = logbeta + pKs_or_logKs(x)
   denom = denom + 10 ^ (logbeta - pH * x)
   If (x = j) Then numerator = 10 ^ (logbeta - pH * x)
   Next
Else            'Calculation using dissociation constants
   For x = N To 1 Step -1
   logbeta = logbeta + pKs_or_logKs(x)
   denom = denom + 10 ^ (logbeta - pH * (N - x + 1))
   If (N - x + 1 = j) Then numerator = 10 ^ (logbeta - pH * (N - x +
1))
   Next
End If
ALPHA = numerator / denom
1000   End Function
```

Figure 18-2. ALPHA custom function

The custom function ALPHA provides a convenient way to calculate alpha values for polyprotic species, and thus to construct distribution diagrams. The macro is shown in Figure 18-2. The syntax of the function is **ALPHA(*j, pH, pKs_or_logKs, pKa_logical*)**. Either pK_a values or $\log K^H$ values can be used. If pK_a values are used, pKa_logical is TRUE; if $\log K^H$ values are used, pKa_logical is FALSE or omitted.

ANALYSIS OF TITRATION DATA

The location of the end-point of a titration by using either the first or second derivative of the titration data was discussed in Chapter 14. These methods use only the data points near the end-point. Another approach, Gran's method, makes use of the complete data set. It is particularly useful either when (i) the inflection at the end-point is poorly defined, or (ii) data at the end-point is missing.

Consider an acid-base titration. At any point before the end-point, the concentration of unreacted H^+ is given by 10^{-pH}. Thus, to estimate the volume required to reach the end-point, it is merely necessary to plot 10^{-pH} versus titrant volume V and extrapolate to $10^{-pH} = 0$. If dilution by the titrant is important, then the function 10^{-pH} should be multiplied by $(V_0 + V)/V_0$, where V_0 is the initial volume.

A similar approach can be used with other electrodes. In the following example illustrating the titration of a chloride sample with standard silver nitrate, the potential of a silver electrode in combination with a saturated calomel reference electrode was used to follow the course of the titration. The potential of the electrode pair is a direct measure of the free chloride ion concentration; as the chloride ion concentration decreases, the potential increases. The titration results are shown in Figure 18-3.

The chloride ion concentration at any point on the titration curve can be calculated from equation 18-5, which can be derived from the Nernst equation.

$$[Cl^-] = antilog \left[\frac{-(E - offset)}{slope} \right] \tag{18-5}$$

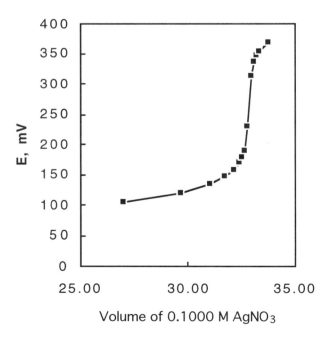

Figure 18-3. Potentiometric titration curve of chloride titrated with silver ion.

Offset can be any potential to scale the *E* values into a more convenient range; *slope* is the Nernst slope, theoretically 59.2 mV at 25°C.

Figure 18-4 shows the titration data *V* and *E*; column C contains the formula =10^(-(E-offset)/slope). Figure 18-5 shows the linear relationship between the Gran function (column C in Figure 18-4) and the volume.

The end-point is considered to be the volume where the straight-line portion of the Gran plot crosses the X-axis. To find the end-point you can make an expanded chart of the titration data near the end-point and estimate the end-point reading visually. From Figure 18-6 the end-point can be estimated to be approximately 32.82 mL.

	A	B	C
1	**Gran's Method** Used to Determine the		
2	Endpoint of a Titration of Chloride with		
3	0.1000 M silver nitrate.		
4	Unreacted chloride calculated from:		
5	=10^-((E-offset)/slope))		
6	(A rearranged form of the Nernst equation.)		
7			
8	offset =	-20	
9	slope =	59	
10			
11	**V**	**E**	**[Cl−], mM**
12	27.00	104	7.91
13	29.70	120	4.24
14	31.1	135	2.36
15	31.73	147	1.48
16	32.13	159	0.92
17	32.40	171	0.58
18	32.54	180	0.41
19	32.67	191	0.27
20	32.81	230	0.06
21	32.94	313	0.00
22	33.08	337	0.00
23	33.21	348	0.00
24	33.35	355	0.00
25	33.75	370	0.00

Figure 18-4. Gran's method calculations.

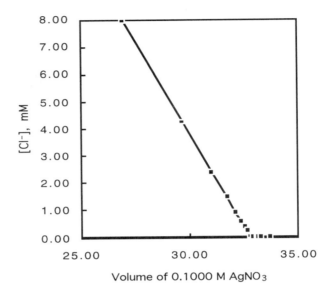

Figure 18-5. Gran plot.

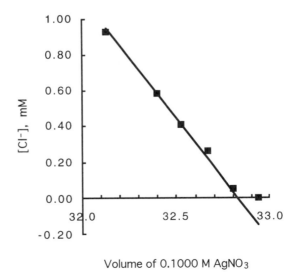

Figure 18-6. Visual estimation of end-point by using a Gran plot.

	E	F
17	*From LINEST:*	
18	-1.3534	44.4350
19	0.01	0.18
20	0.99987	0.03

Figure 18-7. Slope and intercept of Gran plot.

Alternatively the end-point can be obtained algebraically. LINEST was used to obtain the slope and intercept of the straight-line portion of the data, shown in Figure 18-7 (the last 5 rows of data points were not included). The intercept = 44.44 and the slope = –1.35.

To obtain the end-point volume, the value of V where $[Cl^-] = 0$, you need to calculate the *x-intercept*. The *x*-intercept = –intercept/slope = 32.83 mL, the end-point volume for the titration.

A second example illustrates how to obtain an end-point that was missed. To determine the concentration of a weak base compound, a known excess of HCl was added to a solution of the compound, and the excess acid was back-titrated with standard base, using a micrometer syringe buret. Unfortunately the student doing the titration did not take small increments near the inflection point (which was not very pronounced, in any case), and the result shown in Figure 18-8 was obtained.

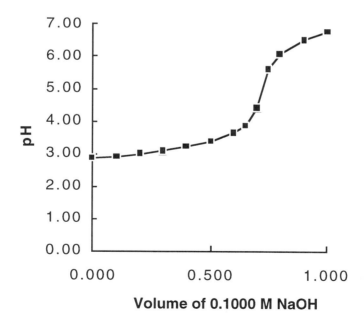

Figure 18-8. Acid-base titration with "missed" end-point.

	A	B	C
3	Y	pH	[H+]
4	0.000	2.85	0.0014
5	0.100	2.91	0.0012
6	0.200	2.99	0.0010
7	0.300	3.08	0.0008
8	0.400	3.20	0.0006
9	0.500	3.36	0.0004
10	0.600	3.62	0.0002
11	0.650	3.86	0.0001
12	0.700	4.36	0.0000
13	0.750	5.60	0.0000
14	0.800	6.08	0.0000
15	0.900	6.50	0.0000
16	1.000	6.75	0.0000

Figure 18-9. Gran's method calculations.

Nonetheless, the end-point can be estimated by using Gran's method. From the pH measurements the concentration of free $[H^+]$ was calculated using the relationship $[H^+] = 10^{-pH}$, as illustrated in Figure 18-9. The data, when plotted (Figure 18-10), gave an excellent straight line with a correlation coefficient (omitting the last 4 data points) of 0.99993. The end-point volume was calculated to be 0.722 mL.

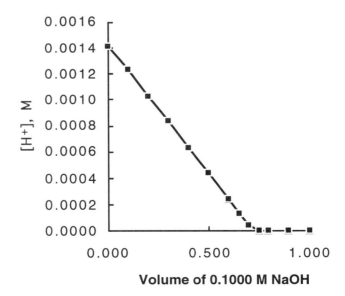

Figure 18-10. Gran plot.

SIMULATION OF TITRATION CURVES
USING A SINGLE MASTER EQUATION

In calculating titration curves, separate equations for different regions of the curve ("before the equivalence point", "at the equivalence point", "after the equivalence point", etc.) are often employed. The following illustrates how to calculate points on a titration curve using a single "master" equation. Instead of calculating pH as a function of the independent variable V, it is convenient to use pH as the independent variable and V as the dependent variable. The species distribution at a particular pH value is calculated from the $[H^+]$, and the volume of titrant required to produce that amount of each species is calculated. For example, in the titration of a weak monoprotic acid HA, we can calculate the concentration of A^- at a particular pH, and then calculate the number of moles of base required to produce that amount of A^-. In general $(J - j)$ moles of base are required to produce the species H_jA from the original acid species H_JA.

Using the relationship

(initial moles of available protons) – (moles of titrant base added) =

(moles of bound protons) + (moles of $[H^+]$)   (18-6)

then in the general case, for C_H moles of strong acid + C_{HA} moles of weak acid H_JA in an initial volume V_0, titrated with V milliliters of standard base of concentration C_{OH}, the relationship is:

$$C_H + J\,C_{HA} + [OH^-]\,(V_0 + V) - C_{OH}V = j\Sigma\,[H_jA] + [H^+]\,(V_0 + V) \quad (18\text{-}7)$$

From equation 18-7, after rearrangement, we obtain equation 18-8. (The $[OH^-]$ is an indicator of one of the sources of protons, namely H_2O.) The $[H_jA]$ are calculated employing the usual α factors, from K's and $[H^+]$.

$$V_{calc} = \frac{C_H + J\,C_{HA} + [OH^-]\,(V_0 + V) - [H^+](V_0 + V) - j\Sigma\,[H_jA]}{C_{OH}} \quad (18\text{-}8)$$

Equation 18-8 permits the calculation of all points on a titration curve using a single equation. As written, it handles strong acids, weak acids or mixtures, and it is readily expanded to handle mixtures of polyprotic acids.

Figure 18-11 illustrates a portion of a spreadsheet for the calculation of the titration curve of 2.500 mmol of a weak acid ($pK_a = 5$) with 0.1000 M strong base. The volume required to obtain a given pH value was calculated for pH values from 3 to 12 in increments of 0.20. The formula used to calculate V in cell C9 is

=(CHA+10^-(pKw-A9)*(V_0) -10^(-A9)*(V_0)-B9*CHA)/COH.

The terms in the formula are in the same order as those in equation 18-8.

The titration curve is shown in Figure 18-12.

	A	B	C	D
1	**Weak Acid–Strong Base Titration Curve**			
2	weak acid, mmol		2.500	(CHA)
3	titrant base, M		0.1000	(COH)
4	total volume, mL		50.0	(V_0)
5	pKa		5.00	(logK)
6	pcKw		14.00	(pcKw)
7				
8	**pH**	**alpha1**	**V(calc)**	
9	3.00	9.9E-01	0	
10	3.50	9.7E-01	0.61	
11	4.00	9.1E-01	2.22	
12	4.50	7.6E-01	5.99	
13	5.00	5.0E-01	12.50	
14	5.50	2.4E-01	18.99	
15	6.00	9.1E-02	22.73	
16	6.50	3.1E-02	24.23	
17	7.00	9.9E-03	24.75	
18	8.00	1.0E-03	24.98	
19	9.00	1.0E-04	25.00	
20	10.00	1.0E-05	25.05	
21	10.50	3.2E-06	25.16	
22	11.00	1.0E-06	25.50	
23	11.50	3.2E-07	26.58	
24	12.00	1.0E-07	30.00	

Figure 18-11. Spreadsheet for weak acid-strong base titration curve.

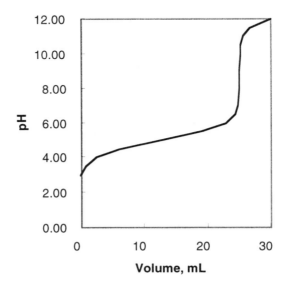

Figure 18-12. Weak acid/strong base titration curve.

19

ANALYSIS OF SPECTROPHOTOMETRIC DATA

In this chapter you'll learn how to handle calibration curves that are not straight lines, how to analyze the spectra of mixtures of components and how to deconvolute a spectrum into its individual absorption bands.

CALIBRATION CURVES FOR SPECTROPHOTOMETRY

Linear calibration lines can be handled quite easily using LINEST. However, when a calibration curve is not linear, the problem is a little more difficult. The calibration curve in Figure 19-1 shows readings taken on a series of sodium standards, using a CIBA-Corning Model 410 flame photometer. The calibration line is noticeably curved.

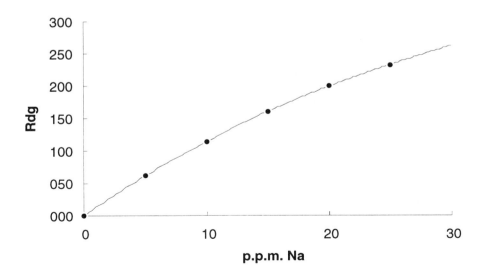

Figure 19-1. Flame photometry calibration curve.

	D	E	F	G
4		Power Series in X		
5	ppm Na	ppm^2	ppm^3	Rdg
6	0	000	000	000
7	5	025	125	062
8	10	100	1000	115
9	15	225	3375	160
10	20	400	8000	200
11	25	625	15625	233
12				
13		From LINEST:		
14	d	c	b	a
15	0.00141	-0.193	13.280	0.079
16	0.00050	0.019	0.193	0.498
17	0.999986	0.50787	#N/A	#N/A

Figure 19-2. Fitting a calibration curve to a cubic equation.

One way to handle a curved calibration line is to fit the line to a power series. A cubic equation ($y = a + bx + cx^2 + dx^3$) is usually sufficient to fit a case such as Figure 19-1. (In any event, since there are only six known points, you couldn't use a polynomial with more than five adjustable parameters.) You can use either LINEST or the Solver to obtain the coefficients of the power series. Figure 19-2 shows a spreadsheet in which LINEST is used to find the regression coefficients for the equation $Rdg = a + b*ppm + c*(ppm)^2 + d*(ppm)^3$.

In the above case either

$$Rdg = a + b*ppm + c*(ppm)^2 + d*(ppm)^3$$

with $a = 0.079$, $b = 13.28$, $c = -0.193$ and $d = 0.00141$

or

$$Rdg = b*ppm + c*(ppm)^2 + d*(ppm)^3$$

with $b = 13.30$, $c = -0.195$ and $d = 0.00144$ give excellent fits to the data points.

To obtain concentration information (x) from a flame photometer reading (y), it is necessary to find the value of x that gives the observed value of y. The **Goal Seek** command in the **Formula** menu (Excel 4.0) or in the **Tools** menu (Excel 5.0) performs this task very conveniently (see "Solving a Problem Using Goal Seek..." in Chapter 15).

An even simpler way to obtain concentration values from flame photometer readings is to fit the data to a power series using LINEST or the Solver, but using the concentration values as the y (dependent) variables and the readings as the independent variables. In this way you will obtain a polynomial such as

$$ppm = a + b*Rdg + c*(Rdg)^2 + d*(Rdg)^3$$

rom which the concentration can be calculated directly. Figure 19-3 illustrates this approach.

	M	N	O	P
3		Power Series in Y		
4	Rdg	Rdg^2	Rdg^3	ppm Na
5	000	000	000	0
6	062	3844	238328	5
7	115	13225	1520875	10
8	160	25600	4096000	15
9	200	40000	8000000	20
10	233	54289	12649337	25
11				
12		From LINEST:		
13	d	c	b	a
14	3.75653E-07	4.09564E-05	0.077250495	-0.00811603
15	1.02243E-07	3.62575E-05	0.003375014	0.080465303
16	0.999969982	0.081032976	#N/A	#N/A

Figure 19-3. Alternate approach.

Thus the equation

$$ppm = -0.00812 + 0.07725 \times Rdg + (4.096 \times 10^{-5}) \times Rdg^2 + (3.7565 \times 10^{-7}) \times Rdg^3$$

yields the concentration directly.

ANALYSIS OF SPECTRA OF MIXTURES

A common analytical problem in spectrophotometry is the analysis of a mixture of components. If the spectra of the pure components are available, the spectrum of a mixture can be analyzed to determine the concentrations of the individual components. If the mixture contains N components, then absorbance measurements at N suitable wavelengths are necessary to solve the set of N linear equations in N unknowns.

APPLYING CRAMER'S RULE TO A SPECTROPHOTOMETRIC PROBLEM

As a simple example of the analysis of mixtures, consider an aqueous solution containing a mixture of Co^{2+}, Ni^{2+} and Cu^{2+}, to be analyzed by spectrophotometric measurements at three different wavelengths. The spectra of the individual ions and of a mixture are shown in Figures 19-4 and 19-5. The most suitable wavelengths for analysis are 394, 510 and 808 nm (determined from an examination of Figure 19-4 and the data table). The molar absorptivities of the three species at these wavelengths are shown in Figure 19-6, together with absorbance readings for a mixture of the three ions, measured in a 1.00-cm cell.

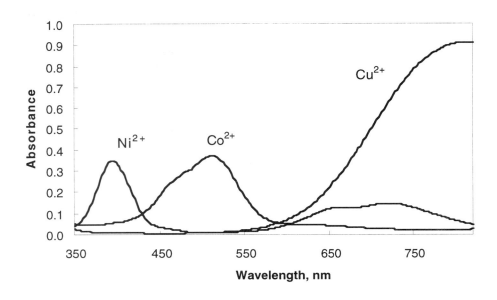

Figure 19-4. Spectra of Co^{2+}, Ni^{2+} and Cu^{2+} ions in aqueous solution (standards).
(Spectrophotometric data provided by Dr. Lev Zompa.)

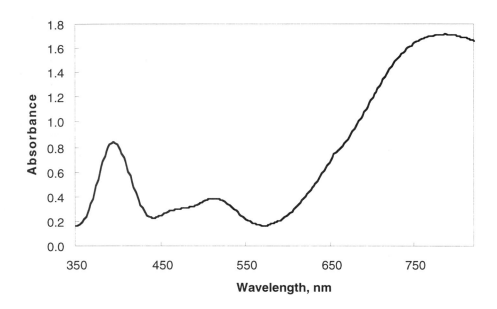

Figure 19-5. Spectrum of a mixture of Co^{2+}, Ni^{2+} and Cu^{2+} ions in aqueous solution.

	N	O	P	Q	R
2			*STANDARDS*		*UNKNOWN*
3		Molar Absorptivity, $M^{-1} cm^{-1}$			*MIXTURE*
4	λ/nm	Co^{2+}	Ni^{2+}	Cu^{2+}	(absorbance)
5	394	0.995	6.868	0.188	0.845
6	510	6.450	0.215	0.198	0.388
7	808	0.469	1.179	15.052	1.696

Figure 19-6. Data table for the determination of a mixture of Co^{2+}, Ni^{2+} and Cu^{2+} ions .

	O	P	Q
15	0.845	6.868	0.188
16	0.388	0.215	0.198
17	1.696	1.179	15.052

Figure 19-7. The determinant for calculating Co^{2+}.

Following the Cramer's rule procedures described in Chapter 15, we obtain the determinant to determine Co^{2+} concentration shown in Figure 19-7.

The formula

=MDETERM(O15:Q17)/MDETERM(O5:Q7)

yields the value 0.05328 M for the Co^{2+} concentration. From similar formulas, $[Ni^{2+}] = 0.1125$ M and $[Cu^{2+}] = 0.1022$ M.

SOLUTION USING MATRIX INVERSION

A set of simultaneous linear equations can also be solved by using matrices, as shown in Chapter 15. The *solution matrix* is obtained by multiplying the matrix of constants by the inverse of the matrix of coefficients. Applying this simple solution to the spectrophotometric data used above, the inverted matrix is obtained by selecting a 3R × 3C array of cells, entering the array formula

=MINVERSE(O5:Q7).

The inverted matrix is shown in Figure 19-8.

The solution matrix is obtained by selecting a 3R × 1C array, then entering the array formula

=MMULT(O38:Q40,R5:R7).

The single array formula

=MMULT(MINVERSE(O5:Q7),R5:R7)

accomplishes the same result. The solution matrix is shown in Figure 19-9.

	O	P	Q
38	-0.0045315	0.15587674	-0.0019965
39	0.14657147	-0.0224973	-0.001539
40	-0.011342	-0.0030951	0.06661918

Figure 19-8. The inverted matrix.

	O	P
46	0.05328	(Co^{2+})
47	0.1125	(Ni^{2+})
48	0.1022	(Cu^{2+})

Figure 19-9. The solution matrix .

DECONVOLUTION OF SPECTRA

The resolution of a complex absorption spectrum into individual absorption bands may be necessary if information about the position, height or width of individual bands is required.   There are a number of computer programs designed for the deconvolution of spectra, but you can do a reasonable job with Excel.

The procedures described below were developed for the deconvolution of electronic absorption spectra (UV-visible spectra) but are equally applicable to the deconvolution of infrared, Raman or NMR spectra. UV-visible spectra differ from vibrational spectra in that the number of bands is much smaller and the bandwidths are much wider. Band shape may also be different. UV-visible spectra are also usually recorded under conditions of high resolution and high signal-to-noise.   Spectra from older instruments usually require manual digitization from a spectrum on chart paper, at e.g., 10 nm intervals. With the widespread use of computer-controlled instruments, it is a simple matter to obtain a file of spectral data at, e.g., 1 nm intervals. In fact, it may be necessary to reduce the size of the data set in order to speed up calculations.

MATHEMATICAL FUNCTIONS FOR SPECTRAL BANDS[*]

A symmetrical spectral band is described by three parameters: position (wavelength or frequency corresponding to the absorption maximum), intensity (absorbance or molar absorptivity at the band maximum) and width (usually the bandwidth at half-height). The band shape functions most commonly used for deconvolution are the Gaussian function and the Lorentzian function. Both are symmetrical functions.   UV-visible spectra generally have a Gaussian band shape. The Lorentzian function is useful for the simulation of NMR spectra. The log-normal band function has been applied to unsymmetrical spectral band shapes.

Many spectral bands can be closely approximated by a Gaussian line shape when the independent variable v is in energy units, e.g., cm^{-1}. The absorbance A at a wavenumber v is given by equation 19-1, where A_{max} is the band maximum, v_{max} is the wavenumber of the band maximum and Δv is the half-width.

$$A = A_{max} \exp\left[-(4 \ln 2)\frac{(v - v_{max})^2}{\Delta v^2} \right] \qquad (19\text{-}1)$$

---

[*]   P. Pelikán, M. Čeppan and M. Liška, *Applications of Numerical Methods in Molecular Spectroscopy*, CRC Press, Boca Raton FL, 1993.

The corresponding equation for a Lorentzian line shape is given by equation 19-2.

$$A = \frac{A_{max}}{1 + 4\dfrac{(\nu - \nu_{max})^2}{\Delta\nu^2}} \tag{19-2}$$

For unsymmetrical bands, the equation for the lognormal line-shape is

$$A = A_{max}\exp\left\{-\frac{\ln 2}{(\ln \rho)^2}\left[\ln\left(\frac{(\nu - \nu_{max})^2}{\Delta x^2}\frac{\rho^2 - 1}{\rho} + 1\right)\right]^2\right\} \tag{19-3}$$

for the region $\nu \geq \nu_{max} - (\Delta x\rho / (\rho^2 - 1))$ and $A = 0$ elsewhere. The asymmetry parameter ρ is given by:

$$\rho = \frac{\nu_R - \nu_{max}}{\nu_{max} - \nu_L} \tag{19-4}$$

A simpler form of the Gaussian band shape, where σ is simply treated as an adjustable parameter, is given in equation 19-5. This is the equation that will be used in the following treatment. By using an embedded chart to compare calculated and experimental data, you can fairly easily find a set of A_{max}, ν_{max} and σ values that approximate the band shape, to use as initial guesses for the deconvolution procedure outlined in the box on the following page.

$$A = A_{max}\exp\left[-\frac{(\nu - \nu_{max})^2}{\sigma^2}\right] \tag{19-5}$$

DECONVOLUTION OF A SPECTRUM: AN EXAMPLE

The spreadsheet shown in Figure 19-10 illustrates the deconvolution of the UV-visible spectrum of a mixed-ligand complex of nickel(II). Four bands are apparent in the spectrum, one a weak shoulder lying between relatively intense bands at approximately 350 and 550 nm. The fourth band appears only as the tail of a fairly intense band lying at longer wavelengths.

The formulas in cells C10 (converting wavelength to wavenumber) and D10 (calculating the Gaussian band profile of band 1) are:

=10000/A10

=band1 A_0*EXP(-(((C10-band1 max)/band1 s)^2)/2)

The Solver was used to vary the values in cells D4:F6 and G4:G5 to make cell I7 a minimum. Because the data did not permit a complete resolution of band 4, cell G6, the bandwidth parameter for band 4, was held constant at the reasonable value of 1.5. The results are shown on the spreadsheet. The resolved spectrum (solid line), with the four bands (broken lines), is shown in Figure 19-11.

The λ_{max} values for bands 2 and 4, from other experimental measurements, are 445 and 880 nm, respectively.

> ### Deconvolution of a Spectrum
>
> 1. Start with a table of wavelength, absorbance data pairs.
> 2. Create a column of wavenumbers.
> 3. Determine the number of bands necessary to describe the spectrum. This can usually be arrived at by inspection: a strongly asymmetric band generally indicates one or more hidden bands; a band with a flat maximum indicates two strongly overlapped bands, etc. Alternatively, you can use the first derivative of the spectrum ($\Delta A / \Delta x$). Except for the most hidden shoulders, each $\Delta A / \Delta x = 0$ value indicates a band maximum.
> 4. Estimate the half-width of the bands by using one or more bands not overlapped by other bands. As first approximation, use this value for all bands in the spectrum
> 4. Set up a table of v_{max}, A_{max} and σ for each band.
> 5. Calculate the band profile for each contributor.
> 6. Sum the individual band contributions.
> 7. Calculate the sum of squares of the residuals $(A_{obsd} - A_{calc})^2$.
> 8. Create an embedded chart, plotting A_{obsd} and A_{calc}.
> 9. Perform some manual adjustment of the parameters, attempting to make the calculated spectrum coincide with the observed. This is especially important if the spectrum is complicated (more than three or four bands, especially if they are overlapped strongly).
> 10. Use the Solver to minimize sum of squares of residuals by varying (ultimately) the $3N$ parameters for the N bands in the spectrum.

TACKLING A COMPLICATED SPECTRUM

For a complicated spectrum, it may be helpful to operate on a reduced-size data set. Many spectrometers record absorbance readings at 1-nm intervals; a complete UV-visible spectrum (200 to 700 nm) contains 500 data points. If the spectrum contains eight bands, you're performing calculations on more than 4000 cells. Start with a data set consisting of every 10th data point, for example. After getting a reasonably good fit to this data set, use these values as initial parameters for the complete data set.

It may be necessary to first minimize portions of the spectrum separately.

	A	B	C	D	E	F	G	H	I
1			**Ni(2,3,2-tet)(en)2+ spectrum**						
2			**(b = 5 cm, c = 0.0252 M, pH = 10.6)**						
3				**band1**	**band2**	**band3**	**band4**		
4	(wavenumber)	n_{max}		29.25	22.72	18.56	11.69		
5	(absorbance)	A_{max}		1.12	0.15	0.87	0.77		
6	(bandwidth)	S		1.60	1.54	1.38	1.5		$\Sigma(\partial^2)$
7	(wavelength)	l_{max}		342	440	539	855		0.0148
8									
9	l,nm	Abs	n,cm^{-1}	band1	band2	band3	band4	Σ	d^2
10	300	0.173	33.33	0.043	0.000	0.000	0.000	0.043	
11	310	0.274	32.26	0.191	0.000	0.000	0.000	0.191	0.007
12	320	0.514	31.25	0.512	0.000	0.000	0.000	0.512	0.000
13	325	0.694	30.77	0.714	0.000	0.000	0.000	0.714	0.000
14	330	0.871	30.30	0.903	0.000	0.000	0.000	0.903	0.001
15	335	1.026	29.85	1.046	0.000	0.000	0.000	1.046	0.000
16	340	1.126	29.41	1.118	0.000	0.000	0.000	1.118	0.000
17	345	1.141	28.99	1.109	0.000	0.000	0.000	1.109	0.001
18	350	1.036	28.57	1.028	0.000	0.000	0.000	1.028	0.000
19	355	0.908	28.17	0.896	0.000	0.000	0.000	0.896	0.000
20	360	0.737	27.78	0.737	0.001	0.000	0.000	0.738	0.000
21	370	0.406	27.03	0.429	0.003	0.000	0.000	0.432	0.001
22	380	0.195	26.32	0.210	0.010	0.000	0.000	0.220	0.001
23	390	0.111	25.64	0.089	0.025	0.000	0.000	0.114	0.000

Figure 19-10. Deconvolution of the UV-visible spectrum.

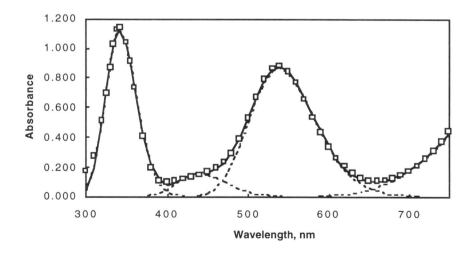

Figure 19-11. Deconvoluted spectrum.

CALCULATION
OF BINDING CONSTANTS

The measurement of binding constants (also called stability constants or formation constants) is of interest in many areas of chemistry. Quantitative information concerning the products of acid–base, metal–ligand or enzyme–substrate interactions is invaluable in analytical chemistry, industrial process chemistry, biochemistry, etc.

A wide range of experimental methods has been applied to the determination of binding constants. Three methods of most common use — potentiometry, spectrophotometry and NMR — are described here. The wide range of experimental methods has produced a wide range of types of experimental data and methods of calculation. Many "canned" computer programs are available, but for the occasional user, or for greater flexibility in tackling a non-standard situation, Microsoft Excel provides an ideal tool for the calculation of binding constants.

The binding constants (illustrated here as metal–ligand formation constants) may be either *stepwise formation constants* K_n (equation 20-1) or *cumulative formation constants* β_n (equation 20-2). Charges have been omitted for clarity.

$$ML_{n-1} + L \rightleftharpoons ML_n \qquad K_n = [ML_n]/\{[ML_{n-1}][L]\} \qquad (20\text{-}1)$$

$$M + nL \rightleftharpoons ML_n \qquad \beta_n = [ML_n]/\{[M][L]^n\} \qquad (20\text{-}2)$$

The relationship between stepwise and cumulative formation constants is

$$\beta_n = \prod_1^n K_n \qquad \text{e.g., } \beta_3 = K_1 K_2 K_3 \qquad (20\text{-}3)$$

Writing the equilibrium constant as a cumulative equilibrium constant does not indicate that two or more ligands are added simultaneously. All association reactions occur in a stepwise manner, although in a few cases the successive reactions overlap extensively.

DETERMINATION OF BINDING CONSTANTS
BY pH MEASUREMENTS

The majority of ligands are weak bases. In aqueous solution the ligand base can be protonated, and thus metal-ligand complex formation involves competition between proton and metal ion for the ligand base. The progress of this competition reaction can be monitored by means of pH measurements. This method was pioneered by Bjerrum.[*] Advantages of the method include its precision and its applicability to a wide range of metal ions and ligands. Because potentiometry measures the activity of a species, three kinds of constant can be identified:

activity-product (thermodynamic) constants, e.g., $K_a = \{H^+\}\{A\}/\{HA\}$

concentration-product constants, e.g., $K_a = [H^+][A]/[HA]$

mixed or Brønsted constants, e.g., $K_a = \{H^+\}[A]/[HA]$.

EXPERIMENTAL TECHNIQUES

The most common procedure is to carry out the measurements in the form of a titration. Most commonly a solution containing metal-ion, ligand and acid is titrated with base. Occasionally, when the rate of attaining equilibrium in the system is slow, a "batch" method is adopted: individual solutions of appropriate concentrations are prepared, sealed, placed in a constant-temperature bath and allowed to reach equilibrium, at which time the final pH measurements are made.

The following provides a brief description of a typical experimental set-up and methods for the determination of metal-ligand binding constants using the pH titration method.

PROCEDURE. Appropriate volumes of ligand solution, metal-ion solution, acid and distilled water are pipetted into a titration cell, usually a double-walled beaker through which thermostatted water is flowed. The solution is stirred magnetically and blanketed with nitrogen to prevent reaction with carbon dioxide (and occasionally oxygen). Increments of titrant (usually standard base) are added and the pH is recorded after the addition of each increment.

ACTIVITY COEFFICIENT CORRECTIONS. To eliminate uncertainties arising from activity constant variations, it is common practice to keep activity coefficients constant by use of a "background electrolyte" or "constant ionic atmosphere" (e.g., 0.10 M $NaClO_4$). Since the glass electrode measures (for practical purposes) hydrogen ion *activity*, i.e., $pH_{meas} = -\log\{H^+\} = -\log[H^+]\gamma_+$, it is necessary to convert activity to concentration in the calculations that follow. The relationship of equation 20-4 may be used, where the activity correction $C = \log\gamma_+$.

$$-\log[H^+] = pH_{meas} + C \qquad\qquad (20\text{-}4)$$

---

[*]   J. Bjerrum, *Metal Ammine Formation in Aqueous Solution*, P. Haase and Son, Copenhagen, 1941.

The correction factor C may be determined from pH measurements on appropriate acid solutions, or calculated using the Debye-Hückel equation. For aqueous solutions at 25°C and ionic strength of 0.10, the correction factor is often taken to be −0.10 (calculated from the Debye-Hückel equation). Experimentally measured values are similar. For the calculation of $[OH^-]$, the concentration-product value of K_W must be used; at 25°C and $I = 0.10$, $p_cK_W = 13.75$.

SEPARATION OF OVERLAPPING PROTONATION CONSTANTS FOR A POLYPROTIC ACID

In the case of a polyprotic acid for which the individual ionizations are well separated (ideally, by at least 3 log units), values for the individual constants can be calculated from data points in the appropriate regions of the titration curve. If the individual ionizations overlap, the Bjerrum $\bar{n}$ (n-bar) method may be used. This mathematical approach was introduced by Bjerrum for the calculation of stability constants of metal-ligand complexes, but can also be applied to the determination of proton-ligand equilibrium constants.

The equilibrium constants to be determined in this example are *protonation constants*, introduced in Chapter 19. The protonation constant K^H of a base L is the reciprocal of the acid dissociation constant K_a for the corresponding conjugate acid HL. For a polyprotic acid, the general expression for the protonation constant is given by equation 20-5.

$$H_{n-1}L + H^+ \rightleftharpoons H_nL \qquad K_n^H = [H_nL]/\{[H_{n-1}L][H^+]\} \qquad (20\text{-}5)$$

Protonation constants lend themselves much more readily to the systematic treatment of equilibria than do dissociation constants. Thus, for example, the definition of $\bar{n}_H$, the average number of protons bound per ligand, leads to the general expression (20-6) for a ligand base L derived from an acid of general formula H_JL. The β's are cumulative protonation constants (see Chapter 18).

$$\bar{n}_H = \frac{[HL] + 2[H_2L] + \cdots}{[L] + [HL] + [H_2L] + \cdots} = \frac{\beta_1^H[H^+] + 2\beta_2^H[H^+]^2 + \cdots + J\beta_J^H[H^+]^J}{1 + \beta_1^H[H^+] + \beta_2^H[H^+]^2 + \cdots + \beta_J^H[H^+]^J} \qquad (20\text{-}6)$$

To apply the $\bar{n}$ method to the determination of protonation constants, the quantity $\bar{n}_H$, the average number of protons bound, can be defined as (total available protons − free hydrogen-ion)/total ligand. At any point in a titration, the stoichiometric concentration of available protons is equal to the sum of the concentrations of dissociable protons from the ligand, from added strong acid and of hydrogen ions arising from the dissociation of water, less the concentration of added strong base. For the titration of an acid of general formula H_JL of concentration C_L plus added strong acid of concentration C_A, titrated with standard sodium hydroxide:

$$\bar{n}_H = \frac{j\,C_L + C_A + [OH^-] - [Na^+] - [H^+]}{C_L} \qquad (20\text{-}7)$$

The quantity $\bar{n}_H$ is a normalized variable; for an acid of stoichiometry H_JL, it can have values $0 \le \bar{n}_H \le J$. A plot of $\bar{n}_H$ vs. pH is termed the formation curve or formation function. If protonation equilibria are well separated (by at least 3 log units), then the titration curve will exhibit a "break" between the two regions and the protonation constants can be calculated separately, using equation 20-8:

$$K_j^H = \frac{\bar{n}_H - j + 1}{(j - \bar{n}_H)\,[H^+]} \qquad (20\text{-}8)$$

If two equilibria overlap, the constants can be obtained from the slope and intercept of a straight-line transformation of the $\bar{n}_H$ expression. If three or more equilibria overlap, either multiple linear regression or the Solver can be used to obtain the constants.

TWO OVERLAPPING PROTONATION CONSTANTS
OF N-(2-AMINOETHYL)-1,4-DIAZACYCLOHEPTANE

The triamine N-(2-aminoethyl)-1,4-diazacycloheptane (aedach) was

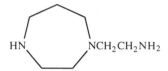

N-(2-aminoethyl)-1,4-diazacycloheptane

synthesized as a potential ligand for complexation of nickel(II) in a square-planar environment. The titration curve of N-(2-aminoethyl)-1,4-diazacycloheptane trihydrobromide with standard NaOH is shown in Figure 20-1. Two protonation equilibria overlap strongly in the pH 9–11 region, while the third protonation constant is much lower, occurring in the pH 3–4 region. The first two protonation constants are normal for aliphatic amines, while the third demonstrates the strong base-weakening effect of charge repulsion by the protonated primary and secondary amines.

It is clear that log K_1^H and log K_2^H cannot be calculated individually using equation 20-8. Instead, the general expression 20-6 for $\bar{n}_H$ can be used to obtain 20-9

$$\bar{n}_H = \frac{\beta_1^H\,[H^+] + 2\beta_2^H\,[H^+]^2}{1 + \beta_1^H\,[H^+] + \beta_2^H\,[H^+]^2} \qquad (20\text{-}9)$$

then, rearranging in the form $y = mx + b$, results in equation 20-10. This equation can be employed, either graphically or by linear regression, to provide the constants β_1^H and β_2^H from the slope and intercept.

$$\frac{\bar{n}_H}{[H^+]\,(1 - \bar{n}_H)} = \beta_2^H\,\frac{[H^+]\,(2 - \bar{n}_H)}{(1 - \bar{n}_H)} + \beta_1^H \qquad (20\text{-}10)$$

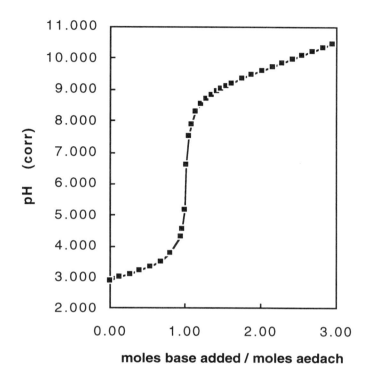

Figure 20-1. Titration curve of aedach·3HBr

THE SPREADSHEET. The constants table from the spreadsheet is shown in Figure 20-2. The names applied to the cell references are shown in column F of the data table; Create Names was used to assign names to the references.

A portion of the data table of the spreadsheet is shown in Figure 20-3. Columns A and B contain the experimental data. The expressions used in columns C, D and E are, respectively:

moles base added/moles aedach   =v*CB/(V_0*CL)

pH_{corr} =pH+C_

$\bar{n}_H$ =(3*CL*V_0/(V_0+v)+10^-(pcKw-

                                 -10^-pH_corr)/(CL*V_0/(V_0+v))

	B	C	D	E	F	G
4	Initial volume, mL				(V_0)	50.00
5	Initial Concentration of aedach.3HBr, M				(CL)	0.002056
6	pH correction factor C				(C_)	-0.11
7	Concentration of titrant NaOH, M				(CB)	0.1381
8	Buret calibration factor				(calib)	0.990
9	pcKw				(pcKw)	13.78

Figure 20-2. Constants table of the spreadsheet.

	A	B	C	D	E	F
	Y, mL	**pH**	**mol b/a**	**pH(corr)**	**n-bar**	**log K**
12	Y, mL	pH	mol b/a	pH(corr)	n-bar	log K
13	0.000	3.008	0.00	2.898	2.385	2.69
14	0.100	3.096	0.13	2.986	2.364	2.74
15	0.200	3.194	0.27	3.084	2.332	2.78
16	0.300	3.310	0.40	3.200	2.292	2.82
17	0.400	3.447	0.54	3.337	2.242	2.84
18	0.500	3.627	0.67	3.517	2.186	2.87
19	0.600	3.890	0.81	3.780	2.120	
20	0.700	4.438	0.94	4.328	2.046	
21	0.720	4.684	0.97	4.574	2.029	
22	0.740	5.270	0.99	5.160	2.012	
23	0.760	6.710	1.02	6.600	1.989	

Figure 20-3. Portion of the spreadsheet for the calculation of the protonation constants of aedach

The plot of $\bar{n}_H$ as a function of pH is shown in Figure 20-4. Since K_3^H does not overlap with the other protonation constants, it can be calculated using the following expression, entered in column F:

=LOG((n_bar-2)/(3-n_bar))+pH_corr

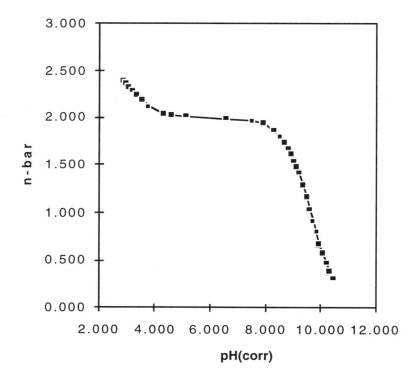

Figure 20-4. $\bar{n}_H$ as a function of pH

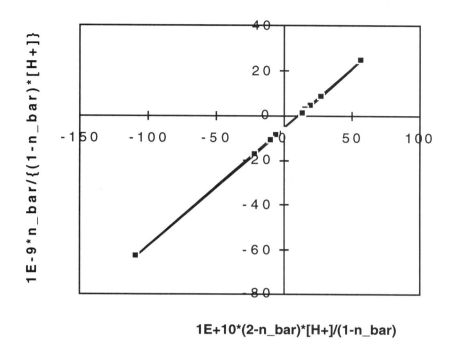

1E+10*(2-n_bar)*[H+]/(1-n_bar)

Figure 20-5. Linear transformation of the formation function equation used to obtain K_1^H and K_2^H from the slope and intercept.

The overlapping protonation constants K_1^H and K_2^H were resolved by using equation 20-10. The following scaled expressions were employed, to eliminate clutter on the y- and x-axis labels.

=1E-9*n_bar/((1-n_bar)*10^-pH_corr)

=1E+10*(2-n_bar)*10^-pH_corr/(1-n_bar)

The chart is shown in Figure 20-5. It is used only to verify that the transformed data fit a linear relationship. LINEST was used to obtain the constants from the slope and intercept of the regression line.

	G	H
45	Slope & Intercept by LINEST	
46	*slope*	*intercept*
47	1.89E+19	1.15E+10
48	6.26E+16	1.13E+08
49	0.99984	438911125

Figure 20-6. Slope and Intercept using LINEST.

From the slope and intercept of Figure 20-6, the values log K_1^H = 10.06 ± 0.005 and log K_2^H = 9.22 ± 0.005 were obtained. The "literature" values, calculated from the same data using a FORTRAN program, are 10.10 ± 0.008 and 9.22 ± 0.01.[*]

THREE OVERLAPPING PROTONATION CONSTANTS OF A POLYAMINE USING LEAST-SQUARES CURVE FITTING AND THE SOLVER

There are a number of computer programs available for the determination of stability constants from pH titration data. The most general of these perform a least-squares fit of the data to a calculated titration curve. The programs are able to handle protonated complexes, polynuclear systems, etc. In this example least-squares curve fitting is applied to a somewhat simpler case, a polyprotic acid in which the equilibria overlap extensively. The method is that used in the computer program SCOGS[*] (Stability Constants of Generalized Species) and described in Chapter 18, namely using pH as the independent variable to calculate the volume of titrant. Then, using the Solver, the sum of squares of residuals ($V_{calc} - V_{obsd}$) is minimized to find the best values of the set of equilibrium constants used to generate the curve.

The example used here is the determination of the protonation constants of 3,2,3-tet (1,5,8,12-tetraazadodecane, $H_2N(CH_2)_3NH(CH_2)_2NH(CH_2)_3NH_2$). The tetraprotonated amine was titrated with standard base. It can be seen from the titration curve shown in Figure 20-7 that while one proton is much more acidic

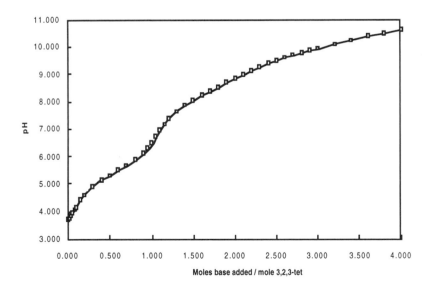

Figure 20-7. Titration curve of 3,2,3-tet. The line is calculated using the four protonation constants found using the **Solver.**

[*]   B. N. Patel and E. J. Billo, *Inorg. Nucl. Chem. Lett.* **1977**, *13*, 335.

[*]   I. G. Sayce, *Talanta* **1968**, *15*, 1397.

	A	B	C	D	E	F
1	**Protonation Constants of 1,5,8,12-tetraazadodecane**					
2	(N,N'-bis(3-aminopropyl)ethylenediamine or 3,2,3-tet)					
3		ligand, mmol		0.1037	(CL)	
4		strong acid, mmol		0.4232	(CA)	
5		titrant base, M		0.1050	(CB)	
6		total volume, mL		55	(V_0)	
7		buret calibration factor		0.990	(calib)	
8		pH correction factor C		-0.1	(C_)	
9		pcKw		13.78	(pcKw)	
10						

Figure 20-8. Data header table for the titration of protonated 3,2,3-tet with standard base

than the others and dissociates in the region from pH 4 to pH 6, the acidities of the other three protons are similar, so that three of the protonation regions overlap.

Using equation 20-11, you can calculate the volume of titrant as a function of the measured pH, then minimize the sum of squares of residuals $V_{exptl} - V_{calc}$, using the Solver.

$$V_{calc} = \frac{C_A + J C_L + [OH^-](V_0 + V) - [H^+](V_0 + V) - j\Sigma H_j L}{C_B} \qquad (20\text{-}11)$$

THE SPREADSHEET. The spreadsheet header table is shown in Figure 20-8. Create Names was used to assign the names in cells D3:D9 to cells C3:C9. Figure 20-9 shows a small portion of the data (entered in columns A and C) and the corrected volume and pH values. V_{corr} is obtained by multiplying the nominal volume by the buret calibration factor, and pH_{corr} by adding the correction factor C to the measured pH. Columns of intermediate calculations are illustrated in Figure 20-10.

Expressions for the denominator of the α expressions, for α_4 and for V_{calc} are as follows:

=1+10^(logK1H-pH)+10^(logK1H+logK2H-2*pH)+10^(logK1H+logK2H
+logK3H-3*pH) +10^(logK1H+logK2H+logK3H+logK4H-4*pH)

=10^(logK1H+logK2H+logK3H+logK4H-4*pH)/denom

=(CA+10^-(pKw-pH)*(V_0+v) -10^(-pH)*(V_0+v)-
(G12+2*H12+3*I12+4*J12) *CL)/CB

	A	B	C	D
11	**V, mL**	**V(corr)**	**pH**	**pH(corr)**
12	0.000	0.000	3.798	3.698
13	0.020	0.020	3.862	3.762
14	0.040	0.040	3.951	3.851

Figure 20-9. Portion of the spreadsheet data table.

10							SUM =	0.0025
11	Denom	alpha0	alpha1	alpha2	alpha3	alpha4	V(calc)	d^2
12	1.1E+19	9.5E-20	5.0E-13	5.5E-07	1.8E-02	9.8E-01	-0.007	5.5E-05
13	5.9E+18	1.7E-19	7.7E-13	7.4E-07	2.1E-02	9.8E-01	0.010	1.0E-04
14	2.6E+18	3.8E-19	1.4E-12	1.1E-06	2.5E-02	9.7E-01	0.031	7.5E-05

Figure 20-10. Portion of the spreadsheet showing intermediate calculations and the sum of squares of residuals.

The standard deviations shown in Figure 20-11 were obtained by using the SOLVSTAT.XLM macro described in Chapter 17.

	G	H	I	J
3	Protonation constants:			± std.dev.
4	logK1H =		10.42	0.007
5	logK2H =		9.74	0.006
6	logK3H =		8.21	0.006
7	logK4H =		5.44	0.006

Figure 20-11. The final values of the protonation constants of 3,2,3-tet.

DETERMINATION OF BINDING CONSTANTS BY SPECTROPHOTOMETRY

The spectrophotometric method is the method of choice for the chemist with an occasional or one-time need to determine a binding constant. The basic concept is obvious, the apparatus is widely available and (usually) the chemist is experienced in the technique. By contrast the potentiometric method requires significant preparation and familiarization before reliable results can be obtained. A further advantage of the spectrophotometric method is that it is applicable to non-basic ligands, such as halide ions.

In the spectrophotometric method the molar absorptivity of the complex is an additional variable to be determined. As well, it is necessary to determine the stoichiometry of the complex before calculations can be performed. The *mole-ratio* and *continuous variations* methods are useful in determining the stoichiometry. The observation (or lack) of isosbestic points is also a useful guide to the complexity of the system.

In the mole-ratio method a series of solutions is prepared in which the concentration of one reactant (usually the metal ion) is held constant while the other reactant (usually the ligand) is varied. In the discussion that follows it is assumed that the ligand concentration is varied.

Absorbance measurements are made at a wavelength where the complex absorbs strongly (it is convenient if neither the metal ion nor the ligand absorbs at that wavelength, but this is not a necessity). The absorbance is plotted versus concentration of the ligand (Figure 20-12). If only one complex of high stability is formed, the graph consists of two linear intersecting parts. The ratio of the concentration of ligand at the intersection point to the (fixed) concentration of metal ion gives the stoichiometry of the complex.

If the stability constant of the complex is high (curve A in Figure 20-12), there will be no appreciable dissociation of the complex at or near the stoichiometric point. If the complex is moderately stable (curve B in Figure 20-12), the plot will consist of two straight-line portions with a central curved portion. Extrapolation of the two straight-line portions yields the intersection point. If a complex of low stability is formed, a large excess of ligand will have to be used to drive the reaction to completion, and there will be no detectable break in the curve from which to obtain the stoichiometry (curves C and D in Figure 20-12).

Once the stoichiometry of the complex has been established, the stability constant(s) can be calculated, provided the data yields a curve showing some dissociation in the neighborhood of the stoichiometric point (curve B in Figure 20-12). Briefly, for any data point in the region of curvature, complex formation did not proceed to completion, as evidenced from the difference between the measured curve and the "theoretical" one. Here there is obviously an equilibrium between metal ion, ligand and complex, and from each data point a value of the stability constant can be calculated.

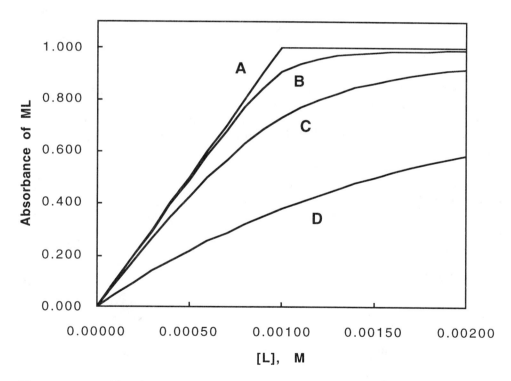

Figure 20-12. Absorbance curves for the formation of a complex ML in solutions containing a fixed stoichiometric concentration of metal ion M (0.00100 M) and varying concentrations of ligand L. Curves are shown for equilibrium constants $K = 1 \times 10^9$ (A), 1 $\times 10^5$ (B), 1×10^4 (C) and 1×10^3 (D).

The following example assumes the simplest case, in which a complex of stoichiometry ML is formed (equation 20-12), and M and L do not absorb at the

$$M + L \rightleftharpoons ML \qquad\qquad K = [M][L]/[ML] \qquad\qquad (20\text{-}12)$$

wavelength used. The concentrations of the three species can readily be calculated from the following relationships:

$$[M] = [M]_T - [ML] \qquad\qquad\qquad (20\text{-}13)$$
$$[L] = [L]_T - [ML] \qquad\qquad\qquad (20\text{-}14)$$
$$[ML] = A/(b\,\varepsilon) \qquad\qquad\qquad (20\text{-}15)$$

where A is the measured absorbance reading, b is the path length (cm), ε is the molar absorptivity of the complex, $[M]_T$ is the analytical concentration of metal and $[L]_T$ is the analytical concentration of ligand.

Note that no valid calculation can be made where the data points are on, or very close to, the straight-line portions of the curve. Here the reaction has proceeded to completion (driven by excess metal ion or ligand), and the concentration of the other species (ligand or metal ion, respectively) is essentially zero.

EXPERIMENTAL TECHNIQUES

For maximum accuracy, the batch method (separate solutions prepared in volumetric flasks) is preferred, but a titration method (aliquots of ligand solution added to a single solution in a spectrophotometer cell) may be necessary in some situations. Often, much lower concentrations of metal ion and ligand can be used than in the pH titration method, if the molar absorptivity of the complex is high.

Buffers can be used to control pH, provided it is determined that the buffer components do not interact, e.g., with the metal ion. However, buffers often absorb in the UV. In any event, a blank consisting of background electrolyte plus buffer should be used in the reference cell. Sodium or potassium nitrate usually can't be used as background electrolyte, because of the absorption band of nitrate centered at 300 nm.

CALCULATIONS

All equations derived for use in spectrophotometric methods are based on four fundamental relationships: (1) Beer's law ($A = \varepsilon b C$), (2) additivity of absorbances ($A = \Sigma A_i$), (3) mass balance ($C_T = \Sigma C_i$), and (4) equilibrium constant expressions; A is the absorbance, ε the molar absorptivity, b the path length and C the concentration.

Absorbance data from different experiments (where concentration and/or path length are different) may be combined by using the effective molar absorptivity $\varepsilon' = A_{obs}/bC_T$.

DETERMINATION OF TWO OVERLAPPING
PROTONATION CONSTANTS OF 4,5-DIHYDROXYACRIDINE

When spectrophotometric methods are used in cases involving multiple, overlapping equilibria, the situation becomes complicated.

The following example deals with the determination of two rather closely-spaced protonation constants (equations 20-16 and 20-17) of 4,5-dihydroxyacridine[*].

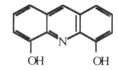

4,5-dihydroxyacridine

$$H^+ + A^{2-} \rightleftharpoons HA^- \qquad\qquad K_1^H \qquad\qquad (20\text{-}16)$$

$$H^+ + HA^- \rightleftharpoons H_2A \qquad\qquad K_2^H \qquad\qquad (20\text{-}17)$$

Because of solubility considerations, the measurements were made in 50% (v/v) dioxane-water. 4,5-Dihydroxyacridine is essentially colorless in 50% (v/v) dioxane-water solution at pH 7, but becomes yellow as the pH is increased and the phenol groups ionize. UV-visible spectrophotometry revealed that the ionized compound absorbed at 450 nm. Solutions were prepared in volumetric flasks and the pH and A_{450} measured, in 1.00-cm cells. A graph of absorbance vs. pH is shown in Figure 20-13.

The pH ranges for the gain or loss of the two protons overlap considerably, and as a result the graph does not exhibit a pH region where only the monoprotonated species HA^- absorbs, although the absorbances corresponding to the species A^{2-} and H_2A can readily be obtained. Estimates of the two protonation constants could be obtained by, for example, calculating log K_1^H from the data at high pH where, presumably, only the species A^{2-} and HA^- absorb. However, you can use the Solver to fit the complete range of absorbance-pH data with three parameters: K_1^H, K_1^H and A_1. The absorbance at any point is given by the equation

$$A_{obs} = \alpha_0 A_0 + \alpha_1 A_1 + \alpha_2 A_2 \qquad\qquad (20\text{-}18)$$

where the α's are the fractions in the forms containing 0, 1 or 2 protons and the A's are the corresponding absorbances. The α's are calculated using the equations given in Chapter 18.

---

[*]   A. Corsini and E. J. Billo, *J. Inorg. Nucl. Chem.* **1970**, *32*, 1241.

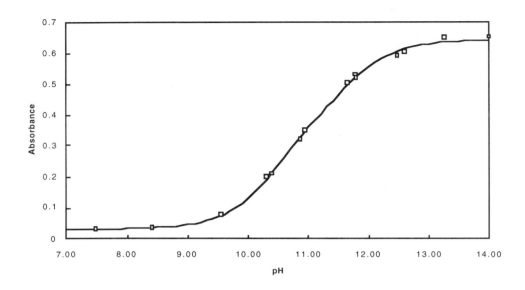

Figure 20-13. Spectrophotometric data for the protonation of 4,5-dihydroxyacridine. The line is calculated using the constants obtained using the Solver.

THE SPREADSHEET. Columns C, D and E of Figure 20-14 contain the expressions for α_0, α_1 and α_2 respectively:

=1/(1+10^(log_K1-A11)+10^(log_K1+log_K2-2*A11))

=10^(log_K1-A11)/(1+10^(log_K1-A11)+10^(log_K1+log_K2-2*A11))

=10^(log_K1+log_K2-2*A11)/(1+10^(log_K1-A11)+10^(log_K1+log_K2-2*A11))

and column F contains the expression for A_{calc}:

=A_0*C11+A_1*D11+A_2*E11

Column G contains the squares of the residuals, which are summed in $G$8. This is the target cell, to be minimized by variation of the changing cells $E$6 and $G$5:$G$6. Examination of the data suggested the following initial values: $A_1 = 0.4$, $\log K_1^H = 12$, $\log K_2^H = 10.5$. $A_0 = 0.650$ and $A_2 = 0.032$ were held constant. Use of these values gives a satisfactory initial fit, as indicated by the residuals. The Solver obtained the solution shown in Figure 20-15.

	A	B	C	D	E	F	G	H
1	\multicolumn{8}{c}{Protonation Constants of 4,5–Dihydroxyacridine}
2				\multicolumn{2}{c}{Using the Solver}			
3	\multicolumn{8}{c}{(solvent 50%v/v dioxane–water, ionic strength 0.1, 25 deg C)}
4		Constants:		A_0 =	0.650	log K1 =	12	
5				A_1 =	0.4	log K2 =	10.5	
6				A_2 =	0.032			
7						SUM(d^2) =	7.7E-03	
8	pH	Abs	alpha0	alpha1	alpha2	Acalc	d^2	
9	7.48	0.032	0.000	0.001	0.999	0.032	1.2E-07	
10	8.42	0.037	0.000	0.008	0.992	0.035	3.9E-06	
11	9.56	0.077	0.000	0.103	0.897	0.070	4.7E-05	
12	10.30	0.200	0.008	0.384	0.608	0.178	4.8E-04	
13	10.39	0.209	0.011	0.432	0.557	0.198	1.3E-04	
14	10.87	0.320	0.049	0.666	0.284	0.308	1.5E-04	
15	10.95	0.349	0.062	0.693	0.246	0.325	5.8E-04	
16	11.65	0.502	0.294	0.659	0.047	0.456	2.1E-03	
17	11.78	0.530	0.364	0.604	0.032	0.479	2.6E-03	
18	11.79	0.520	0.370	0.600	0.031	0.481	1.5E-03	
19	12.48	0.590	0.749	0.248	0.003	0.586	1.3E-05	
20	12.61	0.602	0.802	0.197	0.002	0.600	4.6E-06	
21	13.27	0.648	0.949	0.051	0.000	0.637	1.2E-04	
22	14.00	0.650	0.990	0.010	0.000	0.648	6.1E-06	
23								Acalc
24	7.00		0.000	0.000	1.000			0.032
25	7.10		0.000	0.000	1.000			0.032
26	7.20		0.000	0.001	0.999			0.032

Figure 20-14. Spreadsheet for the determination of the two protonation constants of 4,5-dihydroxyacridine from spectrophotometric data.

	A	B	C	D	E	F	G
4		Constants:		A_0 =	0.650	log K1 =	11.86
5				A_1 =	0.432	log K2 =	10.51
6				A_2 =	0.032		
7						SUM(d^2) =	1.1E-03

Figure 20-15. Changing cells ($D$5, $F$4 and $F$5) and target cell ($G$7) for the determination of the two protonation constants of 4,5-dihydroxyacridine using the Solver.

THE BJERRUM pH-SPECTROPHOTOMETRIC METHOD

The heterocyclic ligand 1,10-phenanthroline forms an orange-red complex

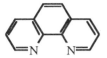

1,10-phenanthroline (phen)

with iron(II), $Fe(phen)_3^{2+}$, which absorbs at 510 nm (molar absorptivity 1.1×10^4 $M^{-1} cm^{-1}$). The stability of the complex is very high.[*] The mole-ratio plot (a plot of absorbance versus ligand concentration, with Fe(II) concentration constant) is a pair of intersecting straight lines, similar to curve A in Figure 20-14, intersecting at n = 3. This permits the determination of the stoichiometry with little uncertainty about the value of n, but does not provide data from which the stability constant can be determined. However, the stability constant can be determined by making use of the fact that the ligand is a weak base. In acidic solution protons compete with Fe^{2+} for the basic nitrogen donors and at low pH, the complex will be partially dissociated. The competition reaction (equation 20-19) can be used to determine the formation constant.

$$Fe^{2+} + 3\,HL^+ \rightleftharpoons FeL_3 + 3\,H^+ \qquad\qquad (20\text{-}19)$$

The procedure used in this example begins with the preparation of a series of solutions, all containing the same concentration of metal ion and the same concentration of ligand, but having different low pH values. The absorbance of each solution is measured; a plot of absorbance versus pH provides a curve that shows the extent of formation of the complex as a function of pH, similar to Figure 20-16. From this data, the stability constant can be calculated, in the following manner.

At a given pH the conditional constant β', so called because it is a constant valid for only a particular pH value, can be determined.

$$\beta' = \frac{[FeL_3]}{[Fe]\,[L']^3} \qquad\qquad (20\text{-}20)$$

where [L'], the total concentration of all forms of L, = [L] + [HL] + [H_2L].

Using the pK_a values of the ligand and the measured pH, the α_0 factor (see Chapter 18) for the ligand can be calculated, where α_0 is defined as the fraction of the total ligand in the unprotonated form, i.e., $\alpha_0 = [L]/[L']$. For 1,10-phenanthroline, which effectively behaves as a mono-base,

$$\alpha_0 = \frac{K_a}{K_a + [H^+]} \qquad\qquad (20\text{-}21)$$

---

[*]   Because of a change in spin state upon addition of the third ligand, the stability of the tris complex is much greater than that of the mono or bis complex. As a result, the reaction essentially yields only the tris complex. The concentration of the mono or bis complexes is close to zero.

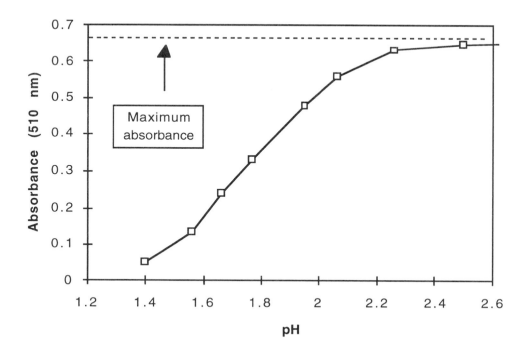

Figure 20-16. Absorbance data for Fe(II)-phen complex as a function of pH.

The equivalent equation using protonation constants instead of pK_a values is given in equation 20-22.

$$\alpha_0 = \frac{1}{1 + K^H [H^+]}\qquad(20\text{-}22)$$

Then $[phen] = \alpha_0 / [phen']$. $\qquad(20\text{-}23)$

From this, the overall formation constant β_3 can be calculated using equation 20-23 or 20-24.

$$\beta_3 = \frac{[Fe(phen)_3]}{[Fe]\,[phen]^3} = \frac{\beta'}{\alpha_0{}^3}\qquad(20\text{-}23)$$

$$\log \beta_3 = \log \beta' - 3 \log \alpha_0\qquad(20\text{-}24)$$

THE SPREADSHEET.   The analytical concentrations of $[Fe^{2+}]_T$, $[phen]_T$, the molar absorptivity of the $Fe(phen)_3{}^{2+}$ complex (determined in a separate experiment), and the pK_a of 1,10-phenanthroline are entered as constants in cells B3, D3, F3 and H3, respectively.   The experimental data (measured pH and absorbance at 525 nm) are in columns A and B. The expressions for $[Fe(phen)_3]$, free $[Fe]$, total $[phen']$, α_0, free $[phen]$ and $\log \beta_3$ are given in Figure 20-17.

	A	B	C	D	E	F	G	H
1			Log beta	for Fe(II)-	1,10-phenanthroline			
2	[Fe]T =	5.98E-05	[phen]T =	2.47E-04	E =	1.11E+04	pKa =	4.98
3	pH	Abs.	[FeL3]	[Fe]	[L']	alpha	[L]	log beta
4	1.56	0.134	1.21E-05	4.77E-05	2.10E-04	0.00038	7.99E-08	20.7
5	1.66	0.239	2.15E-05	3.83E-05	1.82E-04	0.00048	8.70E-08	20.9
6	1.76	0.337	3.04E-05	2.94E-05	1.55E-04	0.00060	9.36E-08	21.1
7	1.77	0.332	2.99E-05	2.99E-05	1.57E-04	0.00062	9.66E-08	21.0
8	1.93	0.480	4.32E-05	1.66E-05	1.17E-04	0.00089	1.04E-07	21.4
9	1.95	0.480	4.32E-05	1.66E-05	1.17E-04	0.00093	1.09E-07	21.3
10	2.06	0.561	5.05E-05	9.28E-06	9.49E-05	0.00120	1.14E-07	21.6
11	2.26	0.632	5.69E-05	2.88E-06	7.57E-05	0.00190	1.44E-07	21.8
12							Mean =	21.2
13	Equations used (in row 4):							
14		[FeL3] =	=B4/$F$3					
15		[Fe] =	=$B$3-C4					
16		[L'] =	=$D$3-3*C4					
17		alpha =	=10^-$H$3/(10^-$H$3+10^-A4)					
18		[L] =	=F7*E4					
19		log beta =	=LOG(C4/(D4*G4^3))					

Figure 20-17. Spreadsheet for the determination of the overall formation constant of tris(1,10-phenanthroline)iron(II). (Student data, from E. J. Billo)

The formation constant of the tris(1,10-phenanthroline)iron(II) complex has been determined by a number of methods, including potentiometry, a competitive spectrophotometric method, and a method involving partition between aqueous and organic solvents. The typical value for log β_3 is about 21.3[*]

DETERMINATION OF BINDING CONSTANTS BY NMR MEASUREMENTS

Nuclear magnetic resonance spectroscopy is a powerful tool for the determination of structural information of complexes in solution. It can also be used for the examination of solution equilibria. In addition to providing quantitative binding constant data, the NMR method can often yield information concerning the site of binding.

The majority of NMR studies of binding constants have employed proton NMR, and discussion will be restricted to that nucleus. Consider an organic ligand with one or more protons, forming a 1:1 complex with a metal ion. A ligand proton will experience different chemical environments when the ligand is free or bound, giving rise to two different chemical shifts, δ_{free} and δ_{bound}. The appearance of the NMR spectrum of the complex will depend on whether chemical exchange of the ligand between the free state and the bound state is slow or fast. If exchange is slow relative to the "NMR time scale" (i.e., lifetimes of minutes or longer) then the spectrum of both species, the bound ligand and the free ligand, will be observable. If the system is truly at equilibrium, then the

---

[*]   T. S. Lee, I. M. Kolthoff and D. L. Leussing, *J. Am. Chem. Soc,* **1948,** *70,* 2348. H. Irving and D. H. Mellor, *J. Chem. Soc.* **1955,** 3457. H. Irving and D. H. Mellor, *J. Chem. Soc.* **1962,** 5222. G. Anderegg, *Helv. Chim. Acta* **1963,** *46,* 2397.

concentration of each of the species can be obtained from the NMR spectrum, and the equilibrium constant determined. This case is rarely encountered.

If the system is at the *fast exchange limit* (i.e., lifetimes of milliseconds or less), then the magnetic environment experienced by the proton will be averaged over the environments of the free and bound states, and an NMR singlet will be observed, at a frequency which is the weighted average of the time spent in the two states. This leads to equation 20-25 for the position of the NMR singlet under conditions of fast exchange, where the α's are the fractions of the total ligand in the free and one or more bound states.

$$\delta_{obsd} = \alpha_{free}\,\delta_{free} + \Sigma\,\alpha_{bound}\,\delta_{bound} \qquad (20\text{-}25)$$

EXPERIMENTAL TECHNIQUES

For studies in aqueous solution, it is necessary to use D_2O as solvent. In this case, exchangeable protons (e.g., protons on O, N, etc.) are almost never observable; the chemical shift of protons on, e.g., carbon are monitored. If the NMR study is carried out using an organic solvent, then protons on N or O can be observed.

Often, reagent additions are made to a single solution in the NMR tube using a microliter pipet. In this case, concentrations should be corrected for dilution. For the most precise work, individual solutions should be made up in e.g., 1-mL volumetric flasks.

Measurement of pH can be made in the NMR tube using a micro combination pH electrode of approximately 3 mm diameter. For adjustment of pH in D_2O solutions, DCl, D_2SO_4, $DClO_4$ and/or NaOD solutions are used. The correction for pH measurements made in D_2O with a glass electrode calibrated against standard buffers solutions is: $pD = pH_{meas} + 0.4$.[*]

CALCULATIONS

Only systems undergoing fast exchange, and only 1:1 binding, will be discussed here. Depending on the magnitude of the binding constant, the following situations may be observed:

Case I: both δ_{free} and δ_{bound} can be measured independently. The calculations in this case are identical to the spectrophotometric determination of the pK_a of an indicator, and will not be discussed further.

Case II: only δ_{free} can be measured independently. There are two unknowns to be determined, K and δ_{bound}, and these can be obtained as the slope and intercept of a linear transformation of the data.

Case III: neither δ_{free} or δ_{bound} can be measured independently. A curve-fitting approach is necessary, using the Solver.

It will be recognized that these situations are the same as for the spectrophotometric method.

---

[*]   P. K. Glasoe and F. A. Long, *J. Phys. Chem.* **1960**, *64*, 188.

MONOMER-DIMER EQUILIBRIUM

As part of a study of host–guest complexation, the hydrogen-bonded dimerization of the substituted urea N-phenyl-N'-(2-pyridyl)urea (U) in 1:1 CH_2Cl_2/toluene was studied. The chemical shift of the high-field urea proton is especially sensitive to concentration (see Figure 20-20).

The variation in chemical shift was analyzed assuming dimer formation. The equations are:

dimerization: $\qquad\qquad\qquad 2\,U \rightleftharpoons U_2 \qquad K = [U_2]\,/\,[U]^2 \quad$ (20-26)

mass balance: $\qquad\qquad [U]_T = [U] + 2\,[U_2]$ $\qquad\qquad\qquad\qquad$ (20-27)

chemical shift: $\qquad\qquad \delta_{obsd} = \alpha_1\,\delta_1 + \alpha_2\,\delta_2$ $\qquad\qquad\qquad\qquad$ (20-28)

where $\alpha_1 = [U]/[U]_T$ and $\alpha_2 = 2\,[U_2]/[U]_T$ are the fractions of U_T in the monomeric and dimeric form respectively.

Combining equations 20-26 and 20-27 yields the following expressions for the concentration of free [U] and the chemical shift:

$$[U] = \frac{\sqrt{8\,K\,[U]_T + 1} - 1}{4\,K} \qquad\qquad (20\text{-}29)$$

$$\delta_{calc} = \frac{[U]\,\delta_1 + 2\,K\,[U]^2\,\delta_2}{[U]_T} \qquad\qquad (20\text{-}30)$$

Using the Solver, the sum of squares of residuals, $\Sigma(\delta_{obsd} - \delta_{calc})^2$, is minimized in order to find the best values of K, δ_1 and δ_2. The results are shown in the spreadsheets of Figures 20-18 and 20-19.

THE SPREADSHEET. Figure 20-18 shows a portion of the spreadsheet and illustrates the layout used for generating the theoretical curve. The experimental data are in A11:A18 and B11:B18. Column C contains the expression for free [U]

=(SQRT(8*K*A11+1)-1)/(4*K)

and column D contains the expression for the calculated chemical shift.

=(C11*delta1+2*K*(C11^2)*delta2)/A11.

Below the data section lies an extensive table (not shown), used to obtain the smooth calculated curve, in rows 20 to 118.

Using the Solver, the sum of squares of residuals (in cell E19) is minimized, the changing cells being D3 (K), D4 (δ_1) and D5 (δ_2).

To plot the theoretical binding curve, generate δ_{calc} for a range of [U], as follows. In cell A20 type 0.000, in A21 type 0.001, then use AutoFill to generate values from 0.000 to 0.100 in steps of 0.001. **Copy** the expression for [U] and **Paste** in cell C20. **Fill Down** to end of table. Transfer the expression for δ_{calc} from any cell in row D to cell F20 (**Copy** to D20, **Cut** and **Paste** to F20 to transfer correctly). **Fill Down** to end of table.

To create the chart, select A11:B118 and G11:G118, then choose the

ChartWizard. Enter labels for the X- and Y-axes. To format the chart into a more presentable form, first double-click on it. The Y-axis begins at zero, so click on the Y-axis to select it, then choose **Scale** and set Minimum = 6. To remove the plotted points from the theoretical line, click on the curve to select the second data series, containing the δ_{calc} values in column F. Choose **Patterns**, then set Line = Custom and choose the solid line; set Marker = None. Then select the first data series (the experimental data points) by clicking on the line connecting them. Choose **Patterns** and set Line = None, Marker = Custom. To change the type of plotting symbol, you may have to alter the Foreground and Background settings until you get the symbol you want.

	A	B	C	D	E
1	Dimerization of N-phenyl-N'-(2-pyridyl)urea (U) Studied by NMR				
2	(Chemical shift ∂ of high field urea proton)				
3		K=	50.0		
4		∂1 =	7.000		
5		∂2 =	10.000		
6	Equations used:				
7	free [U]	=(SQRT(8*K*A11+1)-1)/(4*K)			
8	∂(calc)	=(C11*delta1+2*K*(C11^2)*delta2)/A11			
9					
10	[U]total/M	∂(obsd)	[U] free	∂(calc)	resid^2
11	0.0010	7.074	0.0009	7.252	3.16E-02
12	0.0020	7.298	0.0017	7.438	1.95E-02
13	0.0050	7.718	0.0037	7.804	7.37E-03
14	0.0070	7.929	0.0047	7.966	1.34E-03
15	0.0100	8.128	0.0062	8.146	3.20E-04
16	0.0300	8.736	0.0130	8.697	1.50E-03
17	0.0800	9.358	0.0237	9.110	6.13E-02
18	0.1000	9.503	0.0270	9.190	9.83E-02
19				Σ =	2.21E-01

Figure 20-18. The spreadsheet before using the Solver, with initial estimates of K, $\delta 1$ and $\delta 2$, and showing some of the data used to calculate the theoretical binding curve. (NMR data provided by Dr. Steve Bell.)

	A	B	C	D	E
1	Dimerization of N-phenyl-N'-(2-pyridyl)urea (U) Studied by NMR				
2	(Chemical shift ∂ of high field urea proton)				
3		K=	39.1		
4		∂1 =	6.840		
5		∂2 =	10.588		
6	Equations used:				
7	free [U]	=(SQRT(8*K*A11+1)-1)/(4*K)			
8	∂(calc)	=(C11*delta1+2*K*(C11^2)*delta2)/A11			
9					
10	[U]total/M	∂(obsd)	[U] free	∂(calc)	resid^2
11	0.0010	7.074	0.0009	7.094	4.20E-04
12	0.0020	7.298	0.0018	7.293	2.59E-05
13	0.0050	7.718	0.0038	7.706	1.40E-04
14	0.0070	7.929	0.0050	7.897	1.01E-03
15	0.0100	8.128	0.0066	8.115	1.60E-04
16	0.0300	8.736	0.0142	8.813	5.86E-03
17	0.0800	9.358	0.0262	9.359	1.65E-06
18	0.1000	9.503	0.0299	9.466	1.37E-03
19				Σ =	8.99E-03

Figure 20-19. The spreadsheet after refinement of the constants.

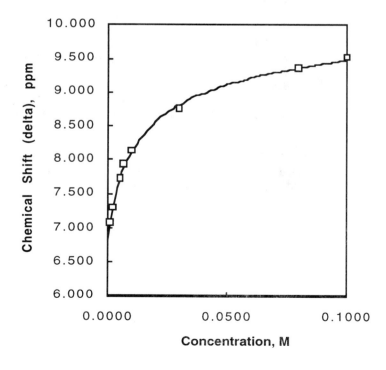

Figure 20-20. The fit of the chemical shift data. The curve is generated using equations and constants found in the text and in Figure 20-21.

ANALYSIS OF KINETICS DATA

In this chapter you'll learn how to extract rate constant information from simple first-order processes, from biphasic processes and from complex rate processes.

EXPERIMENTAL TECHNIQUES

In principle, any measurable property of a reacting system that is proportional to the extent of reaction may be used to monitor the progress of the reaction. The most common techniques are spectrophotometric (UV-visible, fluorescence, IR, polarimetry and NMR) or electrochemical (pH, ion-selective electrodes, conductivity and polarography). Either a "batch" method can be used, in which samples are withdrawn from the reaction mixture and analyzed, or the reaction may be monitored *in situ*. By far the most-used technique involves UV-visible spectrophotometry.

Since reaction rate is sensitive to temperature, the system must be thermostatted. For most reactions in aqueous solution the ionic strength should be controlled at a fixed value (see "Experimental Techniques" in Chapter 20).

ANALYSIS OF MONOPHASIC KINETICS DATA

Most reactions are characterized by a change in reactant or product concentration that can be described by a single exponential. The differential form of the rate equation contains a single term; the integrated form yields a straight line from which the rate constant can be obtained. Some of the more common and useful cases are described below.

FIRST-ORDER KINETICS

First-order reactions are by far the most common. They are also the simplest to study experimentally. For reactions of higher order, experimental conditions can usually be arranged so that they are first-order (see below). This simplifies the situation considerably.

For the reaction of species A to give product B, with rate constant k

$$A \xrightarrow{k} B$$

the rate of disappearance of A is proportional to the amount of A:

$$\frac{d[A]_t}{dt} = -k[A]_t \tag{21-1}$$

Of course the rate of appearance of product can also be used to monitor the reaction, since

$$\frac{-d[A]_t}{dt} = \frac{d[B]_t}{dt}$$

Integration of equation 21-1 leads to the relationship

$$\ln [A]_t = -kt + \ln [A]_0$$

or

$$\log [A]_t = -2.303\, k\, t + \log [A]_0 \tag{21-2}$$

that is, a plot of the logarithm of the concentration of A, plotted vs. time, yields a straight line from which the rate constant k can be obtained. The intercept term is usually of no interest.

An alternative form of equation 21-1 that sometimes is useful is

$$[A]_t = [A]_0\, e^{-k\, t} \tag{21-3}$$

Occasionally a first-order rate constant is obtained by experimental determination of the half-life $t_{1/2}$, the time required for the reactant concentration to decrease to one-half of its original value. From equation 21-2 it follows that $k = \ln(2)/t_{1/2} = 0.693/t_{1/2}$.

If a reaction is monitored by UV-visible spectrophotometry, for example, the concentration may be replaced by the absorbance (A) in equation 21-2. In the general case, both reactant and product may absorb at the monitoring wavelength, and thus the final absorbance is non-zero. Under these conditions the form of equation 21-4 that must be used is

$$\ln |A_t - A_\infty| = -k\, t + \ln |A_i - A_\infty| \tag{21-4}$$

where A_i is the initial absorbance reading and A_∞ is the absorbance value when the reaction is "complete". For first-order reactions the rule of thumb is that 10 half-lives must elapse before the reaction can be considered to be complete. After 10 half-lives a first-order reaction is $(1 - 0.5^{10})$ or 99.9% complete.

Figure 21-1 illustrates the application of equation 21-4 in the determination of the hydrolysis of a substrate by the enzyme thermolysin. The parameters returned by the SLOPE and INTERCEPT functions were used to calculate the theoretical line in column D of Figure 21-1. The formula in cell D8 is

=$B$26+A8*$B$25.

The first-order behavior is verified by the straight-line fit of the data, shown in Figure 21-2.

	A	B	C	D
1	Enzymatic hydrolysis of N–α–furyl acryloylglycyl–L–leucinamide substrate (FAGLA) by Thermolysin			
2	Run No:	XIII–28A	Wavelength:	340 nm
3	Date:	4/12/83	Pathlength:	1 cm
4	Operator:	M. McCarthy	Concentration:	1.00 mM
5			pH:	7.5
7	time, min	A(340)	$-\ln(A - A\infty)$	$-\ln A(\text{calc})$
8	0	1.731	0.4	0.357
9	2	1.581	0.644	0.627
10	4	1.457	0.914	0.897
11	6	1.367	1.168	1.167
12	8	1.292	1.444	1.437
13	10	1.254	1.619	1.707
14	12	1.201	1.931	1.977
15	14	1.158	2.283	2.247
16	16	1.138	2.501	2.517
17	18	1.12	2.749	2.787
18	20	1.103	3.058	3.057
19	22	1.091	3.352	3.327
20	24	1.082	3.650	3.597
21	26	1.077	3.863	3.867
22	28	1.072	4.135	4.137
23	∞	1.056		
24				
25		SLOPE =	0.13499118	
26		INTERCEPT =	0.35708031	

Figure 21-1. Data table for the enzymatic hydrolysis of FAGLA by thermolysin.

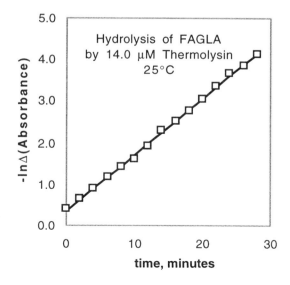

Figure 21-2. First-order plot for the hydrolysis of FAGLA.

REVERSIBLE FIRST-ORDER REACTIONS

If the reaction is reversible, e.g.,

$$A \underset{k_r}{\overset{k_f}{\rightleftharpoons}} B$$

then the rate of approach to equilibrium is a first-order process. If the A_∞ value is denoted by A_{eq}, then the first-order rate expression is simply

$$\ln |A_t - A_{eq}| = -k_{obsd}\, t + \ln |A_i - A_\infty| \qquad (21\text{-}5)$$

and $k_{obsd} = k_f + k_r$. Only the experimental constant k_{obsd} can be obtained from the first-order plot. If the equilibrium constant is known, the values of k_f and k_r can be calculated, since $k_f / k_r = K_{eq}$.

WHEN THE FINAL READING IS UNKNOWN

Occasionally it may not be possible to obtain A_∞ — for example, if the reaction is very slow, if secondary reactions occur toward the end of the primary reaction or if the experiment was terminated before the final reading was obtained. Obviously if a reaction has a half-life of one year it may not be practical to wait for the reaction to be complete.

Several ways have been developed to deal with a reaction for which the A_∞ value is not available. The Guggenheim method, for example, uses paired readings at t and $t + \Delta t$ to calculate the rate constant. By now you probably realize that a much simpler and direct method will be to use the Solver to find both the rate constant k and the A_∞ value by non-linear least-squares curve fitting.

The worksheet in Figure 21-3 illustrates a case of a reaction so slow that it was necessary to use the Solver to find the final absorbance reading. The unstable *cis*-octahedral isomer of the nickel(II) complex of the macrocyclic ligand cyclam (1,4,8,11-tetraazacyclotetradecane) isomerizes to the planar complex [Ni(cyclam)]$^{2+}$, which absorbs at 450 nm.[*] In acidic solution the reaction is slow; a single absorbance reading each day is more than sufficient to monitor the progress of the reaction.

Note the use of date and time arithmetic to calculate the elapsed time between readings. The formula in cell B7 is

=1440*(A7-$A$7).

Because the absorbance of the product is being monitored, the formula in cell D7 is

=Af-Af*EXP(-k_obsd*t).

The Solver was used to minimize the value in the target cell (E18, sum of squares of residuals) by varying the values of the changing cells (C19 and C20, A_∞ and k_{obsd}).

---

[*]   E. J. Billo, *Inorg. Chem.* **1984**, 23, 236.

	A	B	C	D	E
1	Isomerization of cis-[Ni(cyclam)(H₂O)₂]²⁺ in 0.50 M HClO₄				
2	Run No :	XII–21E	Wavelength :	450 nm	
3	Date :	1/15/81	Pathlength :	10 cm	
4	Operator :	J. Billo	Concentration :	2.41 mM	
5			pH :	0.50 M HClO₄	
6	Date & Time	t, min	Absorbance	A(calc)*	∂^2
7	01/15/81 10:05 AM	0	0.040	0.000	1.60E–03
8	01/15/81 12:20 PM	135	0.057	0.027	9.24E–04
9	01/16/81 09:10 AM	1385	0.270	0.240	9.06E–04
10	01/19/81 09:15 AM	5710	0.660	0.661	1.58E–06
11	01/20/81 09:20 AM	7155	0.724	0.735	1.25E–04
12	01/23/81 09:45 AM	11500	0.840	0.859	3.66E–04
13	01/26/81 09:20 AM	15795	0.906	0.908	4.07E–06
14	01/28/81 09:30 AM	18685	0.933	0.923	9.88E–05
15	02/02/81 09:10 AM	25865	0.950	0.937	1.71E–04
16	02/05/81 09:30 AM	30205	0.937	0.939	5.11E–06
17			* =Af–Af*EXP(–k_obsd*t)		
18				Σ(∂²) =	4.20E–03
19		A∞ =	0.941	(Af)	
20		k_obsd/min⁻¹ =	2.13E–04	(k_obsd)	

Figure 21-3. Obtaining the rate constant when the A_∞ value is unknown, by using the Solver.

SECOND-ORDER KINETICS

For the bimolecular reaction of species A and B to give product or products, with rate constant k

$$A + B \xrightarrow{\quad k \quad} C$$

the reaction is second-order and the rate depends on the concentration of both A and B:

$$\frac{d[A]_t}{dt} = -k\,[A]_t\,[B]_t \tag{21-6}$$

Integration of equation 21-6 yields equation 21-7, which can be used to demonstrate that a reaction is second order, and to obtain the rate constant.

$$\frac{1}{[A]_0 - [B]_0}\,\ln\frac{[B]_0\,[A]_t}{[A]_0\,[B]_t} = k\,t$$

or

$$\frac{2.303}{[A]_0 - [B]_0}\,\log\frac{[B]_0\,[A]_t}{[A]_0\,[B]_t} = k\,t \tag{21-7}$$

For the special case [A] = [B] equation 21-7 fails (since the denominator term becomes zero) and the alternate second-order expression 21-8 must be used:

$$\frac{1}{[A]_t} - \frac{1}{[A]_0} = k\,t \qquad (21\text{-}8)$$

The same equation applies if the reaction is second-order in a single reactant, e.g.,

$$\frac{d[A]_t}{dt} = -k\,[A]_t{}^2 \qquad (21\text{-}9)$$

PSEUDO-FIRST-ORDER KINETICS

If the concentration of species B (for example) is large relative to A, it will remain essentially unchanged during the course of the reaction and the rate expression 21-6 is simplified to 21-10, a form of equation 21-1. The reaction is said to be run under *pseudo-first-order* conditions.

$$\frac{d[A]_t}{dt} = -k\,[B]_i\,[A]_t = -k_{obsd}\,[A]_t \qquad (21\text{-}10)$$

and thus $\qquad\qquad\qquad k_{obsd} = k\,[B]_i \qquad\qquad\qquad (21\text{-}11)$

Once the first-order behavior with respect to [A] has been verified, the reaction can be run with varying concentrations of B (B still in large excess over A). A graph of k_{obsd} as a function of [B] should be linear; the slope is the rate constant k. For large variations in [B], resulting in large variations in k_{obsd}, it is often useful to plot log k_{obsd} vs. log [B]. The slope of the plot gives the order of the reaction with respect to [B}, in this case 1.0.

ANALYSIS OF BIPHASIC KINETICS DATA

Often a plot of concentration vs. time, or the monitored parameter vs. time, or the rate plot, will not be monophasic. This can arise from a number of different situations, the more common of which are described below.

CONCURRENT FIRST-ORDER REACTIONS

If, in a mixture of A and B, these components react by parallel first-order processes to give a common product C, and A and B do not interconvert, then a first-order plot of the rate of appearance of P will be curved, having a fast and a slow component.

$$A \xrightarrow{\ k_1\ } C$$
$$B \xrightarrow{\ k_2\ } C$$

This situation is commonly encountered in the measurement of radioactive decay of a mixture of radioisotopes.

CONSECUTIVE FIRST-ORDER REACTIONS

For consecutive first-order processes,

$$A \xrightarrow{\ k_1\ } B \xrightarrow{\ k_2\ } C$$

the rate expressions are

$$\frac{d[A]_t}{dt} = -k_1 [A]_t$$

$$\frac{d[B]_t}{dt} = k_1 [A]_t - k_2 [B]_t$$

$$\frac{d[C]_t}{dt} = k_2 [B]_t$$

which lead to the following expressions for the concentrations:

$$[A]_t = [A]_0 e^{-k_1 t} \tag{21-12}$$

$$[B]_t = [A]_0 \frac{k_1}{k_2 - k_1} \left[e^{-k_1 t} - e^{-k_2 t} \right] \tag{21-13}$$

$$[C]_t = [A]_0 \left[1 - \frac{k_2}{k_2 - k_1} e^{-k_1 t} + \frac{k_1}{k_2 - k_1} e^{-k_2 t} \right] \tag{21-14}$$

The concentrations of A, B and C for a typical series first-order process are shown in Figure 21-4.

The disappearance of A is purely first-order and can be used to determine the rate constant k_1. The species B is formed and then decays in an unmistakable series-first-order manner (Figure17-4 is an example of this). The appearance of C may seem to be pure first-order if the slight deviation from first-order behavior at the beginning of the reaction is missed. In addition, more than one species may

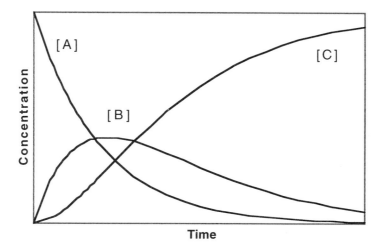

Figure 21-4. Concentration vs. time for consecutive first-order reactions.

absorb at a particular wavelength, complicating and confusing the situation. In the example that follows, both B and C absorb at the same wavelength. This results in behavior that is similar to, and difficult to distinguish from, concurrent first-order reactions.

EXAMPLE. The unstable *cis*-octahedral isomer of the nickel(II) complex of the macrocyclic ligand 13aneN$_4$ (1,4,7,10-tetraazacyclotridecane) isomerizes to an intermediate planar isomer, which then converts to the stable planar isomer [Ni(13aneN$_4$)]$^{2+}$; both planar isomers absorb at 425 nm.[*] The reaction exhibits a fast and a slow component, as illustrated in Figure 21-5.

The rate constants for the fast and slow reactions can be obtained in the following manner: the rate constant for the slow reaction is obtained from the data in the latter part of the reaction, by the usual first-order plot. The intercept of this plot at $t = 0$ is used to obtain A$_\infty$ for the fast reaction; the early-time data is then used to construct a second first-order plot. The first-order plots of ln(A$_\infty$ − A$_t$) vs. t for the data are shown in Figure 11-29.

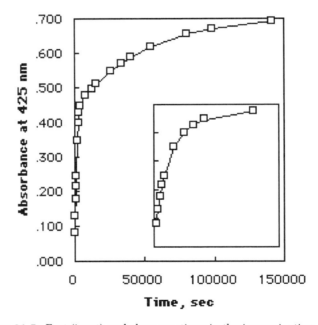

Figure 21-5. Fast (inset) and slow reactions in the isomerization of *cis*-[Ni(13aneN$_4$)(H$_2$O)$_2$]$^{2+}$.

---

[*]   Anne M. Martin, Kenneth J. Grant and E. Joseph Billo, *Inorg. Chem.* **1986**, *25*, 4904.

	A	B	C	D
1	Rate of Isomerization of cis-[Ni(13aneN₄)(H₂O)₂]²⁺			
2	Run No:	XIII-28A	Wavelength:	425 nm
3	Date:	7/19/82	Pathlength:	5 cm
4	Operator:	E. J. Billo	Concentration:	1.43 mM
5			pH:	2.34
7	t, sec	A(obs)	ln(A∞−A)	ln(A'∞−A)
8	160	.080	−0.465	−1.023
9	320	.131	−0.550	−1.176
10	480	.177	−0.633	−1.338
11	640	.216	−0.709	−1.499
12	800	.246	−0.772	−1.643
13	1600	.350	−1.027	−2.414
14	2400	.400	−1.178	
15	3200	.429	−1.277	
16	4000	.449	−1.351	
17	8000	.477	−1.465	
18	12000	.496	−1.551	
19	15200	.510	−1.619	
20	26000	.548	−1.833	
21	33200	.570	−1.981	
22	40400	.590	−2.137	
23	54800	.618	−2.408	
24	80000	.655	−2.937	
25	98000	.670	−3.270	
26	141200	.692	−4.135	
27	∞	0.708		

Figure 21-6. Data table for the isomerization of *cis*-[Ni(13aneN₄)(H₂O)₂]²⁺.

Applying LINEST to the data in the straight-line portion of the slow process (rows 17-26 of Figure 21-6) yields the rate constant for the slow process and permits the calculation of A'_∞ from the intercept value (A_0 for the slow process is A_∞ for the fast process). From ln $(A_\infty - A_0) = -1.315$, $A_\infty - A_0 = 0.269$, from which $A'_\infty = A_0 = 0.439$, as shown in Figure 21-7.

	A	B	C
28	Slow Reaction		
29	(LINEST on Rows 17−26)		
30		−*kobsd*	*intercept*
31		−2.00E−05	−1.315
32		8E−08	0.005
33		0.99988	0.01024
34			
35	A'∞ (from intercept):		0.439

Figure 21-7. Results of first-order plot of the slow part of the isomerization of *cis*-[Ni(13aneN₄)(H₂O)₂]²⁺.

	A	B	C
37		**Fast Reaction**	
38		**(LINEST on Rows 17–13)**	
39		*–kobsd*	*intercept*
40		-9.66E-04	-0.872
41		5E-06	0.004
42		0.99990	0.00554

Figure 21-8. Results of first-order plot of the fast part of the isomerization of cis-[Ni(13aneN$_4$)(H$_2$O)$_2$]$^{2+}$.

Having established the value for A'_∞, the data from the early stages of the reaction can be analyzed as a first-order process, by plotting ln $|A_t - A'_\infty|$ (see the inset in Figure 11-29). The results are shown in Figure 21-8.

CONSECUTIVE REVERSIBLE FIRST-ORDER REACTIONS

The sequence of coupled reversible first-order processes

$$A \underset{k_2}{\overset{k_1}{\rightleftharpoons}} B \underset{k_4}{\overset{k_3}{\rightleftharpoons}} C$$

yields the following set of differential equations:

$$\frac{d[A]_t}{dt} = -k_1[A]_t + k_2[B]_t$$

$$\frac{d[B]_t}{dt} = k_1[A]_t - k_2[B]_t - k_3[B]_t + k_4[C]_t$$

$$\frac{d[C]_t}{dt} = k_3[B]_t - k_4[C]_t.$$

The differential equations can be solved to yield analytical expressions for the concentrations of A, B and C. For the case where $[B]_0 = [C]_0 = 0$ at $t = 0$, the expressions are[*]:

$$[A]_t = [A]_0\left[\frac{k_2\,k_4}{\lambda_2\,\lambda_3} + \frac{k_1\,(\lambda_2 - k_3 - k_4)}{\lambda_2\,(\lambda_2 - \lambda_3)}\,e^{-\lambda_2\,t} + \frac{k_1\,(k_3 + k_4 - \lambda_3)}{\lambda_3\,(\lambda_2 - \lambda_3)}\,e^{-\lambda_3\,t}\right] \tag{21-15}$$

$$[B]_t = [A]_0\left[\frac{k_1\,k_4}{\lambda_2\,\lambda_3} + \frac{k_1\,(k_4 - \lambda_2)}{\lambda_2\,(\lambda_2 - \lambda_3)}\,e^{-\lambda_2\,t} + \frac{k_1\,(\lambda_3 - k_4)}{\lambda_3\,(\lambda_2 - \lambda_3)}\,e^{-\lambda_3\,t}\right] \tag{21-16}$$

$$[C]_t = [A]_0\left[\frac{k_1\,k_3}{\lambda_2\,\lambda_3} + \frac{k_1\,k_3}{\lambda_2\,(\lambda_2 - \lambda_3)}\,e^{-\lambda_2\,t} - \frac{k_1\,k_3}{\lambda_3\,(\lambda_2 - \lambda_3)}\,e^{-\lambda_3\,t}\right] \tag{21-17}$$

---

[*]   John W. Moore and Ralph G. Pearson, *Kinetics and Mechanism*, 3rd edition, Wiley-Interscience, New York, 1982, p. 298.

where $\lambda_2 = (p + q)/2$ (21-18)

              $\lambda_3 = (p - q)/2$ (21-19)

              $p = (k_1 + k_2 + k_3 + k_4)$ (21-20)

              $q = (p^2 + 4(k_1\,k_3 + k_2\,k_4 + k_1\,k_4)^{1/2}$ (21-21)

A system such as this can readily be solved using the Solver. The previous example, the isomerization of cis-[Ni(13aneN$_4$)(H$_2$O)$_2$]$^{2+}$ (Figure 21-5), is actually a case of coupled reversible first-order processes. There are two observable rate processes, k_{fast} and k_{slow}, but each is reversible. Treatment of the data according to the consecutive reversible first-order scheme is shown in Figure 21-6.

The expressions for [A]$_t$, [B]$_t$, [C]$_t$ and Abs$_{calc}$ in cells C14, D14, E14 and F14 are:

```
=C_T*(k_2*k_4/(L_2*L_3)+k_1*(L_2-k_3-k_4)*EXP(-L_2*t)/
    L_2*(L_2-L_3))+k_1*(k_3+k_4-L_3)*EXP(-L_3*t)/
    (L_3*(L_2-L_3)))

=C_T*(k_1*k_4/(L_2*L_3)+k_1*(k_4-L_2)*EXP(-L_2*t)/
    (L_2*(L_2-L_3))+k_1*(L_3-k_4)*EXP(-L_3*t)/(L_3*(L_2-L_3)))

=C_T*(k_1*k_3/(L_2*L_3)+k_1*k_3*EXP(-L_2*t)/(L_2*(L_2-L_3))-
    k_1*k_3*EXP(-L_3*t)/(L_3*(L_2-L_3)))

=b*(E_a*C14+E_b*D14+E_c*E14)
```

The quantities p, q, λ_2 and λ_3 were calculated using the formulas:

```
=k_1+k_2+k_3+k_4

=SQRT(P*P-4*(k_1*k_3+k_2*k_4+k_1*k_4))

=(P+Q)/2

=(P-Q)/2
```

Using the Solver, the sum of squares of residuals (cell G12, Figure 21-9) was minimized by changing the values of k_1, k_2, k_3 and k_4, and the molar absorptivity of the intermediate species B (ε_A and ε_C could be determined separately). The Solver provided a reasonable fit to the data. All five changing cells could be varied at once, although approximate estimates of the parameters had to be provided, before the Solver could proceed to a solution. The five parameters of the Solver solution are shown in bold in Figure 21-9.

	A	B	C	D	E	F	G
1	Rate of Isomerization of cis-[Ni(13aneN$_4$)(H$_2$O)$_2$]$^{2+}$						
2	Run No:		XIII-28A		Wavelength:		425 nm
3	Date:		7/19/82		Pathlength:		5 cm
4	Operator:		E. J. Billo		Concentration:		1.43 mM
5					pH:		2.34
6	(Values in bold are the changing cells, $G$12 is the target cell.)						
7			Rate constants:		Molar Absorptivities:		
8			k_1 =	5.27E-04	E_a =	0	
9			k_2 =	5.14E-04	E_b =	122	
10			k_3 =	3.07E-05	E_c =	115	
11			k_4 =	7.19E-06			
12						$\Sigma\delta^2$ =	4.52E-04
13	t, sec	A(obs)	[A]*	[B]*	[C]*	A(calc)	δ^2
14	160	.080	1.32E-03	1.11E-04	2.80E-07	0.067	1.6E-04
15	320	.131	1.23E-03	2.04E-04	1.06E-06	0.125	4.2E-05
16	480	.177	1.15E-03	2.82E-04	2.26E-06	0.173	1.7E-05
17	640	.216	1.08E-03	3.48E-04	3.81E-06	0.214	4.3E-06
18	800	.246	1.02E-03	4.04E-04	5.65E-06	0.249	7.3E-06
19	1600	.350	8.40E-04	5.73E-04	1.79E-05	0.358	6.8E-05
20	2400	.400	7.57E-04	6.40E-04	3.28E-05	0.408	6.4E-05
21	3200	.429	7.17E-04	6.65E-04	4.86E-05	0.432	8.5E-06
22	4000	.449	6.95E-04	6.71E-04	6.47E-05	0.445	1.8E-05
23	8000	.477	6.45E-04	6.42E-04	1.43E-04	0.472	2.1E-05
24	12000	.496	6.09E-04	6.07E-04	2.14E-04	0.492	1.5E-05
25	15200	.510	5.82E-04	5.81E-04	2.67E-04	0.507	1.2E-05
26	26000	.548	5.05E-04	5.05E-04	4.20E-04	0.548	1.4E-07
27	33200	.570	4.63E-04	4.64E-04	5.03E-04	0.571	1.3E-06
28	40400	.590	4.27E-04	4.29E-04	5.73E-04	0.591	2.7E-07
29	54800	.618	3.71E-04	3.74E-04	6.85E-04	0.621	9.3E-06
30	80000	.655	3.07E-04	3.11E-04	8.11E-04	0.656	4.1E-07
31	98000	.670	2.80E-04	2.84E-04	8.66E-04	0.671	6.0E-07
32	141200	.692	2.45E-04	2.50E-04	9.35E-04	0.690	5.4E-06
33	* (Equations for A–B–C reversible system are from Moore & Pearson, p. 298)						

Figure 21-9. Data table for the isomerization of cis-[Ni(13aneN$_4$)(H$_2$O)$_2$]$^{2+}$.

SIMULATION OF KINETICS
BY NUMERICAL INTEGRATION

Complex systems like that of Figure 21-5 can also be analyzed by Runge-Kutta numerical integration discussed in Chapter 14. The big advantage of direct numerical integration is its generality. The method can be applied to reaction sequences of (in theory, at least) any complexity. The following example applies the Runge-Kutta method to a more complex case.

In the experiment described in the previous section, the reactant species A was prepared from the final product C. The absorbance of the experiment at $t = 0$ showed that the solution contained a small amount of unconverted C. Instead of

$[A]_0 = 1.43$ mM and $[B]_0 = [C]_0 = 0$, the initial conditions were actually $[A]_0 = 1.37$ mM, $[B]_0 = 0$, $[C]_0 = 0.06$ mM. This set of initial condition cannot be accurately treated by using equations 21-15, 21-16 and 21-17, but it can be treated using RK integration.

The worksheet of Figure 21-9 was modified to use the RK method, as described in Chapter 14.

The Runge-Kutta formulas entered in row 18, for TA1, TA2, TA3, TA4 and $[A]_{t+\Delta t}$, and for TB1, etc, are:

=(-k_1*C_A+k_2*C_B)*(A19-A18)

=(-k_1*(C_A+TA1/2)+k_2*(C_B+TB1/2))*(A19-A18)

=(-k_1*(C_A+TA2/2)+k_2*(C_B+TB2/2))*(A19-A18)

=(-k_1*(C_A+TA3)+k_2*(C_B+TB3))*(A19-A18)

=C_A+(TA1+2*TA2+2*TA3+TA4)/6

=(k_1*C_A-(k_2+k_3)*C_B+k_4*C_C)*(A19-A18)

=(k_1*(C_A+TA1/2)-(k_2+k_3)*(C_B+TB1/2)
    +k_4*(C_C+TC1/2))*(A19-A18)

=(k_1*(C_A+TA2/2)-(k_2+k_3)*(C_B+TB2/2)
    +k_4*(C_C+TC2/2))*(A19-A18)

=(k_1*(C_A+TA3)-(k_2+k_3)*(C_B+TB3) +k_4*(C_C+TC3))*(A19-A18)

=C_B+(TB1+2*TB2+2*TB3+TB4)/6

=(k_3*C_B-k_4*C_C)*(A19-A18)

=(k_3*(C_B+TB1/2)-k_4*(C_C+TC1/2))*(A19-A18)

=(k_3*(C_B+TB2/2)-k_4*(C_C+TC2/2))*(A19-A18)

=(k_3*(C_B+TB3)-k_4*(C_C+TC3))*(A19-A18)

=C_C+(TC1+2*TC2+2*TC3+TC4)/6

Following the layout suggested in Chapter 14, the initial concentration of A was entered in cell C18; cell C19 contains the formula =H18. Similar formulas were used for $[B]_t$ and $[C]_t$.

When using the RK method (or any method involving numerical integration) it is important to use sufficiently small time increments to assure accuracy in calculations. In the RK formulas shown above, a calculated time increment, e.g. (A19-A18) in row 18, was used rather than a constant value, so that the interval between successive data points could be varied. In this way larger time increments can be used at the end of the reaction, where concentrations are changing slowly. Only a few cells in column B contain A_{obsd} measurements;

only one is shown in the spreadsheet fragment of Figure 21-10. The data table of time and A_{obsd} values was located elsewhere in the worksheet; VLOOKUP was used in column B to enter the A_{obsd} values for the appropriate time values. The formula is shown below:

=IF(ISERROR(VLOOKUP(t,DataTable,1,0)),"",VLOOKUP(t,DataTable,2,0))

The ISERROR function was used with *range_lookup* = 0; otherwise VLOOKUP returns #N/A! for all values of *t* for which there is no corresponding entry in DataTable.

An IF statement was used to calculate the squares of residuals only for rows that contained an A_{obsd} value.

The Solver was used in the same way as in the previous example, although it was necessary to perform the initial stages of the refinement using subsets of the complete set of parameters. Note the difference in the results when the small amount of product species that was present at the beginning of the reaction is

	A	B	C	D	E	F	G	H
1			Rate of Isomerization of cis-[Ni(13aneN₄)(H₂O)₂]²⁺					
2			(Simulation by RUNGE-KUTTA method)					
3	Run No :		XIII-28A		Wavelength :		425 nm	
4	Date :		7/19/82		Pathlength :	(b)	5 cm	
5	Operator :		E. J. Billo		Concentration :		1.43E-03 M	
6					pH :		2.34	
7		(Values in bold are the changing cells, $H$15 is the target cell.)						
8			Rate constants :		Molar Absorptivities :			
9		k_1 =	4.34E-04		E_a =	0		
10		k_2 =	4.39E-04		E_b =	122		
11		k_3 =	2.98E-05		E_c =	115		
12		k_4 =	6.43E-06					
13								
14		Concentrations corrected for		[A]i =	1.37E-03			
15		small amount of C present :		[C]i =	6.00E-05		Σδ² =	1.38E-04
16								
17	t, sec	Abs	[A](t)	TA1	TA2	TA3	TA4	[A](t+Δt)
18	0		1.37E-03	-5.95E-06	-5.92E-06	-5.92E-06	-5.90E-06	1.36E-03
19	10		1.36E-03	-5.90E-06	-5.87E-06	-5.87E-06	-5.85E-06	1.36E-03
20	20		1.36E-03	-5.85E-06	-5.82E-06	-5.82E-06	-5.79E-06	1.35E-03
21	30		1.35E-03	-5.79E-06	-5.77E-06	-5.77E-06	-5.74E-06	1.35E-03
22	40		1.35E-03	-5.74E-06	-5.72E-06	-5.72E-06	-5.69E-06	1.34E-03
23	50		1.34E-03	-5.69E-06	-5.67E-06	-5.67E-06	-5.64E-06	1.34E-03
24	60		1.34E-03	-5.64E-06	-5.62E-06	-5.62E-06	-5.60E-06	1.33E-03
25	70		1.33E-03	-5.60E-06	-5.57E-06	-5.57E-06	-5.55E-06	1.32E-03
26	80		1.32E-03	-1.11E-05	-1.10E-05	-1.10E-05	-1.09E-05	1.31E-03
27	100		1.31E-03	-1.09E-05	-1.08E-05	-1.08E-05	-1.07E-05	1.30E-03
28	120		1.30E-03	-1.07E-05	-1.06E-05	-1.06E-05	-1.05E-05	1.29E-03
29	140		1.29E-03	-1.05E-05	-1.04E-05	-1.04E-05	-1.03E-05	1.28E-03
30	160	0.080	1.28E-03	-2.07E-05	-2.03E-05	-2.03E-05	-2.00E-05	1.26E-03
31	200		1.26E-03	-2.00E-05	-1.96E-05	-1.96E-05	-1.93E-05	1.24E-03

Figure 21-10. Data table for the isomerization of *cis*-[Ni(13aneN₄)(H₂O)₂]²⁺ analyzed by the RK method. Columns I through Y (not shown) contain formulas for calculation of [B] and [C].

taken into acount: the values of the regression parameters are significantly different, and the sum of squares of residuals (1.38×10^{-4}) is significantly smaller than in the previous treatment (4.52×10^{-4}).

Calculation time will be significantly longer for a solution by direct numerical integration, compared to a solution using analytical expressions.

APPENDICES

A

SELECTED WORKSHEET FUNCTIONS BY CATEGORY

Excel 4.0 provides more than 300 worksheet functions, in 11 categories: Engineering, Financial, Date & Time, Information, Math & Trig, Statistical, Lookup & Reference, Database, Text, Logical, External. Excel 5.0 added a few more worksheet functions, for a total of 320. This book uses functions from seven of those categories. This appendix lists selected worksheet functions by category. In Appendix B, the functions are listed alphabetically, with comments, examples and related functions. Functions followed by ** indicate a new function in Excel 5.0; * indicates a function that is changed in Excel 5.0.

DATABASE & LIST MANAGEMENT FUNCTIONS
(In the following functions a range must be defined as *Database*.)

DAVERAGE	Returns the average of selected values.
DCOUNT	Returns the number of cells that contain numbers in a specified field, according to specified criteria.
DCOUNTA	Returns the number of cells that contain values in a specified field, according to specified criteria.
DGET	Returns a single record that matches the specified criteria.
DMAX	Returns the maximum value in the range *Database*.
DMIN	Returns the minimum value in the range *Database*.
DPRODUCT	Returns the product of values in a specified field that match specified criteria in a database.
DSTDEV	Returns the standard deviation of values in a specified field that match specified criteria in a database.
DSTDEVP	Returns the standard deviation of values in a specified field that match specified criteria in a database.
DSUM	Returns the sum of numbers in the field column of records in the database that match the criteria.
DVAR	Returns the variance of values in a specified field that match specified criteria in a database.
DVARP	Returns the variance of values in a specified field that match specified criteria in a database.
SUBTOTAL**	Returns a subtotal in a database.

DATE & TIME FUNCTIONS

DATE	Returns the serial number of a date.
DATEVALUE	Converts a date in the form of text to a serial number.
DAY	Converts a serial number to a day of the month.
MONTH	Converts a serial number to a month.
NOW	Returns the serial number of the current date and time.
TODAY	Returns the serial number of today's date.
WEEKDAY*	Converts a serial number to a day of the week.
YEAR	Converts a serial number to a year.

INFORMATION FUNCTIONS

CELL	Returns information about the formatting, location or contents of a cell.
COUNTBLANK**	Counts the number of blank cells within a range.
ERROR.TYPE	Returns a number corresponding to an error value.
INFO	Returns information about the current operating environment.
ISBLANK	Returns TRUE if the cell is blank.
ISERR	Returns TRUE if the argument is any error value except #N/A.
ISERROR	Returns TRUE if the argument is an error value.
ISEVEN	Returns TRUE if the argument is even.
ISLOGICAL	Returns TRUE if the argument is a logical value.
ISNA	Returns TRUE if the argument is #N/A.
ISNONTEXT	Returns TRUE if the argument is not text.
ISNUMBER	Returns TRUE if the argument is a number.
ISODD	Returns TRUE if the argument is odd.
ISREF	Returns TRUE if the argument is a reference.
ISTEXT	Returns TRUE if the argument is text.
N	Returns a value converted to a number. Excel usually does this automatically when necessary.
NA	Returns the error value #N/A
TYPE	Returns a number indicating the data type of a value.

LOGICAL FUNCTIONS

AND	Returns TRUE if all arguments are true, otherwise returns FALSE.
FALSE	Returns the logical value FALSE.
IF	Returns one value if *logical_test* is TRUE, another value if *logical_test* is FALSE.
NOT	Reverses the logic of its argument.
OR	Returns TRUE if any argument is TRUE.
TRUE	Returns the logical value TRUE.

LOOKUP & REFERENCE FUNCTIONS

ADDRESS	Returns a reference in the form of text.
AREAS	Returns the number of areas in a multiple selection.
CHOOSE	Chooses a value from a list of values, based on index number.
COLUMN	Returns the column number of a reference.
COLUMNS	Returns the number of columns in a reference.
HLOOKUP*	Finds the value in the first row of an array which is equal to or less than *lookup_value*. Returns the associated value in the *n*th row, as determined by *offset_num*.
INDEX	Returns a value from a reference or array, using a specified index.
INDIRECT	Returns a reference specified by a text value.
LOOKUP	Looks up values in an array.
MATCH	Looks up a value in an array, and returns its relative position.
OFFSET	Returns a reference offset from a base reference by specified number of rows and columns.
ROW	Returns the row number of a reference.
ROWS	Returns the number of rows in a reference.
TRANSPOSE	Returns the transpose of an array.
VLOOKUP*	Finds the value in the first column of an array which is equal to or less than *lookup_value*. Returns the associated value in the *n*th column, as determined by *offset_num*.

MATH & TRIG FUNCTIONS

ABS	Returns the absolute value of a number.
ACOS	Returns the angle corresponding to a cosine value.
ACOSH	Returns the inverse hyperbolic cosine of a number.
ASIN	Returns the angle corresponding to a sine value.
ASINH	Returns the inverse hyperbolic sine of a number.
ATAN	Returns the angle corresponding to a tangent value.
ATAN2	Returns the angle defined by a pair of x- and y-coordinates.
ATANH	Returns the inverse hyperbolic tangent of a number.
COS	Returns the cosine of a given angle.
COSH	Returns the hyperbolic cosine of a number.
COUNTIF**	Returns the number of non-blank cells within a range that meet the given criteria.
DEGREES	Converts a value in radians to degrees.
EXP	Returns the value of *e* raised to the power *number*.
FACT	Returns the factorial of a number, i. e., 1*2*3*...*number*.
INT	Rounds a number down to the nearest integer.

LN	Returns the base-e logarithm of a number.
LOG	Returns the logarithm of a number to the specified base.
LOG10	Returns the base-10 logarithm of a number.
MDETERM	Returns the determinant of an array.
MINVERSE	Returns the inverse of an matrix .
MMULT	Returns the product of two matrices.
MOD	Returns the remainder of the division of number by divisor.
MROUND	Returns a number down to the nearest specified level.
PI	Returns π.
PRODUCT	Returns the product of the specified arguments.
RADIANS	Converts an angle in degrees to radians.
RAND	Returns a random number between 0 and 1.
RANDBETWEEN	Returns a random number within a specified range.
ROMAN**	Converts an Arabic number to Roman numerals.
ROUND	Rounds a number to a specified number of digits.
ROUNDDOWN**	Rounds a number down.
ROUNDUP**	Rounds a number up.
SIGN	Returns the sign of a number.
SIN	Returns the sine of a given angle.
SINH	Returns the hyperbolic sine of a number.
SQRT	Returns the square root of a number.
SUM	Returns the sum of all the numbers in the reference.
SUMIF**	Returns the sum of all the numbers in the reference that satisfy the specified criteria.
SUMPRODUCT	Returns the sum of the products of corresponding array components.
SUMSQ	Returns the sum of the squares of arguments.
TAN	Returns the tangent of a given angle.
TANH	Returns the hyperbolic tangent of a number.
TRUNC	Truncates a number.

STATISTICAL FUNCTIONS

AVEDEV	Returns the average absolute deviation of data points from the mean.
AVERAGE	Returns the average of all the numbers in the reference.
CORREL	Returns the correlation coefficient between two data sets.
COUNT	Returns the number of numbers in the reference.
COUNTA	Returns the number of non-blank values in the reference.
INTERCEPT	Returns the intercept of the linear regression line $y = mx + b$.
LARGE	Returns the kth largest value in a list of values.
LINEST	Returns the parameters of multiple linear regression.
MAX	Returns the maximum value in a list of arguments.

MEDIAN	Returns the median value in a list of arguments.
MIN	Returns the minimum value in a list of arguments.
PEARSON	Returns the Pearson correlation coefficient.
SLOPE	Returns the slope of the linear regression line $y = mx + b$..
SMALL	Returns the kth smallest value in a list of values.
STDEV	Returns the standard deviation of a sample.
VAR	Returns the variance of a sample.

TEXT FUNCTIONS

CHAR	Returns the character corresponding to the character code number.
CLEAN	Removes all non-printable characters from a text string.
CODE	Returns the numeric code corresponding to a character.
CONCATENATE**	Concatenates text values into a single text string.
DOLLAR	Converts a number to a text value in currency format.
EXACT	Compares two text strings; returns TRUE if they are the same.
FIND	Returns the position at which one text string occurs within another text string.
FIXED	Rounds a number to the specified number of decimal places and returns the result as text.
LEFT	Returns the specified number of characters from a text string, beginning at the left.
LEN	Returns the number of characters in a text string.
LOWER	Converts a text string to lowercase.
MID	Returns the specified number of characters from a text string, beginning at the specified position.
PROPER	Capitalizes the first letter in each word of a text string.
REPLACE	Replaces characters at a specified position within a text string.
REPT	Repeats a text value a specified number of times.
RIGHT	Returns the specified number of characters from a text string, beginning at the right.
SEARCH	Finds the position of a text value within a text string.
SUBSTITUTE	Finds and substitutes characters within a text string.
T	Converts a value to text. Excel usually does this automatically when necessary.
TEXT	Formats a number and returns it as text.
TRIM	Removes spaces from a text string, except for single spaces between words.
UPPER	Converts a text string to uppercase.
VALUE	Converts a text argument to a number. Excel usually does this automatically when necessary.

B

ALPHABETICAL LIST OF
SELECTED WORKSHEET FUNCTIONS

This Appendix lists selected worksheet functions that will be useful in creating advanced worksheet formulas for chemical applications. For each function, a description of the function and its syntax are given; in most cases, an example of the use of the function and any related functions are included. Function arguments are in italics; arguments are in boldface type if the argument is required and plain text if the argument is optional. The required data type for an argument is indicated by the suffixes _num, _ref, _logical, _text. If a function can accept an argument of more than one type, then no suffix is attached. Functions followed by ** indicate a new function in Excel 5.0; * indicates a function that is changed in Excel 5.0.

ABS
Returns the absolute value of a number
Syntax: ABS(**number**)
Example: =ABS(-7.3) returns 7.3
Related Function: SIGN

ACOS
Returns the angle corresponding to a cosine value.
Syntax: ACOS(**number**)
Number must be from −1 to +1. The returned angle is in radians, in the range 0 to π. To convert the result to degrees, multiply by $180/\pi$.
Example: =ACOS(0) returns 1.570796327, or 90 degrees.
Related Functions: COS, other trigonometric functions

AND
Returns TRUE if all arguments are TRUE, otherwise returns FALSE.
Syntax: AND(**logical1**,logical2,...)
Up to 30 logical conditions can be tested.
Example: If A1 contains 0 and A2 contains 110, then =AND(A1=0,A2>100) returns TRUE.
Related Functions: NOT, OR

ASIN

Returns the angle corresponding to a sine value.

Syntax: ASIN(*number*)

Number must be from –1 to +1. The returned angle is in radians, in the range $-\pi/2$ to $+\pi/2$. To convert the result to degrees, multiply by $180/\pi$.

Example: If A1 contains 0.7071, then =ASIN(A1) returns 0.785388573, or 45 degrees.

Related Functions: SIN, other trigonometric functions

ATAN

Returns the angle corresponding to a tangent value.

Syntax: ATAN(*number*)

The returned angle is in radians, in the range 0 to π. To convert the result to degrees, multiply by $180/\pi$.

Example: =ATAN(0) returns 0.785388573 or 45 degrees.

Related Functions: ATAN2, TAN, other trigonometric functions

ATAN2

Returns the angle defined by a pair of x- and y-coordinates. The angle is between the x-axis and the line connecting the origin (0,0) and the point (x,y).

Syntax: ATAN2(*x_num,y_num*)

The returned angle is in radians, in the range $-\pi$ to π. To convert the result to degrees, multiply by $180/\pi$. A negative result represents a clockwise angle from the x-axis.

Example: =ATAN2(3,4) returns 0.927295218, or 53.13 degrees.

Related Functions: ATAN, TAN, other trigonometric functions

AVERAGE

Returns the average of all the numbers in the reference.

Syntax: AVERAGE(*number1,number2,...*)

The arguments may be numbers, or names, arrays or references that contain numbers. Up to 30 separate arguments can be listed. Only numbers in the array or range are counted.

Related Function: MEDIAN

CELL

Returns information about location, formatting and contents of a cell.

Syntax: CELL(*info_type_text,reference*)

Info_type_text specifies the cell information to be returned. If *reference* is omitted, information about the active cell is returned. For further information, see *Excel 5 Worksheet Function Reference*.

Example: If the cell format is 0.00E+00, =CELL("format") returns S2.

Related Function: macro function GET.CELL

CHAR
Returns the character corresponding to the character code number, in either the Macintosh character set or the ANSI character set (Windows).
Syntax: CHAR(*number*)
Number must be between 1 and 255.
Example: =CHAR(65) returns A.
Related Function: CODE

CHOOSE
Chooses a value from a list of values, based on index number.
Syntax: CHOOSE(*index_number, value1*,*value2*,...)
Index_number must be a number between 1 and 29, or a formula or reference that evaluates to same. The arguments may be numbers, names, formulas, macro functions or references. If used in a macro, *value* can be GOTO, subroutine or macro call.
Example: If cell H8 contains   1, then
=CHOOSE(H8+1,$F$2,$G$2,$H$2,$I$2)/SUM($F$2:$I$2) calculates the fraction $G$2/SUM($F$2:$I$2).
Related Functions: INDEX, MATCH

CLEAN
Removes all non-printable characters from a text string.
Syntax: CLEAN(*text*)
Use on text imported from other applications that may contain non-printing characters.
Related Function: TRIM

CODE
Returns the numeric code for the first character of text,
Syntax: CODE(*text*)
Example: =CODE ("A") returns 65.
Related Function: CHAR

COLUMN
Returns the column number of reference.
Syntax: COLUMN(*reference*)
If *reference* is a range of cells, returns the column number of the upper left cell of the range. *Reference* cannot be a multiple selection. If *reference* is omitted, it is assumed to be the reference in which the COLUMN function appears.
Example: If C3:N10 is selected, then =COLUMN(SELECTION()) returns 3.
Related Functions: COLUMNS, ROW, ROWS

COLUMNS
Returns the number of columns in reference.
Syntax: COLUMNS(**reference**)
If *reference* is a multiple selection, use INDEX to select a specified area within the selection.
Example: If C3:N10 is selected, then =COLUMNS(SELECTION()) returns 12.
Related Functions: COLUMN, ROW, ROWS

CORREL
Returns the correlation coefficient between two data sets.
Syntax: CORREL(**array1,array2**)
Array1 and *array2* must have the same number of data points.
Related Functions: SLOPE, INTERCEPT, COVAR, PEARSON, RSQ

COS
Returns the cosine of a given angle.
Syntax: COS(**number**)
Number is the angle in radians. To convert an angle in degrees to one in radians, multiply by $\pi/180$.
Related Functions: ACOS, SIN, TAN, other trigonometric functions

COUNT
Returns the number of numbers in the list or range.
Syntax: COUNT(**value1**,*value2*,...)
Up to 30 arguments are allowed. Any value, other than an empty cell, is counted.
Related Functions: COUNTA, COUNTBLANK, COUNTIF

COUNTA
Returns the number of non-blank values in the list or range.
Syntax: COUNTA(**value1**,*value2*,...)
Up to 30 arguments are allowed. Nulls (""), logical values, numbers formatted as dates, or text values that can be converted to numbers, are counted.
Related Functions: COUNT, COUNTBLANK, COUNTIF

COUNTBLANK**
Returns the number of blank cells within a range.
Syntax: COUNTBLANK(**reference**)
Cells containing null values are counted, but cells containing zero values are not counted.
Related Functions: COUNT, COUNTA, COUNTIF

COUNTIF**
Returns the number of cells within a range that meet the specified criteria.
Syntax: COUNTIF(**reference,criteria**)
Example: =COUNTIF(year_of_graduation,<1990)
Related Functions: COUNT, COUNTA, COUNTBLANK, SUMIF

DATE
Returns the serial number of a particular date.

Syntax: DATE(*year, month, day*)

Year is a number from 1900 (Windows) or 1904 (Macintosh) to 2078. *Year* can be specified using 2 digits if in the range 1920 to 2019, otherwise use 4 digits. *Month* or *day* can be greater than 12 or 31 (see example).

Example: DATE(94,7,4) corresponds to July 4, 1994. DATE(94,7,90) corresponds to Sept. 28, 1994.

Related Functions: DATEVALUE, DAY, MONTH, YEAR, NOW, TODAY

DATEVALUE
Converts a date in the form of text to a serial number.

Syntax: DATEVALUE(*text*)

Example: =DATEVALUE("3 Aug 1938") returns 12633 if the 1904 Date System is in effect.

Related Functions: NOW, TODAY

DAY
Converts a serial number to a day of the month.

Syntax: DAY(*serial_number*)

Serial_number can also be given as text, such as "Jul-4-1994". Day is a value in the range 1 to 31.

Related Functions: NOW, TODAY, WEEKDAY, YEAR, MONTH

DEGREES*
Converts an angle in radians to degrees

Syntax: DEGREES(*number*)

Available in Excel 4.0 only as an add-in function.

Related Functions: RADIANS, trigonometric functions

EXACT
Compares two text strings; returns TRUE if they are the same.

Syntax: EXACT(*text1,text2*)

EXACT is case-sensitive.

Related Functions: FIND, SEARCH

EXP
Returns the value of *e* raised to a power.

Syntax: EXP(*number*)

Returns the value of *e* raised to the power *number*.

Related Functions: LN, LOG

FACT
Returns the factorial of a number.

Syntax: FACT(*number*)

Returns the factorial of a number, i.e., 1*2*3*...*number*. *Number* must be positive. If *number* is not an integer, it is truncated.

FIND

Returns the position at which one text string occurs within another text string.

Syntax: FIND(**find_text, within_text**, start_number)

Find_text is the text string you want to find. Within_text is the string you are searching. Start_number is the character at which to start searching. If start_number is omitted, it is assumed to be 1. FIND is case-sensitive.

Example: If cell   A3 contains BILLO, E JOSEPH, the formula =FIND(",",A3) returns 6.

Related Functions: EXACT, SEARCH

FIXED

Rounds a number to the specified number of decimal places and returns the result as text.

Syntax: FIXED(**number**, decimals, no_commas_flag)

Number is the number to be converted. Decimals is the number of decimal places desired; if decimals is omitted, it is assumed to be 2. If decimals is negative, number is rounded to the left of the decimal point. If no_commas_flag is TRUE, commas are not included in the formatted number.

Example: =FIXED(12345,3,1) returns 12345.000. =FIXED(12345) returns 12,345.00. =FIXED(PI(),4) returns 3.1416.

Related Functions: ROUND, TEXT

HLOOKUP*

Finds the value in the first row of an array that is equal to or less than lookup_value. Returns the associated value in the nth row, as determined by offset_num.

Syntax: HLOOKUP(**lookup_value, array, offset_num**, match_logical*)

The values in the first row must be in ascending order. If match_logical is TRUE or omitted, returns the largest value that is less than or equal to lookup_value. If match_logical is FALSE, returns #N/A if an exact match is not found.

Example: See example under VLOOKUP.

Related Functions: INDEX, MATCH, VLOOKUP

IF

Returns one value if logical_test is   TRUE, another value if logical_test is FALSE.

Syntax: IF(**logical_test, value_if_TRUE**, value_if_FALSE)

Value_if_TRUE and/or value_if_FALSE can be IF functions; up to seven IF functions can be nested.

Example: =IF(result=0, 0.08*estimate, 0.08*result)

Related Functions: AND, NOT, OR, macro function IF

INDEX

Chooses a value from an array, based on row and column number pointers.

Syntax: INDEX(**array**, *row_number, column_number*)

If the array is one-dimensional, only one pointer argument need be specified.

Example: If the reference $C$2:$C$102 contains the atomic weights of the elements, arranged in order of atomic number, then =INDEX($C$2:$C$102,51) returns the atomic weight of element number 51.

Related Functions: CHOOSE, HLOOKUP, MATCH, VLOOKUP

INDIRECT

Returns a reference specified by a text string.

Syntax: INDIRECT(**reference_text,***A1_logical*)

Reference_text can be an A1- or R1C1-style reference, or a name. *A1_logical* specifies the form of *reference_text*: if *reference_text* is TRUE or omitted, *reference_text* is interpreted as an A1-style reference; if *reference_text* is FALSE, *reference_text* is interpreted as an R1C1-style reference.

Example: If cell B3 contains "Sheet2!A5" then the formula =INDIRECT(B3) displays the contents of Sheet2!A5.

Related Functions: OFFSET, macro function TEXTREF

INFO

Returns information about the operating environment.

Syntax: INFO(**type_text**)

Type_text specifies what information is to be returned. Some useful values of *type_text*: "directory" returns path of the current directory or folder, "memavail" returns the amount of memory available, "release" returns Microsoft Excel version, "system" returns mac or pcdos. See *Excel 5 Worksheet Function Reference* for details.

Example: The formula =INFO("release") returns 5.0a if Excel 5.0a is being used.

Related Function: CELL

INT

Rounds a number down to the nearest integer.

Syntax: INT(**number**)

INT rounds downward. Use TRUNC to return the integer part of a number, irrespective of sign.

Example: =INT(3.897) returns 3. =INT(-0.05) returns −1.

Related Functions: ROUND, TRUNC

INTERCEPT

Returns the intercept of the linear regression line $y = mx + b$.

Syntax: INTERCEPT(**known_ys, known_xs**)

See LINEST for details.

Related Functions: LINEST, SLOPE

ISBLANK
Returns TRUE if the value is blank.
Syntax: ISBLANK(***value***)
Related Functions: other IS functions

ISERR
Returns TRUE if the value is an error value other than #N/A.
Syntax: ISERR(***value***)
Related Functions: other IS functions

ISERROR
Returns TRUE if the value is any error value.
Syntax: ISERROR(***value***)
Related Functions: other IS functions

ISNA
Returns TRUE if the value is #N/A.
Syntax: ISNA(***value***)
Related Functions: other IS functions

ISNUMBER
Returns TRUE if the value is a number.
Syntax: ISNUMBER(***value***)
Does not convert text representation of a number.
Example: =ISNUMBER(5) returns TRUE. =ISNUMBER("9") returns FALSE.
Related Function: other IS functions

ISTEXT
Returns TRUE if the value is text.
Syntax: ISTEXT(***value***)
Related Functions: other IS functions

LEFT
Returns the specified number of characters from a text string, beginning at the left.
Syntax: LEFT(***text***, *num_chars*)
If *num_chars* is omitted, it is assumed to be 1.
Example: =LEFT("CHEMISTRY",4) returns CHEM.
Related Functions: LEN, MID, RIGHT

LEN
Returns the number of characters in a text string.
Syntax: LEN(***text***)
Example: =LEN("CHEMISTRY") returns 9.
Related Functions: LEFT, MID, RIGHT

LINEST
Returns an array of linear regression parameters.
Syntax: LINEST(**known_ys, known_xs**, const_logical, stats_logical)
See Chapter 16 for details.
Related Functions: CORREL, INTERCEPT, SLOPE

LN
Returns the natural (base-*e*) logarithm of a number.
Syntax: LN(**number**)
Related Functions: EXP, LOG

LOG
Returns the logarithm of a number to the specified base.
Syntax: LOG(**number**, base)
If *base* is omitted, returns the base-10 logarithm.
Related Functions: LN, LOG10

LOG10
Returns the base-10 logarithm of a number.
Syntax: LOG10(**number**)
Related Functions: LN, LOG

LOWER
Converts a text string to lowercase.
Syntax: LOWER(**text**)
Related Functions: PROPER, UPPER

MATCH
Looks up a value in an array and returns its relative position.
Syntax: MATCH(**lookup_value, array**, match_type)
If *match_type* = 0, the function finds the first value that is exactly equal to *lookup_value*; *array* can be in any order. If *match_type* = 1, finds the largest value that is less than or equal to *lookup_value*; *array* must be in ascending order. If *match_type* = –1, finds the smallest value that is greater than or equal to *lookup_value*; *array* must be in descending order.
Example: =MATCH(MAX(Spectrum),Spectrum,0) returns the relative position in the array *Spectrum* of the maximum value in the array.
Related Functions: HLOOKUP, INDEX, VLOOKUP

MAX
Returns the maximum value in a list of arguments.
Syntax: MAX(**number1**, number2,...)
There may be up to 30 arguments. If an argument is a reference, only numbers in the reference are examined.
Related Function: MIN

MDETERM
Returns the determinant of an array.

Syntax: MDETERM(***array***)

Array must have an equal number of rows and columns. If any cells in *array* do not contain numbers, returns #VALUE!.

Example: See Chapter 14 for details.

Related Functions: MINVERSE, MMULT, SUMPRODUCT, TRANSPOSE

MEDIAN
Returns the median value in a list of arguments.

Syntax: MEDIAN(***number1***, *number2*,...)

If there are an even number of numbers in the set, returns the average of the two median values.

Related Function: AVERAGE

MID
Returns the specified number of characters from a text string, beginning at the specified position.

Syntax: MID(***text, start_num, num_chars***)

If *num_chars* extends beyond the end of text, returns characters to the end of text.

Example: If cell A4 contains   H2SO4, =MID(A4,2,1) returns 2.

Related Functions: LEFT, LEN, RIGHT

MIN
Returns the minimum value in a list of arguments.

Syntax: MIN(***number1***, *number2*,...)

There may be up to 30 arguments. If an argument is a reference, only numbers in the reference are examined.

Related Function: MAX

MINVERSE
Returns the inverse of a matrix.

Syntax: MINVERSE(***array***)

Array must have an equal number of rows and columns. If any cells in *array* do not contain numbers, MINVERSE returns #VALUE!. If MDETERM for the array returns 0, the array cannot be inverted; MINVERSE will return #NUM! error.

Example: See Chapter 14 for details.

Related Functions: MDETERM, MMULT, SUMPRODUCT, TRANSPOSE

MMULT
Returns the product of two matrices.

Syntax: MMULT(***array1, array2***)

COLUMNS for *array1* must equal ROWS for *array2*.

Example: See Chapter 14 for details.

Related Functions: MDETERM, MINVERSE, SUMPRODUCT, TRANSPOSE

MOD

Returns the remainder of the division of *number* by *divisor*.

Syntax: MOD(**number, divisor**)

If *divisor* is greater than   *number,* returns *number.*

Use MOD(number,1) to return the decimal part of a floating-point number.

Example: =MOD(2.3333,2) returns 0.3333. =MOD(2.3333,3) returns 2.3333. =MOD(10.37,1) returns 0.37.

Related Functions: INT, ROUND, TRUNC

MONTH

Converts a serial number to a month.

Syntax: MONTH(**serial_number**)

Serial_number can also be given as text, such as "Jul-4-1994". Returns a number between 1 and 12.

Related Functions: YEAR, DAY, WEEKDAY

NOT

Reverses the logical value of its argument.

Syntax: NOT(**logical**)

Example: =IF(NOT(ISERROR(F2/G2)),F2/G2,"")

Related Functions: AND, OR

NOW

Returns the serial number of the current date and time.

Syntax: NOW()

The integer part of the serial number represents the day, month and year; the decimal part represents the time. NOW is recalculated whenever the sheet is recalculated.

Related Functions: DATE, TODAY, YEAR, MONTH, DAY

OFFSET

Returns a reference offset from a base reference by specified number of rows and columns.

Syntax: OFFSET(**reference, rows, cols**, height, width)

Rows or *columns* can be negative. *Height* and *width* must be positive. If *height* and *width* are omitted, they are assumed to be the same as in *reference.*

Example: =OFFSET(table, rows, cols, 1,1) returns a reference to a single cell within the reference *table.*

Related Function: INDEX

OR
Returns TRUE if any argument is TRUE.
Syntax: OR(*logical1*, *logical2*,...)
Up to 30 logical conditions can be tested.
Example: If cell G2 contains   0.75, the formula
=IF(OR(G2<=0,G2>=1),"",F2/G2) returns 1.33333333; if cell G2 contains   1.75,
the formula returns a null.
Related Functions: AND, NOT

PEARSON
Returns the Pearson product moment correlation coefficient between two data
sets.
Syntax: PEARSON(*array1, array2*)
Array1 is the array of independent (x) values; *array2* is the array of dependent
(y) values. Returns a value of R between –1 and 1.
Example: See *Excel 5 Worksheet Function Reference*   for details.
Related Functions: CORREL, RSQ

PI
Returns π.
Syntax: PI()

PRODUCT
Returns the product of the specified arguments.
Syntax: PRODUCT(*number1*,*number2*,...)
Up to 30 numbers can be multiplied.
Related Function: SUM

PROPER
Capitalizes the first letter in each word of a text string.
Syntax: PROPER(*text*)
Example: =PROPER("JOHN Q. PUBLIC") returns John Q. Public.
Related Functions: LOWER, UPPER

RADIANS*
Converts an angle in degrees to radians.
Syntax: RADIANS(*number*)
Available in Excel 4.0 only as an add-in function.
Related Functions: DEGREES, trigonometric functions

RAND
Returns a random number between 0 and 1.
Syntax: RAND()
A new random number is generated each time the worksheet is recalculated.
You may need to **Copy** and **Paste Special** (Values) to prevent a random-number
series from changing each time you change anything on a worksheet.
Example: =0.1*(0.5-RAND()) returns a number between –0.05 and +0.05.

REPLACE

Replaces characters at a specified position within a text string.

Syntax: REPLACE(*old_text, start_num, num_chars, new_text*)

Old_text is the text in which characters are to be replaced. *Start_num* is the position of the first character in *old_text* to be replaced. *Num_chars* is the number of characters to be replaced. *New_text* is the text to replace characters in *old_text.*

Example: If cell A1 contains the text   General Chemistry I Lab, the formula =REPLACE(A1,19,1,"II") returns General Chemistry II Lab.

Related Functions: SEARCH, SUBSTITUTE

REPT

Repeats a text string a specified number of times.

Syntax: REPT(*text, number*)

Example: If *decimals* = 3, the formula ="0."&REPT("0",decimals) returns 0.000 as text.

RIGHT

Returns the specified number of characters from a text string, beginning at the right.

Syntax: RIGHT(*text*, *num_chars*)

If *num_chars* is omitted, it is assumed to be 1.

Example: =RIGHT(303585842,4) returns 5842.

Related Functions: MID, LEFT, LEN

ROMAN**

Converts an Arabic number to Roman numerals.

Syntax: ROMAN(*number*, *form*)

If *form* = 0, TRUE or omitted, the "classic" form is returned. *Form* can also be 1, 2, or 3, in which case successively more "concise" forms are returned. If *form* = 4 or FALSE, a "simplified" form is returned.

Example: =ROMAN(1997,0) returns MCMXCVII.

ROUND

Rounds a number to a specified number of digits.

Syntax: ROUND(*number, digits*)

If the result ends in a zero, the zero is not displayed.

Example: If cell A1 contains 0.001736, the formula =ROUND(A1,5) returns 0.00174. If cell A1 contains 0.007702, the same formula returns 0.0077, i.e., the number is not padded out with trailing zeros.

Related Functions: INT, TRUNC, TEXT

ROW

Returns the row number of reference.

Syntax: ROW(*reference*)

If *reference* is a range of cells, returns the row number of the upper left cell of the range. *Reference* cannot be a multiple selection.

Example: =ROW(B5:C7) returns 5.

Related Functions: COLUMN, ROWS

ROWS

Returns the number of rows in reference.

Syntax: ROWS(**reference**)

If *reference* is a multiple selection, can use INDEX to select a specified area within the selection.

Example: =ROWS(B5:C7) returns 3.

Related Functions: ROW, COLUMNS

RSQ

Returns the correlation coefficient R^2.

Syntax: RSQ(**known_ys, known_xs**)

This is the correlation coefficient returned by LINEST. For more details, see LINEST.

Related Functions: LINEST, CORREL

SEARCH

Finds the position of a text value within a text string.

Syntax: SEARCH(**find_text, within_text**, *start_num*)

Find_text is the text to be found. Include the ? wildcard character in *find_text* to match any single character or * to match any sequence of characters within the source string *within_text*. *Start_num* is the position in *within_text* to begin searching. SEARCH is not case-sensitive, FIND is case-sensitive.

Example: See example under FIND.

Related Functions: FIND, REPLACE, SUBSTITUTE

SIGN

Returns the sign of a number.

Syntax: SIGN(**number**)

Returns 1, 0, –1 if the number is positive, zero or negative, respectively.

Related Function: ABS

SIN

Returns the sine of a given angle.

Syntax: SIN(**number**)

Number is the angle in radians. To convert an angle in degrees to one in radians, multiply by $\pi/180$.

Related Functions: ASIN, COS, TAN, other trigonometric functions

SLOPE

Returns the slope of the linear regression line $y = mx + b$.

Syntax: SLOPE(**known_ys, known_xs**)

See LINEST for details.

Related Functions: INTERCEPT, LINEST

SQRT

Returns the square root of a number.

Syntax: SQRT(**number**)

STDEV

Returns the standard deviation of a sample.

Syntax: STDEV(**value1**, value2,...)

The sample can consist of up to 30 separate arguments or arrays.

Related Functions: AVERAGE, AVDEV

SUBSTITUTE

Finds and substitutes characters within a text string.

Syntax: SUBSTITUTE(**text, old_text, new_text**, instance_num)

Text is the string in which characters are to be substituted. *Old_text* is the text to be replaced. *New_text* is the text to replace *old_text*. *Instance_num* specifies which occurrence of *old_text* is to be replaced; if omitted, all are replaced.

Example: =SUBSTITUTE(A3," ","",2) replaces the second occurrence of a space in the string in cell A3 with a null .

Related Function: REPLACE

SUM

Returns the sum of all the numbers in the reference.

Syntax: SUM(**number1**, number2,...)

The arguments may be numbers, names, arrays or references that contain numbers. Up to 30 separate arguments. Only numbers in the array or range are counted.

Related Functions: AVERAGE, COUNT, COUNTA

SUMIF**

Returns the sum of all the numbers in the reference that satisfy the specified criteria.

Syntax: SUMIF(**range, criteria**, sum_range)

Range is the range of cells in which criteria will be evaluated. *Sum_range* contains the cells that will be summed. If *sum_range* is omitted, cells in *range* are summed.

Example: =SUMIF(B3:B553,>100,D3:D553)

Related Functions: AVERAGE, COUNT, COUNTA

SUMPRODUCT

Returns the sum of the products of corresponding array components.

Syntax: SUMPRODUCT(***array1, array2***,...)

Up to 30 arrays may be included. All arrays must have the same dimensions.

Related Functions: MMULT, TRANSPOSE

TAN

Returns the tangent of a given angle.

Syntax: TAN(***number***)

Number is the angle in radians. To convert an angle in degrees to one in radians, multiply by $\pi/180$.

Related Functions: ATAN, ATAN2, other trigonometric functions

TEXT

Formats a number and returns it as text.

Syntax: TEXT(***value, format_text***)

Format_text is a number format, similar to those in the Number format dialog box. TEXT converts a number to text; the result will sometimes fail to be calculated as a number.

Example: The formula =TEXT(PI(),"0.000") in cell B1 returns 3.142. The formula =2*B1 returns 6.284, but the formula =SUM(B1) returns 0.

Related Function: FIXED

TIME

Returns the serial number of a time.

Syntax: TIME(***hour, minute, second***)

Hour is a number from 0 to 23, *minute* is a number from 0 to 59, and *second* is a number from 0 to 59.

Related Function: NOW

TODAY

Returns the serial number of today's date.

Syntax: TODAY()

TODAY returns the integer part of the serial number returned by NOW.

Related Functions: DATE, DAY, NOW

TRANSPOSE

Returns the transpose of an array

Syntax: TRANSPOSE(***array***)

Must be entered as an array formula in a range with number of columns equal to number of rows in *array*, and number of rows equal to number of columns in *array*.

Example: Does the same thing as **Paste Special** (Transpose) from the **Edit** menu.

Related Functions: MDETERM, MINVERSE, MMULT, SUMPRODUCT

TRIM
Removes spaces from a text string, except for single spaces between words.
Syntax: TRIM(**text**)
Related Function: CLEAN

TRUNC
Truncates a number.
Syntax: TRUNC(**number**, *num_digits*)
Num_digits is the number of digits after the decimal point. If omitted, *num_digits* is taken to be zero. TRUNC does not round up.
Example: The formula =TRUNC(-8.913) returns -8. The formula =TRUNC(PI(),3) returns   3.141.
Related Functions: INT, ROUND

TYPE
Returns a number indicating the data type of a value.
Syntax: TYPE(**value**)
Returns 1 (if *value* is a number), 2 (text), 4 (logical value), 16 (error value) or 64 (array).
Related Functions: macro function GET.CELL, IS functions

UPPER
Converts a text string to uppercase.
Syntax: UPPER(**text**)
Related Functions: LOWER, PROPER

VALUE
Converts a text argument to a number.
Syntax: VALUE(**text**)
Not usually necessary, since Excel automatically converts text to number.
Example: =ISNUMBER(MID("H2SO4",2,1)) returns FALSE but =ISNUMBER(VALUE(MID("H2SO4",2,1))) returns TRUE.
Related Functions: FIXED, TEXT

VAR
Returns the variance of a sample.
Syntax: VAR(**number1**,*number2*,...)
The sample can consist of up to 30 separate arguments or arrays.
Example: See *Excel 5 Worksheet Function Reference*   for details.
Related Function: STDEV

VLOOKUP*

Finds the value in the first column of *array* that is equal to or less than *lookup_value*. Returns the associated value in the nth column, as determined by *offset_num*.

Syntax: VLOOKUP(***lookup_value, array, offset_num***, *match_logical**)

The values in the first column must be in ascending order. If *match_logical* is TRUE or omitted, returns the largest value that is less than or equal to *lookup_value*. If *match_logical* is FALSE, returns an exact match, #N/A if an exact match is not found.

Example: See example under HLOOKUP.

Related Functions: HLOOKUP, INDEX, MATCH

WEEKDAY*

Converts a serial number to a day of the week.

Syntax: WEEKDAY(***serial_number***, *return_type_num**)

Returns a number indicating the day of the week. If *return_type_num* = 1 or omitted, returns 1 (Sunday) to 7 (Saturday). If *return_type_num* = 2, returns 1 (Monday) to 7 (Sunday). If *return_type_num* = 3, returns 0 (Monday) to 6 (Sunday).

Related Functions: DAY, NOW, TODAY

YEAR

Converts a serial number to a year.

Syntax: YEAR(***serial_number***)

Related Function: NOW

SELECTED MACRO FUNCTIONS FOR EXCEL 4.0 MACRO LANGUAGE BY CATEGORY

This appendix lists selected macro functions from the Command-Equivalent, Control, Customizing, Information and Lookup & Reference categories. There are also Database & List Management, DDE/External, Engineering, Statistical and Text functions, which are not discussed in this book. See Excel's On-line Help for a complete list of functions.

Using a command-equivalent function is the same as choosing the command from a menu. For example, using the macro function CLOSE is equivalent to choosing the **Close** command in the **File** menu.

Macro control functions include the commands used to perform looping or branching.

Customizing functions enable a macro to create menus, commands, tools or dialog boxes.

Information functions return information about cells, documents, menu bars, etc.

Reference functions are used to convert a reference to text, or a reference from one form to another.

COMMAND-EQUIVALENT FUNCTIONS

ACTIVATE	Equivalent to choosing an open document in the **Window** menu.
ACTIVE.CELL.FONT**	Equivalent to formatting characters by using Edit Directly In Cell.
ALIGNMENT	Equivalent to choosing the Alignment tab in the **Cells...** command in the **Format** menu (Excel 5.0), or the **Alignment...** command in the **Format** menu (Excel 4.0).
BORDER	Equivalent to choosing the Border tab in the **Cells...** command in the **Format** menu (Excel 5.0), or the **Border...** command in the **Format** menu (Excel 4.0).
CALCULATION	Equivalent to choosing the Calculation tab in the **Options** command in the **Tools** menu (Excel 5.0), or the **Calculation...** command in the **Options** menu (Excel 4.0).

CELL.PROTECTION	Equivalent to choosing the Protection tab in the **Cells...** command in the **Format** menu (Excel 5.0), or the **Cell Protection...** command in the **Format** menu (Excel 4.0).
CLEAR	Equivalent to choosing the **Clear** command in the **Edit** menu.
CLOSE	Equivalent to choosing the **Close** command in the **File** menu.
CLOSE.ALL	Equivalent to choosing the **Close All** command in the **File** menu.
COLUMN.WIDTH	Equivalent to choosing the **Column Width...** command in the **Format** menu.
COPY	Equivalent to choosing the **Copy** command in the **Edit** menu.
CUT	Equivalent to choosing the **Cut** command in the **Edit** menu.
DISPLAY	Equivalent to choosing the Display tab in the **Options** command in the **Tools** menu (Excel 5.0), or the **Display...** command in the **Options** menu (Excel 4.0).
EDIT.DELETE	Equivalent to choosing the **Delete** command in the **Edit** menu.
FILE.CLOSE	Equivalent to choosing the **Close** command in the **File** menu.
FILL.DOWN	Equivalent to choosing the **Fill Down** command in the **Edit** menu.
FILL.RIGHT	Equivalent to choosing the **Fill Right** command in the **Edit** menu.
FONT.PROPERTIES**	Equivalent to formatting characters by choosing **Cells...** from the **Format** menu.
FORMAT.FONT	Equivalent to choosing the Font tab in the **Cells...** command in the **Format** menu (Excel 5.0), or the **Font...** command in the **Format** menu (Excel 4.0).
FORMAT.NUMBER	Equivalent to choosing the Number tab in the **Cells...** command in the **Format** menu (Excel 5.0), or the **Number...** command in the **Format** menu (Excel 4.0).
FORMULA	Enters a number, text, reference or formula in a worksheet.
FULL	Mouse- or keyboard-equivalent to any operation that makes the window full size
HIDE	Equivalent to choosing the **Hide** command in the **Window** menu.
INSERT	Equivalent to choosing **Cells...**, **Rows** or **Columns** in the **Insert** menu (Excel 5.0), or the **Insert** command in the **Edit** menu (Excel 4.0).
NEW*	Equivalent to choosing the **New...** command in the **File** menu.

OPEN*	Equivalent to choosing the **Open...** command in the **File** menu.
PAGE.SETUP*	Equivalent to choosing the **Page Setup...** command in the **File** menu.
PASTE	Equivalent to choosing the **Paste** command in the **Edit** menu.
PASTE.SPECIAL*	Equivalent to choosing the **Paste Special...** command in the **Edit** menu.
PRINT*	Equivalent to choosing the **Print...** command in the **File** menu.
PROTECT.DOCUMENT*	Equivalent to choosing the **Protection** command in the **Tools** menu (Excel 5.0), or the **Protect Document...** command in the **Options** menu (Excel 4.0).
QUIT	Equivalent to choosing the **Quit** command in the **File** menu of Excel for the Macintosh, or the **Exit** command in the **File** menu of Excel for Windows.
ROW.HEIGHT	Equivalent to choosing the **Row Height...** command in the **Format** menu.
SAVE	Equivalent to choosing the **Save** command in the **File** menu.
SAVE.AS*	Equivalent to choosing the **Save As...** command in the **File** menu.
SELECT*	Mouse- or keyboard-equivalent to selecting a cell or range of cells on a worksheet.
SET.PAGE.BREAK	Equivalent to choosing the **Page Break** command in the **Insert** menu (Excel 5.0), or the **Set Page Break** command in the **Options** menu (Excel 4.0).
SET.PRINT.AREA*	Equivalent to setting the Print Area by choosing the Sheet tab in the **Page Setup...** command in the **File** menu (Excel 5.0) or choosing **Set Print Area** in the **Options** menu (Excel 4.0).
SET.PRINT.TITLES	Equivalent to setting the Print Titles by choosing the Sheet tab in the **Page Setup...** command in the **File** menu (Excel 5.0), or choosing **Set Print Titles...** in the **Options** menu (Excel 4.0).
SORT*	Equivalent to choosing the **Sort...** command in the **Data** menu.
UNHIDE	Equivalent to choosing the **Unhide** command in the **Window** menu.
WINDOW.MAXIMIZE	Mouse- or keyboard-equivalent to any operation that makes the window full size or previous size.

MACRO CONTROL FUNCTIONS

ARGUMENT	Describes the arguments used in a function macro or subroutine.
ELSE	Marks the beginning of a group of statements in a macro that will be carried out if the preceding IF or ELSE.IF statements are FALSE.
ELSE.IF	Marks the beginning of a group of statements in a macro that will be carried out if the preceding IF or ELSE.IF statements are FALSE and *logical_test* is TRUE.
END.IF	Marks the end of a group of statements in a macro associated with an IF statement.
FOR	Marks the start of a FOR...NEXT loop.
FOR.CELL	Similar to the FOR function. The statements between FOR.CELL and NEXT are repeated for each cell in a range of cells.
GOTO	Switches macro execution to another reference.
HALT	Stops execution of all macros.
IF	(Syntax 1): Performs one of two actions based on whether *logical_test* is TRUE or FALSE. (Syntax 2): Marks the beginning of a group of statements in a macro that will be carried out if *logical_test* is TRUE.
NEXT	Marks the end of a FOR-NEXT or WHILE...NEXT loop.
RESULT	Specifies the type of data returned by a custom function macro.
RETURN	Terminates a command or function macro and returns control to the caller.
SET.VALUE	Assigns a value to a cell in a macro sheet.
STEP	Stops normal execution of a macro and executes it one cell at a time.
WHILE	Marks the start of a WHILE...NEXT loop.

CUSTOMIZING FUNCTIONS

ADD.BAR	Creates a new menu bar and returns the bar number.
ADD.COMMAND*	Adds a command to a menu and returns the position number of the command in the menu.
ADD.MENU*	Adds a menu to a menu bar and returns the position number of the menu on the menu bar.
ALERT	Displays a dialog box with a message and one or more buttons.
APP.TITLE	Changes the application title "Microsoft Excel" at the top of the application window (Excel for Windows only).
BEEP	Produces a tone.

CHECK.COMMAND*	Adds a check mark beside a command in a menu to indicate that the command has been chosen.
DELETE.BAR	Deletes a menu bar.
DELETE.COMMAND*	Deletes a command from a menu.
DELETE.MENU*	Deletes a menu from a menu bar.
DIALOG.BOX	Displays a dialog box described by a dialog box definition table.
ECHO	Controls screen updating.
ENABLE.COMMAND*	Enables or disables a custom command in a menu.
ERROR	Specifies action to take if an error occurs.
GET.BAR*	Syntax 1: Returns the number of the active menu bar. Syntax 2: Returns the name or position number of a specified command or menu.
INPUT	Displays a dialog box for input.
MESSAGE	Displays a message in the Status Bar.
ON.KEY	Runs a specified macro when a specified key is pressed.
ON.SHEET**	Runs a specified macro when a specified sheet is activated.
ON.WINDOW	Runs a specified macro when a specified window is activated.
RENAME.COMMAND*	Changes the name of a built-in or custom command on a menu.
SHOW.BAR*	Displays the specified menu bar.

INFORMATION FUNCTIONS

ACTIVE.CELL	Returns the reference of the active cell.
CALLER	Returns information about the toolbutton, menu command, button or other object that called a macro.
DIRECTORY	Returns the path to the current drive and directory of folder or sets the path to the specified drive and directory of folder.
DOCUMENTS	Returns a horizontal array of open documents.
ERROR.TYPE	Returns a number indicating what type of error occurred.
FILES	Returns a horizontal array of names of all files in the specified folder or directory.
GET.CELL*	Returns information about a cell (location, contents, formatting, etc.).
GET.DOCUMENT*	Returns information about a document (name, location, type, printer settings, etc.).
GET.FORMULA	Returns the contents of a cell as it appears in the formula bar.
GET.WINDOW*	Returns information about a document window (name, size, position, etc.).
GET.WORKBOOK*	Returns information about a workbook.

| GET.WORKSPACE* | Returns information about the workspace. |
| SELECTION | Returns the external reference of a selected cell or range of cells. |

LOOKUP & REFERENCE FUNCTIONS

| REFTEXT | Converts a reference to text. |
| TEXTREF | Converts text to a reference, in either A1- or R1C1-style. |

ALPHABETICAL LIST OF
SELECTED MACRO FUNCTIONS
FOR EXCEL 4.0 MACRO LANGUAGE

You will find the following listing of selected macro functions for the Excel 4.0 macro language useful when creating your own macros. The list includes 80 of the over 400 macro functions provided in Excel 4.0 and Excel 5.0 For each function, the required syntax is given, along with some comments on the required and optional arguments, one or more examples, and a list of related functions. Functions followed by ** indicate a new function in Excel 5.0; * indicates a function that is changed in Excel 5.0.

Function arguments are in italics; required arguments are in bold. The required data type for an argument is indicated by the suffixes *_num*, *_ref*, *_logical*, *_text*. If a function can accept an argument of more than one type, then no suffix is attached. For example, the syntax of the ADD.COMMAND function is ADD.COMMAND(***bar_num***, ***menu***, ***command_ref***, *position*), indicating that the function will accept either a number or a name for the argument *menu* to indicate the position of a menu on a menu bar. A few functions can accept a larger number of arguments than have been listed here. This has been indicated by the use of an ellipsis. See Excel's On-line Help for a complete list of arguments for these functions.

ACTIVATE
Command-equivalent to choosing an open document in the **Window** menu.
Syntax: ACTIVATE(*window_text, pane_num*)
Specifies the name of the document to display as the active document. It can be given as text or as a variable name.
Example: =ACTIVATE("Worksheet1") makes the document *Worksheet1* the active document.
Related Function: DOCUMENTS

ACTIVE.CELL.FONT**
Equivalent to formatting characters by using Edit Directly In Cell.
Syntax: ACTIVE.CELL.FONT(*font, ..., start_num, num_characters*)
Related Functions: FONT.PROPERTIES?, FONT.PROPERTIES

ADD.BAR

Creates a new menu bar and returns the bar number.

Syntax: ADD.BAR(*bar_num*)

Bar_num can be the number (1-9) of one of the six built-in menu bars or three shortcut menus, or of a custom menu bar (10-). Up to 15 custom menu bars can be defined at one time. (1 = Worksheet and macro sheet, 2 = Chart, 3 = Null, etc.) See Excel's On-line Help for details.

Example: See the example under ADD.MENU.

Related Functions: ADD.COMMAND, ADD.MENU, DELETE.BAR, SHOW.BAR

ADD.COMMAND

Adds a command to a menu and returns the position number of the command in the menu.

Syntax: ADD.COMMAND(**bar_num**, **menu**, **command_ref**, *position*)

Bar_num is the ID number of the menu bar to which the command is to be added. *Menu* can be either the position number or name (as text) of the menu where the command is to be added. *Command_table_ref* is the area on the macro sheet that describes the command; see Chapter 8 for greater detail. *Position* can be either a position number or the name of an existing command (as text) specifying the location of the new command. If *position* is a name, the new command is inserted above the existing one; if omitted, the command is added at the bottom of the menu.

Example: =ADD.COMMAND(1,"Data", $A$23:$D$23, "Sort...") adds the command specified in cells $A$20:$B$20 to the **Data** menu, immediately above the **Sort...** command.

Related Functions: ADD.BAR, ADD.MENU, CHECK.COMMAND,
DELETE.COMMAND, ENABLE.COMMAND, RENAME.COMMAND

ADD.MENU

Adds a menu to a menu bar and returns the position number of the menu on the menu bar.

Syntax: ADD.MENU(**bar_num**, **menu_ref**, *position*)

Bar_num is the ID number of the menu bar to which the menu is to be added. *Menu_ref* is a reference to an area of the macro sheet that describes the new menu (see Chapter 8 for details). *Position* can be a number or name that specifies the placement of the new menu.

Example: =ADD.MENU(A3,B3:D14)
                 =SHOW.BAR(A3)

adds a new menu (specified in the range $B$3:$D$14) to the menu bar that was created by ADD.BAR() in cell A3, and displays the menu bar.

Related Functions: ADD.COMMAND, DELETE.MENU, ENABLE.COMMAND

ALERT

Displays a dialog box with a message and one or more buttons. There are three types of alert boxes.

Syntax: ALERT(*message_text, type_num, help_ref*)

Message_text is displayed in the dialog box. *Type_num* (1-3) determines the type of dialog box displayed: 1 = box with Question or Warning icon, OK, Cancel and optional Help buttons, 2 = box with Info icon and OK button, 3 = box with Warning icon and OK button.

Example: =ALERT("Maximum grade must be greater than zero.",3) displays an Alert box with a Warning icon.

Related Functions: DIALOG.BOX, INPUT

ALIGNMENT

Command-equivalent to choosing the Alignment tab in the **Cells...** command in the **Format** menu (Excel 5.0), or the **Alignment...** command in the **Format** menu (Excel 4.0).

Syntax: ALIGNMENT(*horiz_align_num, wrap_text_logical, vert_align_num, orientation_num*)

Horiz_align_num is a number specifying the type of horizontal alignment, in same order as in dialog box (1 = General 2 = Left, 3 = Center, 4 = Right, 5 = Fill, 6 = Justify, 7 = Center across selection). If *wrap_text_logical* is TRUE, text in the selected cells is wrapped. *Vert_align_num* is a number specifying the type of vertical alignment (1 = Top 2 = Center, 3 = Bottom). *Orientation_num* is a number specifying the orientation of the text (1 = Horizontal 2 = Vertical, 3 = Upward, 4 = Downward). If an argument is omitted, it is not changed from its previous value.

Example: =SELECT("C5")
                    =COLUMN.WIDTH(7)
                    =ALIGNMENT(3,,,)
                    =FORMAT.NUMBER("000")

selects column E of the active worksheet, sets column width to 7, centers text, and formats numbers with leading zeros (13 is displayed as 013).

Related Function: ALIGNMENT?

APP.TITLE

Changes the application title "Microsoft Excel" at the top of the application window (Excel for Windows only).

Syntax: APP.TITLE(text)

Text is the title to replace the original application title. Use when creating a custom application. Use APP.TITLE() to restore the original title.

Related Function: WINDOW.TITLE

ARGUMENT

Describes the arguments used in a function macro or subroutine.

Syntax 1: ARGUMENT(***name_text***, *data_type_num*)

Name_text is the name of the argument used in the macro. *Data_type_num* specifies the type of argument that will be accepted (1 = Number, 2 = Text, 4 = Logical, 8 = Reference, 16 = Error, 64 = Array). If *data_type_num* is omitted, the default value is 7 (number, text or logical is accepted).

Syntax 2: ARGUMENT(*name_text, data_type_num,* ***reference***)

Reference specifies where on the macro sheet to store the argument. It can be a single cell or a range.

Example: The first statements of the function macro to calculate a molecular weight from a chemical formula are

    =ARGUMENT("formula",2)
    =ARGUMENT("precision",1)

Related Function: RESULT

BEEP

Produces a tone. Intended to get the user's attention.

Syntax: BEEP()

Related Function: ALERT

BORDER

Command-equivalent to choosing the **Border** tab in the **Cells...** command in the **Format** menu (Excel 5.0), or the **Border...** command in the **Format** menu (Excel 4.0).

Syntax: BORDER(*outline_num, left_num, right_num, top_num, bottom_num, shade_logical, ...*)

Outline_num, left_num, right_num, top_num, bottom_num, are numbers that can have values from 0 to 7 (0 = No border, 1 = Thin line, 2 = Medium line, 3 = Dashed line, 4 = Dotted line, 5 = Thick line, 6 = Double line, 7 = Hairline). *Shade_logical* specifies whether the selection is shaded.

Example: =SELECT("R2C2:R2C6")
              =BORDER(0,1,1,1,1)

puts a line around all four sides of each cell in the selection.

Related Function: BORDER?

CALCULATION

Command-equivalent to choosing the Calculation tab in the **Options** command in the **Tools** menu (Excel 5.0), or the **Calculation...** command in the **Options** menu (Excel 4.0).

Syntax: CALCULATION(*type_num,...*)

Type_num is a number specifying the type of calculation (1 = Automatic, 2 = Automatic except tables, 3 = Manual). Use *type_num* = 3 to turn off recalculation during the running of a macro that operates on a worksheet.

Example: =CALCULATION(1) turns automatic recalculation back on.

Related Function: CALCULATION?

CHECK.COMMAND

Adds a check mark beside a command in a menu to indicate that the command has been chosen. (The check mark can be toggled on and off by use of this function.)

Syntax: CHECK.COMMAND(*bar_num, menu, command, check_logical*)

Bar_num is the ID number of the menu bar. *Menu* and *command* can be either the ID number or the name, as text. If *check_logical* is TRUE, a check mark is added to the command; if FALSE, the check mark is removed.

Example: =CHECK.COMMAND(A3,"Weak Acid/Base",5,TRUE) puts a check

mark beside the command in position 5 in the **Weak Acid/Base** menu, located on the menu bar whose ID number is stored in cell A3.
Related Functions: ADD.COMMAND, DELETE.COMMAND, ENABLE.COMMAND, RENAME.COMMAND

CLOSE

Command-equivalent to choosing the **Close** command in the **File** menu.
Syntax: CLOSE(*save_logical*)
Save_logical specifies whether to save the file before closing the window. If *save_logical* is TRUE or 1 the file is saved, if **FALSE** or 0 the file is not saved. If *save_logical* is omitted, the Save Changes? dialog box is displayed if changes have been made to the file.
Example: =CLOSE(1) closes the active file after saving.
Related Functions: CLOSE.ALL, FILE.CLOSE, QUIT, SAVE

COLUMN.WIDTH

Command-equivalent to choosing the **Column Width...** command in the **Format** menu.
Syntax: COLUMN.WIDTH(*width_num, reference, standard_width_logical, type_num, standard_width_num*)
See comments for ROW.HEIGHT.
Example: See example for ALIGNMENT.
Related Functions: COLUMN.WIDTH?, ROW.HEIGHT

COPY

Command-equivalent to choosing the **Copy** command in the **Edit** menu. Can automatically paste to another reference.
Syntax: COPY(*from_ref, to_ref*)
From_ref is the cell or range to be copied. If omitted, the current selection is copied. *To_ref* is the cell or range where the copied cell(s) will be pasted. *To_ref* can be a single cell, a selection of the same size as *from_ref* or a multiple of it. If *to_ref* is omitted, then you can subsequently use PASTE or PASTE.SPECIAL.
Example: =SELECT("R1C11:R"&tblsize&"C13")
                =COPY()
                =ACTIVATE(dest.sheet)
                =SELECT("R1C4")
                =PASTE()
copies the range K1:Mx, where x = *tblsize*, activates the document whose name is in cell *dest.sheet* and pastes in the range whose upper left corner is cell D1.
Related Functions: CUT, PASTE, PASTE.SPECIAL

CUT

Command-equivalent to choosing the **Cut** command in the **Edit** menu. Can automatically paste to another reference.

Syntax: CUT(*from_ref, to_ref*)

See the comments about the COPY function.

Example: =CUT(!$J$3:$L$11,!A$1) cuts cells from the active worksheet and pastes them in another location in the active worksheet.

Related Functions: COPY, PASTE, PASTE.SPECIAL

DELETE.BAR

Deletes a menu bar. Only custom menu bars may be deleted.

Syntax: DELETE.BAR(*bar_num*)

Bar_num is the ID number of the custom menu bar to be deleted. Instead of using the ID number, use the reference of the cell containing the ADD.BAR function that created the custom menu bar.

Example: =DELETE.BAR(A3)

Related Functions: ADD.BAR, SHOW.BAR

DELETE.COMMAND

Deletes a command from a menu.

Syntax: DELETE.COMMAND(*bar_num, menu, command*)

Bar_num is the ID number of the built-in or custom menu bar from which the command is to be deleted. *Menu* is the ID number or name of the menu containing the command. *Command* is the ID number or name of the command to be deleted. When a built-in command is deleted, DELETE.COMMAND saves the ID number, which can be used later to restore the command.

Example: =DELETE.COMMAND(10,"File","Find File...")

Related Functions: ADD.COMMAND, CHECK.COMMAND, ENABLE.COMMAND, RENAME.COMMAND

DELETE.MENU

Deletes a menu from a menu bar. Use this to make room on a built-in menu bar for custom menus.

Syntax: DELETE.MENU(*bar_num, menu*)

Bar_num is the ID number of the built-in or custom menu bar from which the menu is to be deleted. *Menu* is the ID number or name of the menu to be deleted.

Example: =DELETE.MENU(1,"Data") deletes the **Data** menu from the Worksheet and Macro Sheet menu bar.

Related Functions: ADD.MENU, DELETE.COMMAND, ENABLE.COMMAND

DIALOG.BOX

Displays a dialog box described by a dialog box definition table.

Syntax: DIALOG.BOX(*dialog_ref*)

See Excel's On-line Help for details.

Related Functions: ALERT, INPUT

DISPLAY
Command-equivalent to choosing the **Display** tab in the **Options** command in the **Tools** menu (Excel 5.0), or the **Display...** command in the **Options** menu (Excel 4.0).
Syntax: DISPLAY(*formulas_logical, gridlines_logical, headings_logical, ...*)
Formulas_logical specifies whether formulas are displayed (usually FALSE for worksheets). *Gridlines_logical* specifies whether gridlines are displayed. *Headings_logical* specifies whether row and column headings are displayed.
Example: =DISPLAY(,0,0) prevents display of gridlines and row and column headings.
Related Function: DISPLAY?

ECHO
Controls screen updating. Improves macro execution speed if the macro incorporates many statements, such as ROW.HEIGHT, that update the screen.
Syntax: ECHO(*display_logical*)
If *display_logical* is FALSE or 0, turns screen updating off. If *display_logical* is omitted, toggles screen updating.
Related Functions: HIDE, UNHIDE

EDIT.DELETE
Command-equivalent to choosing the **Delete** command in the **Edit** menu.
Syntax: EDIT.DELETE(*shift_type_num*)
Shift_type_num specifies how to shift cells after deleting (1 = shift cells left, 2 = shift cells up, 3 = delete entire row, 4 = delete entire column).
Example: =SELECT("R1C2:R10C3")
                =EDIT.DELETE(2)
selects B1:C10 of the active worksheet, deletes it and shifts cells up.
Related Functions: EDIT.DELETE?, COPY, CUT, INSERT, PASTE

ELSE
Marks the beginning of a group of statements in a macro that will be carried out if the preceding IF or ELSE.IF statements are FALSE.
Syntax: ELSE()
Example: See example under IF (Syntax 2).
Related Functions: ELSE.IF, END.IF, IF

ELSE.IF
Marks the beginning of a group of statements in a macro that will be carried out if the preceding IF or ELSE.IF statements are FALSE and *logical_test* is TRUE.
Syntax: ELSE.IF(*logical_test*)
Example: See example under IF (Syntax 2).
Related Functions: ELSE, END.IF, IF

ENABLE.COMMAND

Enables or disables a custom command in a menu. Disabled commands are "dimmed". Built-in commands cannot be disabled.

Syntax: ENABLE.COMMAND(***bar_num, menu, command, enable_logical***)

Bar_num is the ID number of the menu bar. *Menu* can be either the position number or name (as text) of the menu where the command is located. *Command* is the ID number or name of the command. *Enable_logical* =TRUE enables the command.

Example: =ENABLE.COMMAND(10,1,"Print...",FALSE) dims the custom **Print** command in the custom **File** menu, located in position 1 on the custom menu bar with ID number = 10.

Related Functions: ADD.COMMAND, CHECK.COMMAND, DELETE.COMMAND, RENAME.COMMAND

END.IF

Marks the end of a group of statements in a macro associated with an IF statement.

Syntax: END.IF()

Example: See example under IF (Syntax 2).

Related Functions: ELSE, ELSE.IF, IF

ERROR

Specifies action to take if an error occurs.

Syntax: ERROR(***enable_logical***, *macro_ref*)

Enable_logical = FALSE turns off error-checking. If an error occurs during macro execution, it is ignored. *Error_logical* = TRUE or 1 turns on normal error-checking. If *error_logical* = TRUE and *macro_ref* is included, the macro specified in *macro_ref* is run when an error occurs.

Example: =ERROR(1,Handle.Errors) runs the macro *Handle.Errors* if an error is encountered during macro execution.

Related Functions: LAST.ERROR, ERROR.TYPE

ERROR.TYPE

Returns a number indicating what type of error occurred.

Syntax: ERROR.TYPE(*reference*)

Use to determine what kind of error occurred in a cell containing a formula. Values returned are 1 = #NULL!, 2 = #DIV/0!, 3 = #VALUE!, 4 = #REF!, 5 = #NAME?, 6 = #NUM!, 7 = #N/A!; other errors return #N/A!.

Example: =IF(ERROR.TYPE($B$12)=2, GOTO($B$120)) checks cell B12 in the macro sheet; if a division-by-zero error has occurred, control is transferred to cell B120.

Related Functions: ISERR, ISERROR, ISNA

FILE.CLOSE

Command-equivalent to choosing the **Close** command in the **File** menu. If the document is an Excel 4.0 workbook, closes all documents in the workbook.

Syntax: **FILE.CLOSE**(*save_logical*)

Save_logical specifies whether to save the file before closing the window. If *save_logical* is TRUE or 1, changes to all documents in the workbook are saved, if FALSE or 0, changes are not saved. If *save_logical* is omitted, the Save Changes? dialog box is displayed if changes have been made to any document in the workbook.

Example: =CLOSE(1) closes the active file after saving.

Related Functions: CLOSE, CLOSE.ALL, QUIT, SAVE

FILES

Returns a horizontal array of names of all files in the specified folder or directory.

Syntax: FILES(*name_text*)

Name_text specifies where to look for files. If *name_text* is omitted, filenames from the current directory are returned. If *name_text* is a filename, FILES returns an error if the file does not exist.

Example: =IF(ISNA(FILES(sourcename)),ALERT("The document " &
                        sourcename&" does not exist.",2))

Related Function: DOCUMENTS

FILL DOWN

Command-equivalent to choosing the **Fill Down** command in the **Edit** menu. There is also a FILL.UP function.

Syntax: FILL DOWN()

Related Functions: COPY, PASTE, FILL.RIGHT, FILL.UP, FILL.LEFT

FILL RIGHT

Command-equivalent to choosing the **Fill Right** command in the **Edit** menu. There is also a FILL.LEFT function.

Syntax: FILL RIGHT()

Related Functions: COPY, PASTE, FILL.DOWN, FILL.UP, FILL.LEFT

FONT.PROPERTIES**

Equivalent to formatting characters by choosing **Cells...** in the **Format** menu.

Syntax: FONT.PROPERTIES(*font, ..., start_num, num_characters*)

Related Function: ACTIVE.CELL.FONT

FOR

Marks the start of a FOR...NEXT loop. The statements within the loop are repeated until the loop counter exceeds the *end_num* value.

Syntax: FOR(***counter_text, start_num, end_num,*** *step_num*)

Counter_text is the loop counter; it must be in the form of text. *Start_num* is a number or variable specifying the initial value of *counter_text*. Execution of the loop terminates when *counter_text* becomes greater than *end_num* . *Step_num* is the value by which *counter_text* is incremented; it can be negative. If *counter_text* is omitted, it is set to 1.

Example: =FOR("x",start.row,start.row+rows-1)
                =SELECT("R"&x&"C"&start.col&":R"&x&"C"&start.col+ cols-1)
                =SORT(2,"RC",2)
                =NEXT()

carries out the **Sort** operation on successive cell ranges *RxCy:RxCz*, beginning with *x* = *start.row*, and incrementing the row number *x* by 1 until it is greater than (*start.row* + *rows* −1).

Related Functions: NEXT, WHILE

FOR.CELL

Similar to the FOR function. The statements between FOR.CELL and NEXT are repeated for each cell in a range of cells.

Syntax: FOR.CELL(***name_text***,*area_ref*,*skip_blanks_logical*)

Name_text is the name, as text, given to the single cell that is operated on during each loop; it corresponds to the loop counter *counter_text* in a FOR...NEXT loop. Cells are selected from left to right in successive rows. *Area_ref* is the range of cells to be operated on; if omitted, the current selection is used. If *skip_blanks_logical* = TRUE, blank cells are not examined; if FALSE or omitted, all cells in the selection are examined

Example: =FOR.CELL("cell","R1C2:R125C2")
                =SELECT(cell)
                =IF(ISNUMBER(GET.CELL(5)),IF(GET.CELL(48)=FALSE,
                    FORMAT.FONT(,,1)),)
                =NEXT()

examines a range of cells in column B; if a cell contains a number and if it does not contain a formula, then the cell is formatted Bold.

Related Functions: FOR, NEXT, WHILE

FORMAT.FONT

Command-equivalent to choosing the Font tab in the **Cells...** command in the **Format** menu (Excel 5.0), or the **Font...** command in the **Format** menu (Excel 4.0).

Syntax: FORMAT.FONT(*font_name_text, font_size, bold_logical, italic_logical,...*)

Example: =FORMAT.FONT("times",10,1,0) formats the selected cell for 10 point Times font, bold.

Related Functions: FORMAT.FONT?, FORMAT.NUMBER

FORMAT.NUMBER

Command-equivalent to choosing the Number tab in the **Cells...** command in the **Format** menu (Excel 5.0), or the **Number...** command in the **Format** menu (Excel 4.0). Cells can be formatted to display values as numbers, dates or times.

Syntax: FORMAT.NUMBER(***format_text***)

Format_text is a text string, exemplified by the built-in Format Codes displayed in the Number Format dialog box.

Example: =FORMAT.NUMBER("0.00E+00") displays numbers in scientific format.
Related Functions: FORMAT.NUMBER?, FORMAT.FONT

FORMULA

Enters a number, text, reference or formula in a worksheet.
Syntax: FORMULA(***formula_text***, *reference*)
Formula_text can be text, or a number or formula in the form of text. *Reference* specifies where *formula_text* is to be entered; if omitted, *formula_text* is entered in the active cell. For further information see Chapter 5.
Examples: =FORMULA("ID NUMBER") enters the text in the active cell.
    =FORMULA(B26,"R"&row+x&"C"&col) enters the contents of cell B26 of the macro sheet in the location RrCc, where r = row + x and   c = col.
    =SELECT("R9C1")
    =FORMULA("=R[-1]C-"&$C$3)
selects cell A9 and enters the formula =A8-0.005 (cell C3 of the macro sheet contains the value 0.005).
Related Functions: FORMULA.ARRAY, FORMULA.FILL, SET.VALUE

FULL

Mouse- or keyboard-equivalent to any operation that makes the window full size or previous size (e.g., clicking on the Zoom box in Excel for the Macintosh).
Syntax: FULL(*logical*)
A function from Excel 3.0 or earlier. Replaced by WINDOW.MAXIMIZE in Excel 4.0.
Example: =FULL(TRUE) maximizes the active window.
Related Function: WINDOW.MAXIMIZE

GET.BAR

Returns the ID number of the active menu bar or the name or position number of a specified command or menu.
Syntax 1: GET.BAR()
Returns the ID number of the active menu bar
Syntax 2: GET.BAR(***bar_num***, ***menu***, ***command***)
Returns the name or position number of a specified command or menu. *Bar_num* is the ID number of a built-in or custom menu bar about which you want information. *Menu* and *command* can be specified as either ID number or name, as text. If *command* is given as a name, the position number of the command is returned, and vice versa. If *command* is 0 or omitted, the name or position number of the menu is returned.
Examples: =GET.BAR(1,"File", "Print...") returns 13, the position of the **Print** command in the **File** menu; separator bars are included in the position count. =GET.BAR(1,1, "Print...") is equivalent. =GET.BAR(1,"Data",1) returns the name of the first command in the **Data** menu.
=GET.BAR(1,"Data",) returns 5, the position of the   **Data** menu on the menu bar.
Related Functions: ADD.BAR, SHOW.BAR, ADD.COMMAND,
CHECK.COMMAND, DELETE.COMMAND, RENAME.COMMAND

GET.CELL

Returns information about the location, contents or formatting of a cell.

Syntax: GET.CELL(**info_type_num**, *reference*)

Info_type_num specifies what information will be returned. It can have values from 1 to 53. For example, 1 = absolute reference of the upper-left cell of *reference*, 5 = contents of the cell, 6 = formula in the cell, as text, 14 = TRUE if the cell is locked, 53 = contents of the cell as currently displayed.

Example: =GET.CELL(7) returns the number format of the cell, e.g., General, 000-00-0000, mmm dd, yyyy.

Related Functions: worksheet function CELL, GET.FORMULA

GET.DOCUMENT

Returns information about a document (name, location, type, printer settings, etc.)

Syntax: GET.DOCUMENT(**info_type_num**, *name_text*)

Info_type_num specifies what information will be returned. It can have values from 1 to 68. For example, 1 = name of the document (as text), 3 = number indicating the type of document, 50 = total number of pages, based on current print settings, 53 = a number indicating portrait or landscape, 64 = array of row numbers that are immediately below a page break.

Example: =SET.VALUE(sourcename, GET.DOCUMENT(1)) saves the name of the active document in the cell named *sourcename*.

Related Function: GET.CELL

GET.FORMULA

Returns the contents of a cell as it appears in the formula bar.

Syntax: GET.FORMULA(**reference**)

Returns the cell contents as text. Cell references are in R1C1-style.

Examples: =GET.FORMULA() returns the contents of the current selection in the active worksheet, =GET.FORMULA(!$C$10) returns the contents of cell C10 of the active worksheet; if C10 contains the formula =-LOG(B10) then GET.FORMULA returns -LOG(RC[-1]).

Related Functions: worksheet function CELL, GET.CELL

GOTO

Switches macro execution to *reference*.

Syntax: GOTO(**reference**)

Reference can be a cell reference or a name. It can be an external reference to another macro sheet. Except in IF statements, the use of GOTO should be avoided.

Example: =GOTO($B$113))

HALT

Stops execution of all macros.

Syntax: HALT()

Related Functions: BREAK, RETURN

HIDE
Command-equivalent to choosing the **Hide** command in the **Window** menu.
Syntax: HIDE()
Hides the active window. A macro can refer to a sheet even though it is hidden.
Related Function: UNHIDE

IF
Performs one of two actions based on whether *logical_test* is TRUE or FALSE.
Syntax 1: IF(***logical_test, action_if_true,*** *action_if_false*)
Example: =IF(total>100, GOTO($B$113))
Syntax 2: IF(***logical_test***)
Use this form of the IF statement when more than one action must be performed
if *logical_test* is TRUE.
Example: =IF(logical_test1)
                (several statements)
            =ELSE.IF(logical_test2)
                (several statements)
            =ELSE()
                (several statements)
            =END.IF()
Related Functions: AND, ELSE, ELSE.IF, END.IF, NOT, OR

INPUT
Displays a dialog box for input, with title bar, text message, input box, OK and
Cancel buttons.
Syntax: INPUT(***message_text***, *type_num, title_text, ...*)
Message_text is the text to be displayed in the box. *Type_num* specifies the type
of data to be entered (0 = Formula, 1 = Number, 2 = Text, 4 = Logical, 8 =
Reference, 16 = Error, 64 = Array). *Title_text* is text to be displayed in the title
bar of the box; if omitted, "Input" will be displayed.
Example: =INPUT("Enter range or select with mouse ",8,"FIRST ROW OF
                DATA TO BE SORTED")
Related Functions: ALERT, DIALOG.BOX

INSERT
Command-equivalent to choosing **Cells...**, **Rows** or **Columns** in the **Insert** menu
(Excel 5.0) or the **Insert** command in the **Edit** menu (Excel 4.0).
Syntax: INSERT(*shift_type_num*)
Shift_type_num specifies the direction to shift cells before inserting (1 = shift
cells right, 2 = shift cells down, 3 = shift entire row, 4 = shift entire column). If an
entire row or column is selected, *shift_type_num* is not required and, if specified,
is ignored.
Example: =SELECT("C2:C3")
                =INSERT()
selects columns B and C of the active worksheet, then performs an **Insert**.
Related Functions: INSERT?, COPY, CUT, EDIT.DELETE, PASTE

NEW*
Command-equivalent to choosing the **New...** command in the **File** menu.
Syntax: NEW(*document_type_num*, *xy_series*)
Document_type_num specifies the type of document to create (1 = worksheet, 2 = chart, 3 = macro sheet, etc.); if omitted, a document of the same type as the active document is created. *xy_series* is a number from 0 to 3 that specifies how data is arranged in a chart. See Excel's On-line Help for details.
Example: =NEW(1) opens a new worksheet.
Related Functions: NEW?, OPEN

NEXT
Marks the end of a FOR...NEXT or WHILE...NEXT loop.
Syntax: NEXT()
Related Functions: FOR, WHILE

OPEN*
Command-equivalent to choosing the **Open...** command in the **File** menu.
Syntax: OPEN(**file_text**, *update_links_num*)
File_text is the name of the file to be opened. *Update_links_num* is a number (0-3) specifying what external references are to be updated (0 = do not update, 1 = update external references only, 2 = update remote references only, 3 = update external and remote references). If the document has external references and *update_links_num* is omitted, Excel displays the message "Update references to unopened documents?".
Example: =OPEN(*source.name*, 1) opens the file whose filename is stored in the cell *source.name* and updates external references.
Related Functions: OPEN?, CLOSE, NEW

PAGE.SETUP*
Command-equivalent to choosing the **Page Setup...** command in the **File** menu.
Syntax: PAGE.SETUP(*header_text*, *footer_text*, *left_margin_num*, *right_margin_num*, *top_margin_num*, *bottom_margin_num*, *R&C_headings_logical*, *gridlines_logical*, *H_center_logical*, *V_center_logical*, *orientation_num*, ...)
Values of *orientation* are 1 = portrait, 2 = landscape.
Example: =PAGE.SETUP("","",0,0,0,0) removes header and footer text and sets all margins to zero. All other parameters are unchanged.
Related Functions: PAGE.SETUP?, GET.DOCUMENT, PRINT

PASTE
Command-equivalent to choosing the **Paste** command in the **Edit** menu.
Syntax: PASTE(*to_ref*)
See the comments about the CUT function.
Example: See the examples at the COPY and CUT functions.
Related Functions: COPY, CUT, PASTE.SPECIAL

PASTE.SPECIAL*
Command-equivalent to choosing the **Paste Special...** command in the **Edit** menu. There are four syntax forms: syntax 1, from a worksheet to a worksheet; syntax 2, from a worksheet to a chart; syntax 3, from a chart to a chart; syntax 4, from another application.
Syntax 1: PASTE.SPECIAL(***paste_type_num, operation_type_num, skip_blanks_logical, transpose_logical***)
Paste_type_num is a number specifying what to paste, in the same order as in dialog box (1 = All, 2 = Formulas, 3 = Values, 4 = Formats, 5 = Notes). *Operation_type_num* is a number specifying what operation to perform (in same order as in dialog box). For the other syntax forms, see On-line Help.
Example: =PASTE.SPECIAL(3,1,FALSE,FALSE) pastes the contents of the Clipboard as values, performs no operations, does not skip blanks and does not transpose rows and columns.
Related Functions: PASTE.SPECIAL?, COPY, CUT, PASTE

PRINT*
Command-equivalent to choosing the **Print...** command in the **File** menu.
Syntax: PRINT(*range_num, from_num, to_num, copies_num, ...*)
Range_num specifies what to print (1 = Print all pages, 2 = Print a range of pages). *From_num* and *to_num* specify what pages to print when *range_num* = 2. *Copies_num* specifies the number of copies to print; if omitted, the default value is 1.
Example: =PRINT(2,1,2,1) prints one copy of pages 1-2.
Related Functions: PRINT?, PAGE.SETUP

PROTECT.DOCUMENT*
Command-equivalent to choosing the **Protection** command in the **Tools** menu (Excel 5.0), or the **Protect Document...** command in the **Options** menu (Excel 4.0).
Syntax: PROTECT.DOCUMENT(*contents_logical, windows_logical, password, objects_logical*)
Example: =PROTECT.DOCUMENT(1) protects the contents of the active document.
Related Functions: PROTECT.DOCUMENT?, CELL.PROTECTION

QUIT
Command-equivalent to choosing the **Quit** command in the **File** menu of Excel for the Macintosh, or the **Exit** command in the **File** menu of Excel for Windows.
Syntax: QUIT()
Related Function: FILE.CLOSE

REFTEXT
Converts a reference to text.
Syntax: REFTEXT(***reference***,*a1_logical*)
If *a1_logical* is TRUE, the reference will be in A1-style. If *a1_logical* is FALSE or omitted, the reference will be in R1C1-style.
Example: If cell C10 is the active cell in Sheet1 of Workbook2, then

=REFTEXT(SELECTION()) returns [Workbook2]Sheet1!R10C3,
=REFTEXT(SELECTION(),1) returns[Workbook2]Sheet1!$C$10
Related Functions: INDIRECT, TEXTREF

RENAME.COMMAND
Changes the name of a built-in or custom command on a menu.
Syntax: RENAME.COMMAND(***bar_num, menu, command, name_text***)
See DELETE.COMMAND for a description of the arguments.
Related Functions: ADD.COMMAND, CHECK.COMMAND, DELETE.COMMAND

RESULT
Specifies the type of data returned by a custom function macro.
Syntax: RESULT(*type_num*)
For possible values of *type_num* (1-64) see the entry under ARGUMENT. RESULT
is required if you want the function to return an array or a reference.
Related Functions: ARGUMENT, RETURN

RETURN
Terminates a command or function macro and returns control to the caller.
Syntax: RETURN(*value*)
If the macro is a function macro, *value* specifies the value to return.
Example: =RETURN(B33)
Related Functions: ARGUMENT, RESULT

ROW.HEIGHT
Command-equivalent to choosing the **Row Height...** command in the **Format**
menu.
Syntax: ROW.HEIGHT(*height_num, reference, standard_height_logical,*
type_num)
Height_num specifies the height in points. *Reference,* as an external reference to
the active worksheet or an R1C1-style reference or a name, is the row or rows for
which the height is to be specified. *Standard_height_logical,* if TRUE, sets row
height to the standard height determined by the particular fonts used in the row.
Type_num takes the following actions: 1 = Hide the row by setting row height to
zero, 2 = Unhide the row, 3 = Use best-fit height.
Example: =SELECT("R2C1")
            =FORMULA("(Assignments in italics are changes from last
               listing.)")
            =ROW.HEIGHT(12)
            =FORMAT.FONT(,9,0,1)
selects cell A2 in the active worksheet, inserts a text message, makes the row
height 12 and formats the text as 9 point italic.
Related Functions: ROW.HEIGHT?, COLUMN.WIDTH

SAVE
Command-equivalent to choosing the **Save** command in the **File** menu.
Syntax: SAVE()
Related Function: SAVE.AS

SAVE AS*
Command-equivalent to choosing the **Save As...** command in the **File** menu. Use SAVE.AS if you want to save a file to a different directory or folder.
Syntax: SAVE AS(*document_name_text*, ...)
Document_name_text is the name to be given to the document. See Excel's On-line Help for the other arguments for this function.
Related Functions: SAVE.AS?, SAVE

SELECT*
Mouse- or keyboard-equivalent to selecting a cell or range of cells on a worksheet.
Syntax: SELECT(*selection_ref*)
Selection_ref can be an A1-style reference to the active worksheet or an R1C1-style reference. If *selection_ref* is omitted, the current selection is used.
Example: =SELECT(!$A$1:$E$2) selects the range A1:E2 in the active worksheet.
=SELECT("R1C1:R2C5") selects the range A1:E2 in the active worksheet.
Related Function: SELECTION

SELECTION
Returns the reference of a selected cell, range of cells or object.
Syntax: SELECTION()
If a cell is selected, SELECTION returns the value contained within the cell. To obtain the reference, use REFTEXT(SELECTION()). If an object is selected, the identifier of the object (Button n, Chart n, etc.) is returned.
Example: =IF(SELECTION()="", RETURN())
Related Functions: ACTIVE.CELL, OFFSET, SELECT

SET.PAGE.BREAK
Equivalent to choosing the **Page Break** command in the **Insert** menu (Excel 5.0), or the **Set Page Break** command in the **Options** menu (Excel 4.0).
Syntax: SET.PAGE.BREAK()
Related Functions: REMOVE.PAGE.BREAK, SET.PRINT.AREA

SET.PRINT.AREA*
Equivalent to setting Print Area by choosing the Sheet tab in the **Page Setup...** command in the **File** menu (Excel 5.0) or by choosing **Set Print Area** in the **Options** menu (Excel 4.0).
Syntax: SET.PRINT.AREA()
Related Functions: SET.PAGE.BREAK, SET.PRINT.TITLES

SET.PRINT.TITLES

Equivalent to setting Print Titles by choosing the **Sheet** tab in the **Page Setup...** command in the **File** menu (Excel 5.0) or by choosing **Set Print Titles...** in the **Options** menu (Excel 4.0).

Syntax: SET.PRINT.TITLES(*columns_reference, rows_reference*)

Columns_reference, rows_reference are the references of the titles. To remove print titles, use "".

Related Functions: SET.PRINT.AREA, SET.PRINT.TITLES?

SET.VALUE

Assigns a value to a cell in a macro sheet. Use to store values during execution of a macro.

Syntax: SET.VALUE(*reference, values*)

Reference is the cell on the macro sheet where *value* will be stored.

Example: =SET.VALUE(y,y+1)

Related Function: FORMULA

SHOW.BAR

Displays the specified menu bar.

Syntax: SHOW.BAR(*bar_num*)

Example: See the example under ADD.MENU.

Related Functions: ADD.BAR, DELETE.BAR

SORT

Command-equivalent to choosing the **Sort...** command in the **Data** menu.

Syntax: SORT(***sort_by_num, sortkey_ref1, order_num1, ...***)

Sort_by_num specifies whether to sort by rows (1) or columns (2). *Sortkey_ref* is an external A1-style or R1C1-style reference to a cell identifying which row or column to sort by. *Order_num* specifies whether to sort in ascending (1) or descending (2) order. Two additional sets of sortkeys and orders can be specified.

Example: =SELECT("R5C6:R25C17")
                =SORT(1,$F$6,1)

selects F5:Q25 and sorts by rows in ascending order according to the values in column F.

Related Function: SORT?

STEP

Stops normal execution of a macro and executes it one cell at a time.

Syntax: STEP()

Insert in a macro to debug the macro without stepping through the whole macro sequence.

Related Function: HALT

TEXTREF

Converts text to an absolute reference in either A1- or R1C1-style.

Syntax: TEXTREF(**text**,*a1_logical*)

If *a1_logical* is TRUE, *text* must be an A1-style reference. If *a1_logical* is FALSE or omitted, *text* must be an R1C1-style reference.

Example: If cell   B6 contains the value 11, then

   =TEXTREF("[Workbook2]Sheet2!A"&B6,1)

returns the value contained in cell A11 of Sheet2 of Workbook2.

Related Functions: INDIRECT, REFTEXT

UNHIDE

Command-equivalent to choosing the **Unhide** command in the **Window** menu.

Syntax: UNHIDE(**window_text**)

Window_text is the name of the window to unhide. If a document is not open, UNHIDE returns #N/A. Use GET.WINDOW to see if a document is hidden.

Related Functions: GET.WINDOW, HIDE

WHILE

Marks the start of a WHILE...NEXT loop. The statements within the loop are repeated until *logical_test* is FALSE. Use if you don't know how many times a loop will be executed.

Syntax: WHILE(**logical_test**)

Example: =WHILE(OFFSET(!A1,y,)<>"")

              (perform some operation)

              (increment y)

          =NEXT()

continues to perform the operation as long as the contents of the cell whose reference is OFFSET(!A1,y,) is not empty.

Related Functions: FOR, NEXT

WINDOW.MAXIMIZE

Mouse- or keyboard-equivalent to any operation that makes the window full size or previous size (e.g., clicking on the Zoom box in Excel for the Macintosh).

Syntax: WINDOW.MAXIMIZE(*window_name_text*)

Window_name_text specifies the name, as text or as a reference to a cell containing the name as text, of the window to activate and maximize. If *window_name_text*   is omitted, the active window is maximized.

Example: =WINDOW.MAXIMIZE(dest.sheet) switches to the window whose name is stored in the cell *dest.sheet* and makes the window full size.

Related Function: FULL

SELECTED
VISUAL BASIC KEYWORDS
BY CATEGORY

This appendix lists selected VBA keywords (reserved words) for functions, statements, methods and properties. See Excel's On-line Help for a complete list of keywords.

FUNCTIONS

Abs	Returns the absolute value of a number.
Asc	Returns the numeric code for the first character of text.
Atn	Returns the angle corresponding to a tangent value.
Chr	Returns the character corresponding to a code.
Cos	Returns the cosine of an angle.
Exp	Returns *e* raised to a power.
Fix	Truncates a number to an integer.
Format	Formats a value according to a formatting code expression.
InputBox	Displays an input dialog box and waits for user input.
Int	Rounds a number to an integer.
IsArray	Returns **True** if the variable is an array.
IsDate	Returns **True** if the expression can be converted to a date.
IsEmpty	Returns **True** if the variable has been initialized.
IsMissing	Returns **True** if an optional argument has not been passed to a procedure.
IsNull	Returns **True** if the expression is null (i.e., contains no valid data).
IsNumeric	Returns **True** if the expression can be evaluated to a number.
LBound	Returns the lower limit of an array dimension.
LTrim	Returns a string without leading spaces.
LCase	Converts a string into lower case letters.
Left	Returns the leftmost characters of a string.
Len	Returns the length (number of characters) in a string.
Log	Returns the natural (base-*e*) logarithm of a number.
Mid	Returns a specified number of characters from a text string, beginning at a specified position.

MsgBox	Displays a message box.
Now	Returns the current date and time.
Right	Returns the rightmost characters of a string.
Rnd	Returns a random number between 0 and 1.
RTrim	Returns a string without trailing spaces.
Sgn	Returns the sign of a number.
Sin	Returns the sine of an angle.
Sqr	Returns the square root of a number.
Str	Converts a number to a string.
Tan	Returns the tangent of an angle.
Trim	Returns a string without leading or trailing spaces.
UBound	Returns the upper limit of an array dimension.
UCase	Converts a string into upper case letters.
Val	Converts a string to a number.

STATEMENTS (COMMANDS)

Beep	Makes a "beep" sound.
Call	Transfers control to a **Function** or **Sub** procedure.
Dim	Declares an array and allocates storage for it.
Do...Loop	Delineates a block of statements to be repeated.
Else	Optional part of **If...Then** structure.
ElseIf	Optional part of **If...Then** structure.
End	Terminates a procedure or block.
Exit	Exits a **Do**..., **For**..., **Function**... or **Sub**... structure.
For Each...Next	Delineates a block of statements to be repeated.
For...Next	Delineates a block of statements to be repeated.
Function	Marks the beginning of a **Function** procedure.
GoSub	Branches to a subroutine within a procedure.
GoTo	Unconditional branch within a procedure.
If...Then...End If	Delineates a block of conditional statements.
On...GoSub	Branches to one of several specified subroutines, depending on the value of an expression.
On...GoTo	Branches to one of several specified lines, depending on the value of an expression.
ReDim	Allocates or re-allocates dynamic array storage.
Return	Delineates the end of a subroutine within a procedure.
Select Case	Executes one of several blocks of statements, depending on the value of an expression.
Stop	Stops execution, but does not close files or clear variables.
Sub	Marks the beginning of a **Sub** procedure.
Until	Optional part of **Do...Loop** structure
While	Optional part of **Do...Loop** structure
With...End With	Delineates a block of statements to be executed on a single object.

METHODS

Activate	Activates an object.
Cells	Returns a single cell by specifying the row and column.
Clear	Clears formulas and formatting from a range of cells.
Close	Closes a window, workbook or workbooks.
Columns	Returns a Range object that represents a single column or multiple columns.
Copy	Copies the selected object to the Clipboard or to another location.
Cut	Cuts the selected object to the Clipboard or to another location.
Delete	Deletes the selected object.
FillDown	Copies the contents and format(s) of the top cell(s) of a specified range into the remaining rows.
FillRight	Copies the contents and format(s) of the leftmost cell(s) of a specified range into the remaining columns.
InputBox	Displays an input dialog box and waits for user input.
Insert	Inserts a range of cells in a worksheet.
Paste	Pastes the contents of the Clipboard onto a worksheet.
Quit	Quits Microsoft Excel.
Range	Returns a Range object that represents a cell or range of cells.
Rows	Returns a Range object that represents a single row or multiple rows.
Save	Saves changes to active workbook.
SaveAs	Saves changes to active workbook or other document with a different filename.
Select	Selects an object.
Sort	Sorts a range of cells.

PROPERTIES

ActiveCell	Returns the active cell of the active window.
ActiveSheet	Returns the active sheet of the active workbook.
Bold	Returns **True** if the font is bold. Sets the bold font.
Column	Returns a number corresponding to the first column in the range.
Count	Returns the number of items in the collection.
Font	Returns the font of the object.
FontStyle	Returns or sets the font of the object.
Italic	Returns **True** if the font is italic. Sets the italic font.
Row	Returns a number corresponding to the first row in the range.
Value	Returns the value of an object.

OTHER KEYWORDS AND OPERATORS

False	Boolean keyword.
True	Boolean keyword.
And	Logical operator.
O r	Logical operator.

ALPHABETICAL LIST
OF SELECTED
VISUAL BASIC KEYWORDS

You will find the following listing of VBA functions, statements, methods and properties useful when creating your own macros. For each VBA keyword, the required syntax is given, along with some comments on the required and optional arguments, one or more examples and a list of related keywords. See Excel's On-line Help for further information.

Abs Function
Returns the absolute value of a number.
Syntax: **Abs(*number*)**
Example: **Abs**(-7.3) returns 7.3
See Also: Sgn

Activate Method
Activates an object.
*Syntax: object.***Activate**
Object can be **Chart, Worksheet** or **Window**.
Example: **Workbooks**("BOOK1.XLS").**Worksheets**("Sheet1").**Activate**
See Also: Select

ActiveCell Property
Returns the active cell of the active window. Read-only.
Syntax: **ActiveCell** and   **Application.ActiveCell** are equivalent.
See Also: Activate, Select

ActiveSheet Property
Returns the active sheet of the active workbook. Read-only.
*Syntax: object.***ActiveSheet**
Object can be **Application, Window** or **Workbook**.
Example: **Application.ActiveSheet.Name** returns the name of the active sheet of the active workbook. Returns **None** if no sheet is active.
See Also: Activate, Select

Asc Function
Returns the numeric code for the first character of text.
Syntax: **Asc(*character*)**
Example: **Asc** ("A") returns 65.
See Also: **Chr**

Atn Function
Returns the angle corresponding to a tangent value.
Syntax: **Atn(*number*)**
Number can be in the range $-\infty$ to $+\infty$. The returned angle is in radians, in the range $-\pi/2$ to $+\pi/2$ ($-90°$ to $90°$). To convert the result to degrees, multiply by $180/\pi$.
Example: **Atn**(1) returns 0.785388573 or 45 degrees.
See Also: **Cos, Sin, Tan**

Beep Command
Makes a "beep" sound.
Syntax: **Beep**

Bold Property
Returns **True** if the font is bold. Sets the bold font. Read-write.
*Syntax: object.***Bold**
Object must be **Font**.
Example: **Range("A1:E1").Font.Bold = True** makes the cells bold.
See Also: **Italic**

Call Command
Transfers control to a **Function** or   **Sub** procedure.
Syntax: **Call *name* (*argument1, ...*)**
Name is the name of the procedure. *Argument1*, etc., are the names assigned to the arguments passed to the procedure.   **Call** is optional; if omitted, the parentheses around the argument list must also be omitted.
Example: **Call** Task1(argument1,argument2)
See Also: **Sub, Function**

Cells Method
Returns a single cell by specifying the row and column.
*Syntax: object.***Cells(*row, column*)**
Object is optional; if not specified, **Cells** refers to the active sheet.
Example: **Cells(2,1).Value** = 5 enters the value 5 in cell A2.
See Also: **Range**

Chr Function
Returns the character corresponding to a code.
Syntax: **Chr(*number*)**
Number must be between 1 and 255.
Example: **Chr**(65) returns A.
See Also: **Asc**

Clear Method
Clears formulas and formatting from a range of cells.
*Syntax: object.***Clear**
Object can be **Range** (or **ChartArea**).
Example: **Range**("A1:C10").**Clear**
See Also: **ClearContents, ClearFormats** in Excel's On-line Help.

Close Method
Closes a window, workbook or workbooks.
Syntax: For workbooks, use *object.***Close**. For a workbook or window, use
*object.***Close**(*SaveChangesLogical, FileName*).
Object can be **Window, Workbook** or **Workbooks**. If *SaveChangesLogical* is
False, does not save changes; if omitted, displays a "Save Changes?" dialog box.
Example: **Workbooks**("BOOK1.XLS").**Close**

Column Property
Returns a number corresponding to the first column in the range. Read-only.
*Syntax: object.***Column**
Object must be **Range**.
See Also: **Columns, Row, Rows**

Columns Method
Returns a Range object that represents a single column or multiple columns
*Syntax: object.***Columns**(*index*)
Object can be **Worksheet** or **Range**. *Index* is the name or number (column A =
1, etc.) of the column.
Example: **Selection.Columns.Count** returns the number of columns in the
selection.
See Also: **Range, Rows**

Copy Method
Copies the selected object to the Clipboard or to another location.
*Syntax: object.***Copy**(*destination*)
Object can be **Range, Worksheet, Chart** and many other objects. *Destination*
specifies the range where the copy will be pasted. If omitted, copy goes to the
Clipboard.
Example: **Worksheets**("Sheet1").**Range**("A1:C50").**Copy**
See Also: **Cut, Paste**

Cos Function
Returns the cosine of an angle.
Syntax: **Cos(*number*)**
Number is the angle in radians; it can be in the range $-\infty$ to $+\infty$. To convert an angle in degrees to one in radians, multiply by $\pi/180$. Returns a value between -1 and 1.
See Also: **Atn, Sin, Tan**

Count Property
Returns the number of items in the collection. Read-only.
*Syntax: object.***Count**
Object can be any collection.
Example: The statement N = array.**Count** counts the number of values in the range array.

Cut Method
Cuts the selected object to the Clipboard or to another location.
*Syntax: object.***Cut**(*destination*)
Object can be **Range, Worksheet, Chart** or one of many other objects. *Destination* specifies the range where the copy will be pasted. If omitted, copy goes to the Clipboard.
Example: **Worksheets("Sheet1").Range("A1:C50").Cut**
See Also: **Copy, Paste**

Delete Method
Deletes the selected object.
*Syntax: object.***Delete***(shift)*
Object can be **Range, Worksheet, Chart** and many other objects. *Shift* specifies how to shift cells when a range is deleted from a worksheet (**xlToLeft** or **xlUp**). Can also use *shift* = 1 or 2, respectively. If *shift* is omitted, Excel moves the cells without displaying the "Shift Cells?" dialog box.
Example: **Worksheets**("Sheet12").**Range**("A1:A10").**Delete** (**xlToLeft**) deletes the indicated range and shifts cells to left.

Dim Command
Declares an array and allocates storage for it.
Syntax: **Dim *variable** (subscripts)*
Variable is the name assigned to the array. *Subscripts* are the dimensions of the array; an array can have up to 60 dimensions. Each dimension has a default lower value of zero; a single number for a dimension is taken as the upper limit. Use *lower* **To** *upper* to specify a range that does not begin at zero. Use **Dim** with empty parentheses to specify an array whose dimensions are defined within a procedure by means of the **ReDim** statement.
Example: **Dim** Matrix (5,5) creates a 6×6 array.
See Also: **ReDim**

Do...Loop Command
Delineates a block of statements to be repeated.
Syntax: The beginning of the loop is delineated by **Do** or **Do Until** *condition* or **Do While** *condition*. The end of the loop is delineated by **Loop** or **Loop Until** *condition* or **Loop While** *condition*. *Condition* must evaluate to **True** or **False**.
Example: See examples of   **Do...Loop** structures in Chapter 7.
See Also: **Exit, For, Next, Wend, While**

Else Command
Optional part of **If...Then** structure.

ElseIf Command
Optional part of **If...Then** structure.

End Command
Terminates a procedure or block.
Syntax: **End** terminates a procedure. **End Function** is required to terminate a **Function** procedure. **End If** is required to terminate a block **If** structure. **End Select** is required to terminate a **Select Case** structure. **End Sub** is required to terminate a **Sub** procedure. **End With** is required to terminate a **With** structure.
Example: See examples under **Select Case**.
See Also: **Exit, Function, If, Then, Else, Select Case, Sub, With**

Exit Command
Exits a **Do**..., **For**..., **Function**... or   **Sub**... structure.
Syntax: **Exit Do, Exit For, Exit Function, Exit Sub**
From a **Do** or **For** loop, control is transferred to the statement following the **Loop** or **Next** statement, or, in the case of nested loops, to the loop that is one level above the loop containing the **Exit** statement. From a **Function** or **Sub** procedure, control is transferred to the statement following the one that called the procedure.
Example: See examples of **Exit** procedures in Chapter 7.
See Also: **Do, For...Next, Function, Stop, Sub**

Exp Function
Returns *e* raised to a power.
Syntax: **Exp(*number*)**
Returns the value of *e* raised to the power *number*.
See Also: **Log**

False Keyword
Use the keywords **True** or **False** to assign the value **True** or **False** to Boolean (logical) variables.
When other numeric data types are converted to Boolean values, 0 becomes **False** while all other values become **True** . When Boolean values are converted to other data types, **False** becomes 0 while **True** becomes -1.
Example: **If** SubFlag = **False Then**...
See Also: **True**

FillDown Method
Copies the contents and format(s) of the top cell(s) of a specified range into the remaining rows.
*Syntax: object.***FillDown**
Object must be **Range**.
Example: **Worksheets**("Sheet12").**Range**("A1:A10").**FillDown**
See Also: **FillLeft, FillRight, FillUp** in Excel's On-line Help.

FillRight Method
Copies the contents and format(s) of the leftmost cell(s) of a specified range into the remaining columns.
*Syntax: object.***FillDown**
Object must be **Range**.
Example: **Worksheets**("Sheet12").**Range**("A1:A10").**FillRight**
See Also: **FillDown, FillLeft, FillUp** in Excel's On-line Help.

Fix Function
Truncates a number to an integer.
Syntax: **Fix(***number***)**
If *number* is negative, **Fix** returns the first negative integer greater than or equal to *number*.
Example: **Fix**(-2.5) returns –2.
See Also: **Int**

Font Property
Returns the font of the object. Read-only.
*Syntax: object.***Font**
Example: ActiveCell.Font.Bold = **True** makes the characters in the active cell bold.
See Also: **FontStyle**

FontStyle Property
Returns or sets the font of the object. Read-write.
*Syntax: object.***FontStyle**
Example: **Range**("A1:E1").**Font.FontStyle** = "Bold"
See Also: **Font**

For...Next Command
Delineates a block of statements to be repeated.
Syntax: **For** *counter* = *start* **To** *end* **Step** *increment*
                (statements)
                Next *counter*
Step *increment* is optional; if not included, the default value 1 is used.
Increment can be negative, in which case *start* should be greater than *end*.
Example: See examples of **For...Next** procedures in Chapter 7.
See Also: **Do...Loop, Exit, For Each...Next, While...Wend**

For Each...Next Command
Delineates a block of statements to be repeated.
Syntax: **For** *Each element* **In** *group*
                (statements)
                Next *element*
Group must be a collection or array. *Element* is the name assigned to the variable
used to step through the collection or array. *Group* must be a collection or array.
Example: See examples of **For Each...Next** procedures in Chapter 7.
See Also: **Do...Loop, Exit, For...Next, While...Wend**

Format Function
Formats a value according to a formatting code expression.
Syntax: **Format(*expression,formattext*)**
Expression is usually a number, although strings can also be formatted.
Formattext is a built-in or custom format. Additional information can be found
in *Microsoft Excel/Visual Basic Reference*, or VBA On-line Help.
Example: **Format**(TelNumber,"(###) ###-####") formats the value
TelNumber in the form of a telephone number.

Function Command
Marks the beginning of a **Function** procedure.
Syntax: **Function** *name argument1, ...*
Name is the name of the variable whose value is passed back to the caller.
Argument1, etc., are the names assigned to the arguments passed from the caller
to the procedure.
Example: See examples of **Function** procedures in Chapter 7.
See Also: **Call, Sub**

GoSub Command
Branches to a subroutine within a procedure.
Syntax: **GoSub** *label*
The beginning of the subroutine is delineated by *label*, which can be a name or
a line number. The end of the subroutine is delineated by one or more **Return**
statements. The subroutine must be within the calling procedure.
Example: See examples of subroutines in Chapter 7.
See Also: **GoTo, On...GoSub, On...GoTo, Return, Sub**

GoTo Command
Unconditional branch within a procedure.
Syntax: **GoTo** *label*
Label can be a name or a line number.
See Also: **GoSub**

If...Then...Else...End If Command
Delineates a block of conditional statements.
Syntax: **If** *condition* **Then** ... **Else** ... **End If**
The statement can be all on one line (e.g., **If** *condition* **Then** *statement*).
Alternatively, a block **If** structure can be used, in which case the first line consists
of **If** *condition* **Then**; the end of the structure is delineated by **End If**.
Condition must evaluate to **True** or **False**. The ellipsis following **Then** and
Else can represent a single statement or several statements separated by colons;
these are executed if *condition* is **True** or **False**, respectively.
Examples: If Char = "." **Then GoTo** 2000
                     If (Char >= "0" **And** Char <= "9") **Then**
                     (statements)
                     End If
See Also: **ElseIf, End**

InputBox Function
Displays an input dialog box and waits for user input.
Syntax: **InputBox(***prompt,title,default,xpos,ypos,helpfile,context***)**
See *Microsoft Excel/Visual Basic Reference* or On-line Help for details.
See Also: **InputBox** Method, **MsgBox**

InputBox Method
Displays an input dialog box and waits for user input.
*Syntax: object.***InputBox(***prompt,title,default,left,top,helpfile,context,*
type)
Object must be **Application**. See *Microsoft Excel/Visual Basic Reference* or On-
line Help for details.
See Also: **InputBox** Function, **MsgBox**

Insert Method
Inserts a range of cells in a worksheet.
*Syntax: object.***Insert***(shift)*
Object is a **Range** object. *Shift* specifies how to shift cells when a range is
inserted in a worksheet (**xlToRight** or **xlDown**). Can also use *shift* = 1 or 2,
respectively. If *shift* is omitted, the "Shift Cells?" dialog box is not displayed.
Examples: **Worksheets**("Sheet12").**Range**("A1:A10").**Insert** (1) inserts
the indicated range and shifts cells to right.
Worksheets("Sheet1").**Columns**(4).**Insert** inserts a new column to the left of
column D.
See Also: **Delete**

Int Function
Rounds a number to an integer.
Syntax: **Int**(*number*)
If *number* is negative, **Int** returns the first negative integer less than or equal to *number*.
Example: **Int**(-2.5) returns –3.
See Also: **Fix**

IsArray Function
Returns **True** if the variable is an array.
Syntax: **IsArray**(*name*)
See Also: Other **Is** functions

IsDate Function
Returns **True** if the expression can be converted to a date.
Syntax: **IsDate**(*expression*)
See Also: Other **Is** functions

IsEmpty Function
Returns **True** if the variable has been initialized.
Syntax: **IsEmpty**(*expression*)
See Also: Other **Is** functions

IsMissing Function
Returns **True** if an optional argument has not been passed to a procedure.
Syntax: **IsMissing**(*name*)
See Also: Other **Is** functions

IsNull Function
Returns **True** if the expression is null (i.e., contains no valid data).
Syntax: **IsNull**(*expression*)
See Also: Other **Is** functions

IsNumeric Function
Returns **True** if the expression can be evaluated to a number.
Syntax: **IsNumeric**(*expression*)
See Also: Other **Is** functions

Italic Property
Returns **True** if the font is italic. Sets the italic font. Read-write.
Syntax: object.**Italic**
Object must be **Font**.
Example: **Range**("A1:E1").**Font.Italic = True** makes the cells italic.
See Also: **Bold**

LBound Function
Returns the lower limit of an array dimension.
Syntax: **LBound(*array,dimension*)**
Array is the name of the array. *Dimension* is an integer (1, 2, 3, etc.) specifying the dimension to be returned; if omitted, the value 1 is used.
Example: If the array table was dimensioned using the statement **Dim** table (1 **To** 3, 1000), **LBound**(table,1) returns 1, **LBound**(table,2) returns 0.
See Also: **Dim, UBound**

LCase Function
Converts a string into lower case letters.
Syntax: **LCase(*string*)**
See Also: **UCase**

LTrim Function
Returns a string without leading spaces.
Syntax: **LTrim(*string*)**
See Also: **RTrim**

Left Function
Returns the leftmost characters of a string.
Syntax: **Left(*string,number*)**
If *number* is zero, a null string is returned. If *number* is greater than the number of characters in *string*, the entire string is returned.
Example: **Left("CHEMISTRY",4)** returns CHEM
See Also: **Len, Mid, Right**

Len Function
Returns the length (number of characters) in a string.
Syntax: **Len(*string*)**
Example: **Len("CHEMISTRY")** returns 9.
See Also: **Left, Mid, Right**

Log Function
Returns the natural (base-e) logarithm of a number.
Syntax: **Log(*number*)**
Number must be a value or expression greater than zero. VBA does not provide base-10 logarithms; use **Log(value)/Log(10)**.
See Also: **Exp**

Mid Function
Returns the specified number of characters from a text string, beginning at the specified position.
Syntax: **Mid(*string,start,number*)**
If *start* is greater than the number of characters in *string*, returns a null string. If *number* is omitted, all characters from *start* to the end of the string are returned.

Example: **Mid**("H2SO4",2,1) returns 2.
See Also: **Left, Len, Right**

MsgBox Function
Displays a message box.
Syntax: **MsgBox(*prompt*,*buttons*,*title*,*helpfile*,*context*)**
See *Microsoft Excel/Visual Basic Reference* or On-line Help for details.
See Also: **InputBox**

Now Function
Returns the current date and time.
Syntax: **Now**
See Also: other date and time functions.

On...GoSub Command
Branches to one of several specified subroutines, depending on the value of an expression.
Syntax: **On *expression* GoSub *label1*, ...**
Control is transferred to the label (line number or label) determined by the value of *expression*. *Expression* is any expression that evaluates to a number; if not an integer, it is rounded to the nearest whole number. The number must be between 0 and 255.
Example: See examples of **On...GoSub** procedures in Chapter 7.
See Also: **GoSub, GoTo, Return, Select Case**

On...GoTo Command
Branches to one of several specified lines, depending on the value of an expression.
Syntax: **On *expression* GoTo *label1*, ...**
See explanation under **On...GoSub**   Command.
Example: See examples of **On...GoTo** procedures in Chapter 7.
See Also: **GoSub, GoTo, Return, Select Case**

Open Method
Opens a workbook.
Syntax: ***object*.Open(*filename*, ...)**
Object must be **Workbooks**. *Filename* is required. See On-line Help for the remaining arguments.
Example: **Workbooks.Open("SOLVSTAT.XLS")**
See Also: **Close, Save, SaveAs**

Paste Method

Pastes the contents of the Clipboard onto a worksheet.

Syntax: object.Paste(destination)

Object must be **Worksheet.** The are other **Paste** methods, with different syntax, for **Chart** and many other objects. *Destination* specifies the range where the copy will be pasted. If omitted, copy is pasted to the current selection.

Example: **Worksheets("Sheet1").Range("A1:C50").Copy**
        ActiveSheet.Paste

See Also: **Copy, Cut**

Quit Method

Quits Microsoft Excel.

Syntax: object.Quit

Object must be **Application**.

Example: **Application.Quit**

See Also: **Close, Save**

Range Method

Returns a Range object that represents a cell or range of cells.

Syntax: object.Range(reference)

Object is required if it is **Worksheet**. *Reference* must be an A1-style reference, in quotes, or the name of the reference.

Example: **Worksheets("Sheet12").Range("A1").Value** = 5

See Also: **Cells**

ReDim Command

Allocates or re-allocates dynamic array storage.

Syntax: **ReDim** *variable (subscripts)*

For discussion of *variable* and *subscripts*, see comments under the entry for **Dim**. You can use **ReDim** repeatedly to change the number of elements in an array, or the number or dimensions.

Example: **Dim** Matrix()
        (statements)
        ReDim Matrix (5,5)
        (statements)
        ReDim Matrix (15,25)

See Also: **Dim**

Return Command

Delineates the end of a subroutine within a procedure.

See Also: **GoSub**

Right Function

Returns the rightmost characters of a string.

Syntax: **Right(***string,number)*

If *number* is zero, a null string is returned. If *number* is greater than the number

of characters in *string,* the entire string is returned.
Example: **Right**(303585842,4) returns 5842.
See Also: **Left, Len, Mid**

Rnd Function
Returns a random number between 0 and 1.
Syntax: **Rnd**

Row Property
Returns a number corresponding to the first row in the range. Read-only.
*Syntax: object.***Row**
Object must be **Range**.
Example: **If** ActiveCell.Row = 10 **Then** ActiveCell.Interior.ColorIndex = 27
changes the interior color of the active cell to yellow if it is in row 10.
See Also: **Column, Columns, Rows**

Rows Method
Returns a Range object that represents a single row or multiple rows.
*Syntax: object.***Rows**(*index*)
Object can be **Worksheet** or   **Range**. *Index* is the name or number of the row.
Example: **Selection.Rows.Count** returns the number of rows in the
selection.
See Also: **Columns, Range**

RTrim Function
Returns a string without trailing spaces.
Syntax: **RTrim(string)**
See Also: **LTrim, Trim**

Save Method
Saves changes to active workbook.
*Syntax: object.***Save**(*filename*)
Object must be **Workbook**. If *filename* is omitted, uses a default name.
Example: **ActiveWorkbook.Save**
See Also: **Close, Open, SaveAs**

SaveAs Method
Saves changes to active workbook or other document with a different filename.
*Syntax: object.***SaveAs**(*filename, ...*)
Object can be **Worksheet,   Workbook,   Chart** or other document types. See
Microsoft Excel/Visual Basic Reference or On-line Help for details.
Example: NewChart.**SaveAs**("New Chart")
See Also: **Close, Open, Save**

Select Method
Selects an object.
*Syntax: object.**Select***
Object can be **Chart, Worksheet** or one of many other objects.
Example: **Range("A1:C50").Select**
See Also: **Activate**

Select Case Command
Executes one of several blocks of statements, depending on the value of an expression.

Syntax: **Select Case *expression***
              Case *expression1*
              (statements)
              Case *expression2*
              (statements)
              End Select

You can also use the **To** keyword in *expression* , e.g., **Case** "A" **To** "M".
Expression can also be a logical expression. Use **Case Else** (not required) to handle all cases not covered by the preceding **Case** statements.
Example: See examples of **Select Case** procedures in Chapter 7.
See Also: **If...Then...Else, On...GoSub, On...GoTo**

Sgn Function
Returns the sign of a number.
Syntax: **Sgn(*number*)**
Returns 1, 0 or −1 if *number* is positive, zero or negative, respectively.
Example: **Sgn(-7.3)** returns −1
See Also: **Abs**

Sin Function
Returns the sine of an angle.
Syntax: **Sin(*number*)**
Number is the angle in radians; it can be in the range $-\infty$ to $+\infty$. To convert an angle in degrees to one in radians, multiply by $\pi/180$. Returns a value between −1 and 1.
See Also: **Atn, Cos, Tan**

Sort Method
Sorts a range of cells.
Syntax: **object.Sort(*sortkey1*,*order1*,*sortkey2*,*order2*, ...)**
Object must be **Range**. See *Microsoft Excel/Visual Basic Reference* or On-line Help for details.

Sqr Function
Returns the square root of a number.
Syntax: **Sqr(*number*)**
Number must be greater than or equal to zero.

Stop Command
Stops execution, but does not close files or clear variables.
See Also: **End**

Str Function
Converts a number to a string.
Syntax: **Str(*number*)**
A leading space is reserved for the sign of the number; if the number is positive, the string will contain a leading space.
See Also: **Format**

Sub Command
Marks the beginning of a Sub procedure.
Syntax: **Sub *name* argument1, ...**
Name is the name of the procedure. *Argument1*, etc., are the names assigned to the arguments passed from the caller to the procedure. The end of the procedure is delineated by **End Sub**
Example: See examples of **Sub** procedures in Chapter 7.
See Also: **Call, Function**

Tan Function
Returns the tangent of an angle.
Syntax: **Tan(*number*)**
Number is the angle in radians; it can be in the range $-\infty$ to $+\infty$. To convert an angle in degrees to one in radians, multiply by $\pi/180$. Returns a value between $-\infty$ to $+\infty$.
See Also: **Atn, Cos, Sin**
Trim Function
Returns a string without leading or trailing spaces.
Syntax: **Trim(*string*)**
See Also: **LTrim, RTrim**

True Keyword
Use the keywords **True** or **False** to assign the value **True** or **False** to Boolean (logical) variables.
When other numeric data types are converted to Boolean values, 0 becomes **False** while all other values become **True** . When Boolean values are converted to other data types, **False** becomes 0 while **True** becomes -1.
Example: **If** FirstFlag = **True Then GoTo** 2000
See Also: **False**

UBound Function
Returns the upper limit of an array dimension.
Syntax: **UBound(*array*, *dimension*)**
Array is the name of the array. *Dimension* is an integer (1, 2, 3, etc.) specifying the dimension to be returned; if omitted, the value 1 is used.
Example: If the array   table was dimensioned using the statement **Dim** table (1 **To** 3, 1000), **UBound**(table,3) returns 1, **UBound**(table,2) returns 1000.
See Also: **Dim, LBound**

UCase Function
Converts a string into upper case letters.
Syntax: **UCase(*string*)**
See Also: **LCase**

Until Command
Optional part of **Do...Loop** structure
Syntax: See explanation under **Do...Loop**

Val Function
Converts a string to a number.
Syntax: **Val(*string*)**
Val stops at the first non-numeric character other than the period.
Example: **Val**("21 Lawrence Avenue") returns 21.
See Also: **Str**

Value Property
Returns the value of an object.
*Syntax: object.***Value**
If *object* is **Range**, returns or sets the value(s) of the cell(s). Read-write.
If **Range** contains more than one cell, returns an array of values.
Example: **Worksheets**("Sheet12").**Range**("A1").**Value** = "Volume, mL"

Wend Command
Delineates the end of a **While...Wend** procedure.
Syntax: See explanation under **Do...Loop**
See Also: **Do...Loop, While...Wend**

While Command
Executes a series of statements as long as a specified condition is true.
Syntax: See explanation under **Do...Loop**
See Also: **Do...Loop, Wend**

With...End With Command

Delineates a block of statements to be executed on a single object.

Syntax: **With *object***
          (statements)
          End With

Example: **With** ActiveSheet.PageSetup
          .CenterHeader = "DRAFT VERSION"
          .CenterFooter = ""
          End With

See Also: **Do...Loop, While...Wend**

G

SELECTED SHORTCUT KEYS
FOR WINDOWS AND MACINTOSH

	Windows	Macintosh
Entering or examining formulas		
To paste placeholder <u>A</u>rguments of a function (after typing the function and the opening parenthesis)	CTRL+A	CONTROL+A
To <u>T</u>oggle between relative, absolute and mixed references	F4	COMMAND+T
To enter an array formula	CTRL+SHIFT+ENTER	COMMAND+ENTER
To display the value of a selected reference, variable or function	F9	COMMAND+=
To toggle between displaying values and formulas	CTRL+SHIFT+~	COMMAND+~ or CONTROL+~
Inserting		
To insert a line break	ALT+ENTER	COMMAND+OPTION+ ENTER
To insert a tab	CTRL+TAB	COMMAND+OPTION+tab
To insert the current date	CTRL+(semicolon)	COMMAND+(hyphen)
To insert the current time	CTRL+(colon)	COMMAND+(semicolon)
Applying Formats		
Apply currency format	CTRL+SHIFT+$	CONTROL+SHIFT+$
Apply percentage format	CTRL+SHIFT+%	CONTROL+SHIFT+%
Apply exponential format	CTRL+SHIFT+^	CONTROL+SHIFT+^
Apply general format	CTRL+SHIFT+~	CONTROL+SHIFT+~

Editing in the formula bar

To enter Edit mode	F2	COMMAND+U
To exit Edit mode and save changes	ENTER, or click Check box	RETURN, or click Check box
To exit Edit mode without COMMAND+(period), saving changes	ESC, or click Cancel box	ESC, or click Cancel box
To copy the value in the cell above the active cell into the formula bar	CTRL+SHIFT+"	COMMAND+SHIFT+"
To copy the formula in the cell above the active cell into the formula bar	CTRL+SHIFT+'	COMMAND+SHIFT+'

ABOUT THE DISKETTE
THAT ACCOMPANIES THIS BOOK

This appendix describes the worksheets and macro sheets that are on the diskette that accompanies this book. Each document is provided as an Excel 4.0 worksheet, so that it can be accessed by users of Excel 4.0, 5.0 or 7.0. The only documents not accessible to users of Excel 4.0 are the VBA modules; these are in Excel 5.0 Workbook format.

Read the installation instructions of the Before You Begin section in the frontmatter for information about loading the files on the diskette to your computer. The diskette is in PC format, but since almost all Macintosh computers can read and write PC format diskettes using the PC Exchange utility, you should have no trouble using this diskette with either a PC or Macintosh. If you have problems with the diskette, please contact John Wiley's tech support system at (212) 850-6194.

The filenames assigned to the documents are names compatible with Excel for Windows; that is, they end with an .xls or .xlm extension. If you are using a Macintosh, you can rename the documents if you wish.

Chapter 3   *Creating Advanced Worksheet Formulas*

IFdemo.xls illustrates the use of the **IF** function to prevent the display of error values.

MegaForm.xls illustrates the use of "megaformulas" (in this case involving text functions). Unhide columns C through F to view the separate parts of the megaformula.

ArrayDem.xls illustrates three different array formulas for the calculation of the sum of squares of deviations.

Chapter 5   *Creating Command Macros in Excel 4.0 Macro Language*

Demos.xlm illustrates several ways to implement a simple macro.

Sort.xlm is a simple version of an Excel 4.0 Macro Language command macro that performs a row-by-row sort.

Chapter 6   *Creating Custom Functions in Excel 4.0 Macro Language*

DegF.xlm is a simple Excel 4.0 Macro Language custom function macro that converts Fahrenheit to Celsius.

MolWt.xlm is an Excel 4.0 Macro Language custom function macro that calculates molecular weight from a chemical formula.

Chapter 7   *Excel 5.0 Macro Language: Visual Basic for Applications*

ChemForm.xls is a VBA procedure that formats text as a chemical formula. The macro can be easily assigned to a toolbutton.

Chapter 9   *Creating Custom Tools and Toolbars*

NumFmt.xlm is a simple Excel 4.0 Macro Language command macro that toggles between floating-point and scientific number formats. The macro can be easily assigned to a toolbutton.

FullPage.xlm is a simple Excel 4.0 Macro Language command macro that can be used to obtain the maximum amount of space on a page for printing a worksheet. It sets either portrait or landscape orientation, sets margins to zero and removes header and footer text. The macro can be easily assigned to a toolbutton.

Chapter 11   *Advanced Charting Techniques*

ErrBarMkr.xlm is an Excel 4.0 Macro Language command macro that creates a table of error bar data on a worksheet, tosimplify the plotting of error bars in an X-Y chart.

Chapter 12   *Using Excel as a Database*

Database.xls is a sample database to illustrate Excel's database capabilities.

Chapter 13   *Getting Experimental Data into Excel*

NISPEC.DAT is a comma-delimited text file to be used with Excel's Data Parse or Text to Columns menu commands.

Nth.xls illustrates four different worksheet formulas to select every *N*th data point from a data table.

Chapter 14   *Some Mathematical Tools for Spreadsheet Calculations*

Derivs.xls illustrates how to obtain the first and second derivatives of a data set.

NumDiff.xls illustrates how to calculate the first derivative of a function.

CurvArea.xls illustrates three worksheet formulas that can be used to obtain the area under a curve.

RK4.xls illustrates the Euler and fourth-order Runge-Kutta methods for the solution of differential equations.

Polar.xls illustrates how to convert from polar to Cartesian coordinates.

Chapter 15   Graphical and Numerical Methods of Analysis

NewtRaph.xlm is a simple Excel 4.0 Macro Language command macro that finds roots of a polynomial by the Newton-Raphson method.

Chapter 16   Linear Regression

CalCurv.xls is a example of linear regression applied to a linear calibration curve.

Oxygen.xls illustrates the use of the LINEST function to perform multiple linear regression.

Chapter 17   Non-Linear Regression Using the Solver

NonLin.xls illustrates the use of the Solver to perform multiple non-linear regression analysis.

SOLVSTAT.xlm is an Excel 4.0 Macro Language command macro that returns the standard deviations for non-linear regression analysis performed by the Solver. To the best of my knowledge, there is no other Excel macro or Add-in that provides regression statistics for non-linear least-squares regression performed by using the Solver. SOLVSTAT.XLM is a shareware program. Feel free to distribute it to others.

See "Instructions for Using SOLVSTAT" at the end of this appendix.

Chapter 18   Analysis of Solution Equilibria

Alpha.xlm is an Excel 4.0 Macro Language custom function macro that returns α values for a polyprotic acid species.

Alpha.xls is the VBA implementation of the same custom function macro.

Gran.xls illustrates the use of Gran's method to find the equivalence point of a titration.

Chapter 19   Analysis of Spectrophotometric Data

FlameCal.xls illustrates two methods for treating a curved spectrophotometric calibration curve.

3CmpSpec.xls illustrates methods for the analysis of the spectrum of a mixture of Cu^{2+}, Co^{2+} and Ni^{2+} ions.

Deconv.xls illustrates the deconvolution of a UV-visible spectrum into its component Gaussian bands.

Chapter 20   Calculation of Binding Constants

Titrat.xls is an example of the calculation of the pK_a values of a polyprotic acid from titration data, using the Solver.

NMRdata.xls is an example of the calculation of a binding constant from NMR data, using the Solver.

Chapter 21   Analysis of Kinetics Data

XII-21E.xls illustrates the use of the Solver to obtain a first-order rate constant when the final reading is unavailable.

XIII-28A.xls illustrates the use of the fourth-order Runge-Kutta method to obtain four coupled rate constants from a complex kinetic process.

Instructions for Using SOLVSTAT

This command macro returns the standard deviations of regression coefficients obtained by using the **Solver**, plus the correlation coefficient and the RMSD; these statistical parameters are not available from the **Solver**. The array of values returned is in a format similar to that returned by LINEST.

The sheet must contain Y_{calc} values. The Y_{obsd} and Y_{calc} values must each be in either a single column or row. The regression coefficients returned by the Solver do not have to be in adjacent cells.

To use the macro, simply **Open** SOLVSTAT.XLM; it will appear on screen and then **Hide** itself. It installs a new menu command, **Solver Statistics...**, immediately under the **Solver...** command in either the **Formula** menu (Excel 4.0) or **Tools** menu (Excel 5.0 or 7.0). If the Solver Add-in has not been loaded, the **Solver Statistics...** command will be at the top of the menu. The command will remain in the menu until you exit from Excel.

Activate the document in which the **Solver** has already been used to obtain regression coefficients by minimizing the sum of squares of deviations between observed and calculated Y values. Choose **Solver Statistics...** from the menu. Follow the directions in the dialog boxes.

I

NEW FEATURES IN EXCEL 97

Excel for Windows 97 is slated to appear in 1997. There are a number of changes, some cosmetic, some substantive. On the basis of information I've been able to obtain, the changes that will most interest users of this book are listed below.

- Excel 97 will use a *new file format*. Older versions of Excel won't be able to read Excel 97 documents.

- Worksheets will have *65,536 rows* instead of 16,384 (2^{16} instead of 2^{14}) in case you ever need them.

- You can have *32,767 characters in a cell* instead of 255 ($2^{15} - 1$ instead of $2^8 - 1$) in case you ever need them.

- You can have *32,000 points in a chart* instead of 4000 (except 3D charts, still at 4000 points) in case you ever need to plot that many points.

- Menus and toolbuttons can be mixed on the same "bar".

- The Formula Wizard has been replaced by the *Formula Palette*; the Formula Palette is said to be not as helpful as the Formula Wizard, which in my opinion only got in the way.

- *Data validation*: you can specify the type of data entered into a specified cell.

- *Conditional formatting*: you can have a cell's format be bold or italic, for example, depending on the cell's value.

- The Cell Note feature has been replaced by *Cell Comments*.

- *Multilevel Undo*: instead of being able to Undo just the last action you performed, you can now step back through 16 levels of previous actions, if necessary, in order to correct an error.

- There are *six new chart types*, none of which have any serious scientific use. We're still waiting for a true 3D chart type.

- The *Chart Wizard* has been improved, and the improvements are said to be good ones.

- The *Chart Element Selector*: a new toolbutton to help to select closely spaced chart elements. But see the Excel Tip on page 196 of this book for a way to do this already in Excel 4.0, Excel 5.0 or Excel 7.0.
- There is now an *easier way to change a chart* from an embedded chart to a separate chart sheet or vice versa.
- Several *Internet features* have been introduced.
- A *new version of VBA* has been introduced.

INDEX